Proceedings of SUT International Conference on Subsea Technology and Deepwater Engineering(2012)

Chief Editor: Menglan Duan
Associate Editor: Yu Zhang and Yi Wang

Pretroleum Industry Press

Proceedings of SUT International Conference on Subsea Technology and Deepwater Engineering(2012)

by Menglan Duan et al.

ISBN 978-7-5021-9293-8

Copyright © 2013 by

Petroleum Industry Press

(Building 1, Block 2, Anhuali, Andingmenwai St., Beijing 100011, P. R. China)

All rights reserved. No part of this publication may be reproduced, stored in a retrieval system or transmitted in any form or by any means: electronic, electrostatic, magnetic tapes, mechanical, photocopying, recording or otherwise, without permission in writing from the publisher.

Printed in Beijing, China

图书在版编目(CIP)数据

SUT 水下技术和深水工程国际会议论文集. 2012:英文/段梦兰主编.
北京:石油工业出版社,2013.9
ISBN 978-7-5021-9293-8

Ⅰ. S…

Ⅱ. 段…

Ⅲ. 海上油气田－油气田开发－国际学术会议－文集－英语

Ⅳ. TE5-53

中国版本图书馆 CIP 数据核字(2012)第 229603 号

责任编辑:何莉　李熹蓉
责任校对:黄京萍
封面设计:赛维钰

出版发行:石油工业出版社
　　　　(北京安定门外安华里2区1号　100011)
　　　　网　　址:www.petropub.com.cn
　　　　编辑部:(010)64523535　发行部:(010)64523620
经　销:全国新华书店
印　刷:北京晨旭印刷厂

2013年9月第1版　2013年9月第1次印刷
787×1092 毫米　开本:1/16　印张:31
字数:784 千字
定价:156.00 元
(如出现印装质量问题,我社发行部负责调换)
版权所有,翻印必究

Society for Underwater Technology

The Society for Underwater Technology (SUT) is an international learned society of marine technology with headquarters in London, England that was founded in 1966. There are branches in Aberdeen (UK), Houston (USA), Rio de Janeiro (Brazil), Newcastle (UK), Perth (Australia), Melbourne (Australia), Kuala Lumpur (Malaysia), Lagos (Nigeria), Bergen (Norway), and Beijing (China), and new branches in development in Egypt, India, Nova Scotia, and London.

Goals

SUT promotes the further understanding of the underwater environment and encourages:

- Cross-fertilisation and dissemination of ideas, experience and information between workers in academic research, applied research and technology, industry and government.
- Development of techniques and tools to explore, study and exploit the oceans.
- Proper economic and sociological usage of resources in and beneath the oceans.
- Further education of scientists and technologists to maintain high standards in marine science and technology.

SUT covers all aspects of technology applied to diving technology and physiology, submersible design and operation, naval architecture, underwater acoustics, subsea systems, geophysics, marine resource exploitation, oceanography, environmental studies, pollution and marine biology. Evening lectures are organised by most of the branches. Training courses are offered to industry, in particular "Subsea Awareness" courses.

Education about the marine world, in particular marine industry, is a strong focus for the society and it supports a number of studentships through awards and an "Educational Support Fund".

Journal

SUT publishes a peer-reviewed scientific journal, *Underwater Technology*.

History

The Society for Underwater Technology was founded in 1966 following the demise of the Underwater Equipment Research Society the previous year. This precursor society had been set up to facilitate the "interchange of information between users and suppliers of undersea equipment". Many of its members went on to become early members of the SUT. In 1966 a Steering Committee was put in place to form the society, leading directly to the first general meeting on 2 March 1967, hosted by Lord Wakefield of Kendall in the House of Lords. Lord Wakefield was elected as president, with Rear Admiral Sir Edmund Irving as the first chairman of council, Nic Flemming as honorary secretary, and V. Grimoldby as honorary treasurer. The original technical committees were "Biological Technology", "Earth Science", and "General Technology".

The first annual meeting was held 7 December 1967 at the Institute of Mechanical Engineers, where the Society had also found a home through the institute's "daughter society" scheme, paying £ 500 a year for office space and administrative assistance. At this time the Society's association with what was to become Oceanology International was initiated with plans to run a major international conference in 1969 alongside an existing exhibition series in Brighton.

In the early 1970s, branches were developed, mainly in the United Kingdom, while tie-ups with overseas organisations such as the Marine Technology Society in the US and the Engineering Committee for Ocean Engineering were also being sought. Branches were established in East Scotland, West Scotland, East Anglia, Southern England, and Southwest England, while overseas branch possibilities were looked at in Europe.

At the beginning of the 1980s, the branch structure was reduced to a group of regional organisers who tried to keep activities going across the United Kingdom in the Southwest, Southern England, the Midlands and Northwest, Scotland East, Scotland Northeast, and Scotland West. In the 1990s the Aberdeen Branch, Southern, and Northeastern Branches were formed.

In 1983 the Educational Support Fund was launched.

In 1990 SUT moved home to the Institute of Marine Engineers and in 2011 moved to new premises located in Fetter Lane.

Organizing Committee

Chair	Laibin Zhang	China University of Petroleum, Beijing
Co-chair	Weiliang Dong	China Oil Services Ltd.
	Yuhong Gu	Kingdream Public Limited Company
	Xiaojian Jin	CNOOC Engineering Department
	Fahua Ji	Sinochem Petróleo Brasil Limitada
	Baoping Lu	Sinopec Research Institute for Petroleum Engineering
	Wanghong Li	The National Natural Science Fund Committee, Engineering and Material Science
	Yu Liang	CNOOC Shenzhen Branch East of Nanhai Petroleum Administration Bureau of CNOOC
	Zhenping Weng	China Ship Scientific Research Center
Secretary	Menglan Duan	China University of Petroleum, Beijing
Members	Gilles Demerliac	Technip
	Jijun Gu	Universidade Federal do Rio de Janeiro (UFRJ), Rio de Janeiro, Brazil
	Xuehui Jiang	Federal Mogul Deva Shanghai Office
	Guolin Jin	Yong Wei Valve
	Guangquan Li	Sinopec Research Institute of Petroleum Engineering
	Shulin Li	Suzhou Douson Drilling & Production Equipment Co., Ltd.
	Chengjian Liu	JiangSu Rudong Jinyou Machinery Co., Ltd.
	Dongfeng Mao	China University of Petroleum, Beijing
	Ruiyong Ruan	Jiangsu Guo Rui Hydraulic Machinery Co., Ltd.
	Jianjun Wang	China Oilfield Services Limited
	Youhua Wang	CNPC Offshore Engineering Co., Ltd.
	Zhe Wang	Offshore Oil Engineering Co., Ltd.
	Mike Read	TELEDYNE Oil&Gas, United States
	Chonggao Zeng	National Torch Plan JianHu Province Oil Equipment Characteristic Industry Base
	Guiji Zhang	South Sea Deepwater Gas and Oil Development Project
	Jian Zhang	Shengli Engineering & Consulting Co., Ltd.
	Shihua Zhang	Shengli Oil Field Explotion and Design Institute Co.
	Yandong Zhou	Offshore Oil Engineering Co., Ltd.
	Jianguo Zhu	Sinopec international petroleum exploration and development company
	Xianzhong Zhu	Jiangsu Xianzhong Petroleum Machinery Co., Ltd.
	Xin Zuo	China University of Petroleum, Beijing

Technical Committee

Chair	Shouwei Zhou	Chinese Academy of Engineering
	Hengyi Zeng	Chinese Academy of Engineering
Co-Chair	Frank Lim	2H Offshore
	Jim Neffgen	JM Consultants & Engineers Construction & Engineering Services
	Guohua Li	Sinopec International Petroleum Exploration and Production Corporation
	Xinzhong Li	CNOOC Research Institute
	Zhigang Li	Offshore Oil Engineering Company Limited
	Chao Wu	China Oilfield Services Limited
Members	Jing Cao	CNOOC Research Institute
	An Chen	Universidade Federal do Rio de Janeiro (UFRJ), Rio de Janeiro, Brazil
	Jack Chen	Exxon Mobile Corp.
	Gennadiy V. EGOROV	Marine Engineering Gureau
	Yuan Gao	COOEC Subsea Technology Co., Ltd., Shenzhen, China
	Yanqing He	CNPC Economics & Technology Research Institude
	Weiping Huang	Ocean University of China
	Yvan Jacobzone	Technip Malaysia
	Jinkun Liu	Shengli Engineering & Consulting Co., Ltd.
	Xiaolan Luo	China University of Petroleum, Beijing
	Jakob Pinkster	STC-Group
	Baojiang Sun	China University of Petroleum, Huadong
	Qin Su	Sinopec International Petroleum Exploration and Development Company
	Hongping Tian	Kingdream Public Limited Company
	P. Temarel	University of Southampton
	Decheng Wan	Shanghai Jiao Tong University
	Alan Wang	Offshore Oil Engineering Co., Ltd.
	Liquan Wang	Harbin Engineering University
	Jiaming Wu	South China University of Technology
	Qingxia Yang	China Classification Society
	Young-Soon Yang	Seoul National University
	Bo Zhang	FMC Tehnologies
	Changzhi Zhang	CNOOC, Shenzhen branch
	Dagang Zhang	DMAR Engineering, Inc.
	Tianfeng Zhao	China University of Petroleum, Beijing
	Xiaohuan Zhu	Offshore Oil Engineering Co., Ltd.

Editorial Committee

Editor in Chief	Menglan Duan	China University of Petroleum, Beijing
Associate Editors in Chief	Yu Zhang	China University of Petroleum, Beijing
	Yi Wang	China University of Petroleum, Beijing
Editors	Biao Tang	China University of Petroleum, Beijing
	Shuangni Tan	China University of Petroleum, Beijing
	Penghui Bai	China University of Petroleum, Beijing
	Yiqian Pang	China University of Petroleum, Beijing
	Chenggong Sun	China University of Petroleum, Beijing
	Weiwei Gao	China University of Petroleum, Beijing
	Meiqiu Li	Yangzi University
	Changwei Shi	CNPC Offshore Engineering Company Limited
	Zhihui Hu	CNPC Offshore Engineering Company Limited
	Lina Li	Offshore Oil Engineering Co., Ltd.
	Zhihui Xu	Offshore Oil Engineering Co., Ltd.

President Laibin Zhang's Opening Ceremony Speech

Ladies and Gentlemen,

Good morning! Today is a date of milestone in China's Deepwater Oil and Gas Developing History. Why we say so? Because the first international conference on subsea technology and deepwater engineering hosted by the Society for Underwater Technology is held in China, which will have a historic influence on the deepwater natural resource development especially the oil and gas in South China Sea. The Society for Underwater Technology is a first international institution in underwater engineering which has a history of around 50 years. Mr. Gallett will make an introduction to the SUT.

The SUT has over 10 branches all over the world, and the China branch was established 2 years ago in China University of Petroleum. Soon after, the China Branch, together with China Offshore Oil Engineering Corporation, organized a five days training course last year in Tanggu, Tianjin. More than 80 participants took that training. Today, as a first event of the SUT in China, this conference, the SUTTC 2012, registered participants of over 160 experts which is the double of the expected number.

As we know, the global increase of oil and gas production is from ocean in the recent decade, and the increase mainly from deepwaters. And, more and more other resources such as hydrates have been discovered in the ocean. The SUT plays an important and special role in the rapid development of techniques and technologies for such resources, by bringing together organizations and individuals with a common interest in underwater technology, ocean science and offshore engineering. As Dr. Zhou will mention in the opening presentation of this conference, China is making efforts in developing deepwater oil and gas in South China Sea. The SUTTC is at the right time to promote such efforts and makes it an exchange opportunity of latest technology advances between China and Western countries.

Shenzhen is the first city to open to the West in the 1970s, and witnesses the rapid development of China in the past 30 years. It is an innovative city of sciences and technologies, and is the CNOOC base for exploring and developing deepwater oil and gas in South China Sea. We are now getting together in such a beautiful city to have a dream that the SUTTC 2012 will make a milestone in promoting the advances of technologies in China's deepwater oil and gas field, and it will also promote the closer collaboration between the SUT and the Chinese community of both academic and industrial fields.

We invite 14 keynote presentations and attract 60 technical papers, which will be published by Petroleum Industry Press in the proceedings of 2012 SUT International Conference on Subsea Technologies and Deepwater Engineering. Some papers with permanent interest will be recommended for publication to international SCI-indexed journals such as Ships and Offshore Structures, China Ocean Engineering, etc. The SUT journal of Underwater Technology and the International Journal of Energy Engineering will also select some of the papers for publication. After the conference, the SUT China Branch will host a one day training course on subsea technologies by the SUT experts from United Kingdom, and the CNOOC Shenzhen will arrange a technical visit to its deepwater base. These two activities will greatly enrich the SUTTC 2012. The China University of Petroleum has made corresponding progresses in a wide range of subsea and deepwater engineering in the past years by strong supports from CNOOC and the government. As the President of the CUP, I will definitely support the SUT China Branch in its every effort in collaboration with the SUT and Chinese industries. I also suggest the SUT technical conference could be held in China every year or every two years, attracting more and more participants from all over the world. I do believe that the SUTTC will be one of the most influential events of its kind in the world.

Although it is only the first SUT event in China, the SUTTC 2012 receives strong support from industries and government. About 20 enterprises and institutions are involved in the organization, and over 10 companies make financial supports. On behalf of the Organizing and Technical Committees, I would like to thank them all for their important contributions to the success of this conference. Besides, I'm especially grateful to K. C. Wong Education Foundation, Hong Kong, which is the first support to the conference, and to the CNOOC Shenzhen for arranging the technical site visit. Ian, Jim, I would like to thank you two for taking so long a trip to Shenzhen for this conference and training, and for your insistent support to the China Branch.

Finally, I wish the conference a great success. And I hope you all enjoy your stay in Shenzhen.

Thank you!

ACKNOWLEGDEMENTS

The conference and its proceedings are financially supported by K. C. Wong Education Foundation, Hong Kong, the National Basic Research Program of China (973 Program Grant No. 2011CB013702), the National High-tech Development Program of China (863 Program, 2012AA09A205), and the National Science and Technology Major Projects of Prototype Development of Subsea Connector for Manifolds (2011ZX05026 – 003 – 02), Research on Fabrication and Test Equipment for Subsea Production System (2011ZX05027 – 004 – 01) and Deepwater Equipment and Technology for Emergency Maintenance of Subsea Production System (2011ZX05027 – 005).

Financial support from companies below is also greatly appreciated.

  K.C.Wong Education Foundation, Hong Kong

 Kingdream Public Limited Company

  Yong Wei Valve

  Suzhou Douson Drilling & Production Equipment Co., Ltd.

  第四石油机械厂 SJ PETROLEUM MACHINERY CO.

  Jiangsu Xianzhong Petroleum Machinery Co., Ltd.

 Jiangsu Yangbiao Petroleum Machinery Co., Ltd.

  Jianhu Xian Petroleum Machinery Institute

  Jiangsu Guo Rui Hydraulic Machinery Co.,Ltd.

  RG Petro-Machinery (Group) Co., Ltd.

Contents

ADVANCED OFFSHORE TECHNOLOGY

A New Adding Slots Method for Marginal Field Development ······ (3)

Applications of Virtual Reality Technology in Offshore Engineering ······ (10)

Characteristics and Adaptability of the Deepwater Oil and Gas Field Development Mode ······ (18)

Development & Application of Analysis Program for Offshore Structure Skid shoe engineering
······ (24)

Diving Engineering —Analyses and Challenges of Air Surface-Supplied Diving on DP Vessel
······ (33)

Economic Analysis on Reelwell Drilling Method In Deep Water ······ (39)

Key Technologies and Future Trend for Deepwater Platform ······ (48)

Full Mission Bridge Simulations -a Must for (Complex) Offshore Projects ······ (59)

Motion Analysis and Weather Window Selection for Towing of "Nan Hai Tiao Zhan" Semi-submersible Platform ······ (72)

Numerical Simulation on SPAR platform's Cable Tension in Deep Sea under Wave and Current
······ (80)

Research on Hydrodynamic Characteristics of Heave Damping Plate on Deepwater SPAR Platform ······ (93)

Technique Design of Model Tests for a Typical Wave Energy Converter ······ (104)

The Process Design of Liwan Gas Central Platform ······ (115)

Unified Model of Heat Transfer for Gas-Liquid Flow in Upward Vertical Annuli ······ (120)

Wave Loading Uncertainties and Structural Fatigue Reliability Researches for Semi-Submersible ······ (133)

INSTALLATION AND RISKS

Development of Virtual Simulation System for Subsea Hardware Installation Procedure
······ (149)

Gas Hydrate Problems during Deep Water Gas Well Test ······ (158)

Installation Technology on Riser Clamps for Platform Upgrading ······ (169)

Investigation on Active Model Test Methods Based on the Deepwater Riser Installation Process ······ (175)

Latest Progress in Deepwater Installation Technologies ······ (189)

Pendulous Method to Install Manifolds in Ultra Deep Water ……………………… (211)

Optimization Research of Jetting Parameters for Conductor Installation in Deepwater …… (217)

The Mechanical Analysis of Subsea Manifold Lowered into Deep Water ……………… (223)

PIPELINES & RISERS

A Parametric Study Method for J-lay Installation of Deepwater Pipelines ……………… (243)

A Theoretical Analysis Method for Sandwich Pipes Under Combined Internal-External Pressure and Thermal Load ……………………………………………………… (262)

An Experimental Study of Dynamic Response of a Top Tensioned Riser Model Subject to VIV ……………………………………………………………………………… (275)

On Buckling of Subsea Pipe-In-Pipe System ……………………………………… (289)

Buckling and Collapse of Sandwich Pipes under External Pressure: A Review …………… (301)

Investigation on Failure Mechanism and Prevention Methods of Steel Catenary Riser …… (312)

Numerical Simulation of Submarine Pipeline Upheaval Buckling with Different Imperfection Styles ……………………………………………………………………………… (319)

Research about the Riser Selection Scheme in the South China Sea Deep Water Oil and Gas Field Development ………………………………………………………… (326)

Simulation of the Scour Around a Submarine Pipeline Under Uniform Flows …………… (333)

Study on the Burst Limit States of Submarine Pipeline under Internal Corrosion ………… (345)

VIV Prediction of a Long Tension-Dominated Riser using a Wake Oscillator Model …… (352)

UNDERWATER SYSTEMS

An Overview of Deepwater Subsea Manifold Vertical Connectors …………………… (367)

Analysis of Specification on Offshore Drilling Unit ……………………………… (373)

Application of Explicit Quasi-Static Analysis in the Design of Subsea Connector ………… (382)

Calculation Method of Gas Hydrate Formation Inhibitor Concentration for Qiong Dongnan Deepwatwer Drilling ……………………………………………………… (386)

Casing Program Design and Optimization for Deepwater Drilling and Application in JDZ Block of West Africa …………………………………………………………… (399)

Experimental Study on Tethered Underwater Robot ……………………………… (413)

Maintenance and Repair Methods of Deepwater Equipments ……………………… (422)

Quantitative Risk Analysis of Simplified Subsea Chemical Injection System …………… (427)

Influencing Factors on Subsea Tree Selection …………………………………… (435)

Research on the Replacement of Subsea Christmas Tree Connector Sealing Ring ………… (441)

Selection Method for Subsea X-tree ……………………………………………………… (449)

Simple Analysis of Application Prospect For Several Deepwater Wellhead Unit ………… (457)

Subsea Gate Valve Design ………………………………………………………………… (464)

The Effect of Locking Surface Angles on the Mechanical Advantage of Subsea Connector

……………………………………………………………………………………………… (473)

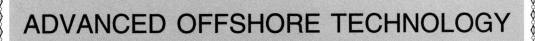

ADVANCED OFFSHORE TECHNOLOGY

A New Adding Slots Method for Marginal Field Development

Xinzhong Li[1], Yi Wang[2], Menglan Duan[2]

[1] CNOOC Research Institute, Chaoyang, Beijing, China
[2] Corresponding author, Offshore Oil/Gas Research Center, China University of Petroleum, Beijing, China

Abstract A new adding slots method, which is different from the traditional "external adding slots" method and "internal adding slots" method, is used in the SZ36 – 1 WHPJ platform. Two new legs are added for supporting the new decks, and special connections are designed to avoid underwater operation during the whole process of installation. This paper presents the design considerations, including structure transportation and installation. The analysis shows that new structure and the existing platform meet the minimum requirements for yielding, fatigue criteria during the service period of another 20 years.

Key Words Adding slot; Docking; Jacket

INTRODUCTION

SZ36 – 1 oil field was found in 1987, it is the first big oil field found in Bohai Gulf which has more than one billion tons of oil reserves. And lots of the jacket platforms were built in eighties last century in SZ36 – 1 oil field. After over twenty years of production, these facilities have created enormous wealth. But, the production drop problem is severer with the increase in production years. As the oil prices keep rising, the China National Offshore Oil Corporation (CNOOC) decided to adjust the phase I of SZ36 – 1 oil field and add eight slots to the WHPJ platform, so as to digging the potential of these platforms. The function expanding and transforming of the active service platforms will revitalize old oil fields.

There are two methods for offshore jacket platform to adding slots: "external adding slots" method and "internal adding slots" method (as shown in Fig. 1). There are not strictly distinctions between the two methods, usually determined by the location relative to the legs of original platform structure. If new slots placed in the rectangular area surrounded by the legs of platform, it is called "internal adding slots". On the other hand, if new slots placed outside the rectangular area surrounded by the legs of platform, it is called "external adding

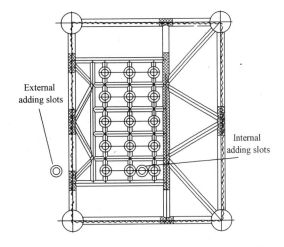

Fig. 1 The methods of adding slots for jacket platform

slots". SZ36 – 1 WHPB platform and WC 13 – 2 platform used "internal adding slots" method, WZ 11 – 4D – A platform used the "external adding slots" method.

Fig. 2 Original situation of WHPJ platform

The WHPJ is a four legs jacket platform with 16 well slots (as shown in Fig. 2). It went to operation in 1997. The water depth is 31.4 m. The platform has four decks (upper deck EL. +19.0m, sub-lower deck EL. +16.0m, lower deck EL. +13.0m and bottom deck EL. +10.0m) and four bracing levels on jacket (EL. +4.3 m, EL. –3.0m, EL. –15.0m and EL. –31.4m). New additional production wells must be added to the platform. But the new wells face the problem of inadequate space on the old platform.

In this paper, a new adding slot method is proposed, based on the combination factors including the basic designing data, on-site conditions and construction technology for SZ36 – 1 WH-PJ platform. And the relevant structural mechanics analysis, transport and installation analysis were conducted. This method was first introduced in China, and will provide a new development mode for the redevelopment of large old oil fields.

1 NEW ADDING SLOTS METHOD

1.1 The Consideration Factors for Adding Slots

WHPJ has serviced for more than a decade and experienced several transformations. Living quarter has been added to the main deck, and a number of equipments have been added to the wellhead areas. Eight slots will be added in this adjustment to meet the requirement of production. And necessary ancillary facilities will also be increased. To meet the requirement of adding eight slots to the WHPJ platform, following factors must be considered:

(1) The effect of structural characteristics of the existing platform.

Because it is adding slots to the existing platform, the layout and form selection of new structure will be limited by the existing platform itself. Especially when the platform has serviced for a long time, after repeated transformation during service, the platform space is small and facilities are crowded. These create tremendous restrictions to the location of the new slots.

(2) The number of new slots.

The number of new slots is an important consideration factor when selecting method of adding slots. "Internal adding slots" method is preferred when the number of new slots is small, while larger number of new slots should be given priority to the "external adding slots" method.

(3) Meet the requirements of drilling and completion.

New adding well slots will use the drilling rig of jack-up platform for drilling if the jacket platform has no drilling rig. So the location of new slots must be placed within the space where can be easily reached by the cantilever of jack-up platform. The workover function is generally completed by the workover rig of jacket, so the adding slots also need to be kept in the space where the track of workover rig can be easily covered.

(4) The difficulty of construction and installation.

The connection problem between new structure and original platform is a challenge, especially in the construction and installation. Underwater construction is inescapable when involving in the connecting between underwater parts of jacket. The operational difficulties and risks are great and the construction costs are high, whether using underwater welding or clamp connection.

Considering the crowded situation of WHPJ and the planning number of slots to be added, "internal adding slots" method can't meet the requirement. And the traditional "external adding slots" method can hardly meet the requirement of adding eight slots. The paper proposes a new creative adding slots method (shown in Fig. 3) to solve the problem of great number of adding slots and the need of expanding platform space.

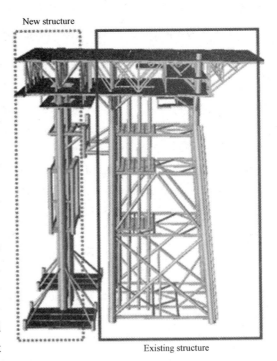

Fig. 3　The new adding slots method

1.2　The Adding Slots Method of WHPJ

The new adding slots method of WHPJ consists of two parts, topsides structure and two piles jacket structure.

Fig. 4　The new added jacket structure

The new added topsides structure will be an integrated deck with equipment, electrical and instrument bulks, utility rooms. The new topsides will consist of three decks. These will be upper deck, lower deck and sub-lower deck. The upper deck elevation will be EL. +19.0 m. The plan area of the deck will be approximately 13.0m × 24.0m. The lower deck elevation will be EL. +13.0 m. The plan area of the deck will be approximately 13.0m × 24.0m. The sub-lower deck elevation will be EL.+ 10.0m. This deck will have a plan area of approximately 9.5 × 16.5m. The well bay area on the central of the topsides shall have a spacing of 4.4m × 8.0m at each of the three deck levels.

The new added jacket structure (as shown in Fig. 4) will be a braced frame structure constructed of steel tubular members. The jacket will have 2 plies. The jacket legs are straight. The jacket leg spacing will consist of two rows of 1 leg (Row A and Row B) spaced at 11.0m on the plan of the working point. The spacing between the new legs and old legs will be 10.0m (Row 1, Row 3).

The jacket will support 8 conductors, the outside diameter of which is 610mm. The conductors will be arranged in one grid of 2 ×4 conductors.

1.3 The Advantages of the New Method

The new adding slots method which has been used on WHPJ platform do not occupy the deck space of existing platform, and solve the crowdedness problem of WHPJ. Meet the requirements of better layout of the new facilities. And the connections between new and existing structures are all above the water surface, avoiding the difficulties of underwater cutting and welding, making it more convenient for quick installation. By application of precision docking GUID technology (as shown in Fig. 5) and the casing of driven piles used horn shape which expanded the area of contact, reducing the accuracy requirements of installation. The heights of two driven piles are different, further reducing installation difficulty. The mudmat of the new jacket structure is larger than the conventional design, prevents the new jacket having a greater settlement, guaranteeing the docking process of connection. The construction process has little effect on existing platform's production and living facilities. The open-hole operation on the deck is unnecessary and the relocation of production equipment is avoided. The workover rig of the existing platform can be applied to new wells workover operations, by extending the original track (as shown in Fig. 6), so the workover rig can be slipped through the track to new slots area. And the drilling ship only needs to be in place once to drilling and completing all new eight wells.

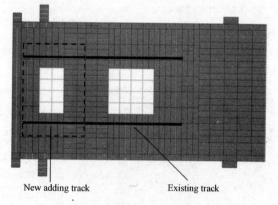

Fig. 5 The connection structure of new and existing platform

Fig. 6 The track of workover rig for WHPJ platform

2 ANALYSIS SUMMARY

The WHPJ platform is an existing platform. In the new adding slots method, two legs jacket and deck are new added structure. To meeting the need of the service period of another 20 years, lots of analysis needs to be carrying out to guarantee the safety of the existing and new structure. In all these analysis, the existing and new structures are modeled as an integer.

2.1 Static Analysis

Static analysis includes the structural strength check and foundation design of platform under the actions of extreme storm, extreme ice and operating environmental events. Environmental events of 50-year return period are selected as the extreme storm condition. Extreme ice events of 50-year return period with environmental events (current and wind) of 10-year return period are selected as the extreme ice condition, while the environmental events of 1-year return period are selected as the operating conditions.

Eight load approaching directions, 0°, 42°, 90°, 138°, 180°, 222°, 270°, and 336° are selected for static analysis. Different water level should be analyzed to determine the specific water level corresponding to the particular load condition that will be critical for platform. And both the heavy platform and light platform conditions are analyzed to get the most critical cases for foundation design. A one-third increase of allowable stresses is used for extreme storm and extreme ice condition. The ice condition check is done according to "Regulations for Offshore Ice Condition & Application in China Sea".

2.2 Seismic Analysis

The seismic analysis for the platform is based on API RP 2A section 2.3.6. A response spectrum analysis was done by using the spectrum which is specified by company. The structural damping ratio is 5%. Dynamic characteristics (frequencies and mode shapes) were calculated which will be used in response spectrum analysis (as shown in Table 1 and Fig. 7). The ductility and strength requirements are considered respectively. The strength level ground acceleration is 0.094g while ductility requirement analysis acceleration is 0.188g.

The responses of the first twenty five natural modes are combined using the complete quadratic combination (CQC) method for the spectrum in three individual directions. The responses of the above three directions are then combined using the Square Root of the Sum of the Square (SRSS) method.

Table 1 Natural frequencies of the WHPJ

Mode number	Frequencies after modification (Hz)	Frequencies of original (Hz)
1	0.795	0.697
2	0.813	0.751
3	0.939	0.877

2.3 Fatigue Analysis

The fatigue life is calculated by combination of wave fatigue analysis and ice fatigue analysis.

A detail wave dynamic spectrum fatigue analysis is performed by SACS. The wave climate information is provided by company. Short-term sea-state is defined by Pierson-Moskowitz spectrum, and the long-term sea-state is characterized by wave scatter diagram and wave directional frequencies. To be conservative, corrosion allowance on the wall thickness of tubular member is taken as much as in-place analysis.

The procedure of detail wave dynamic spectrum fatigue analysis is:

(1) A SEASTATE program of SACS was done to account for the structural self-weight, buoyancy, equipment and live load.

(2) Equivalent linear foundation was created based on the 'center-of-damage sea-state, i. e. the sea-state estimated to contribute most significantly to fatigue damage.

(3) Dynamic characteristics were calculated by DYNPAC program of SACS.

(4) Determine transfer functions using SACS WAVE RESPONSE.

(5) Select fatigue wave and create equivalent static load for each direction.

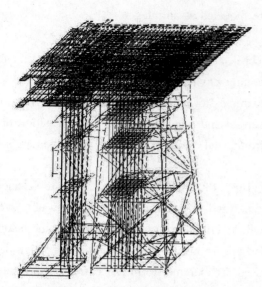

Fig. 7　First natural model of WHPJ after modification

(6) Perform spectrum fatigue analysis.

A detail ice fatigue analysis is performed by ABAQUS. The ice climate information is provided by company. The fatigue analysis is based on fatigue damage accumulation using Miner's law in connection with API X' S-N curves. Geometric nonlinearities lead to stress concentrations that need to be considered within the fatigue assessment. To produce ice fatigue analysis, the ice thicknesses are considerate from 5cm to 40cm in order to cover the main probability of ice (considerate co-liner with the main current direction) in Bohai Gulf. Using these as input, the response of the structure is calculated and the sectional forces are analyzed with the rain-flow method and evaluated the fatigue life. This procedure is very time-consuming since several seeds of 1 minute's long time histories have to be generated per bin to achieve statistically acceptable results.

The safety factor is 2 in the estimation of fatigue life. And the results are shown that the WHPJ meet the requirement of another 20 years of service life.

2.4　Transportation Analysis

For transportation analysis, the jacket and deck will be towed from fabrication yard to installation site by the cargo barge BH306. The analysis contains only the analysis for the jacket and the deck not for the barge and the sea-fasten structure. A one third increase in API RP 2A allowable stress was considered. The analysis result indicates the jacket and deck are safe in transportation condition.

2.5　Lifting Analysis

For lifting analysis, the jacket and deck will be lifted by one derrick barge. In order to make sure all members can meet the strength requirements on lifting condition, the lifting analysis is carried out. For the jacket, the four padeyes locate on the legs at EL. +5.2 m. For the deck, the four padeyes locate on the deck at EL.+19.0 m and the minimum sling angle is 60°. The analysis result indicates the jacket and deck are safe in lifting condition.

2.6 Docking Analysis

For docking analysis, the SZ36 – 1 WHPJ new added jacket structure will be lifted and connected with the existing platform by docking GUID. During installation the clearance between the docking GUID is very small, there is a possibility of collision. It is necessary to check the magnitude of lateral forces that can be applied to the new jacket structure and existing jacket legs at docking GUID for the jacket structure to remain undamaged. 10% of new added jacket structure self-weight is applied to the jacket docking analysis. The analysis result indicates that the jacket is safe under the impact of jacket when installing the new added jacket.

CONCLUSIONS

The paper presents a new adding slots method. This method needs no underwater construction such as underwater cutting and welding, and is convenient during the whole process of installation by using the GUID structure. The workover rig of the existing platform can be applied to new wells workover operations by extending the track. Two new legs are added for supporting the new decks so as to expand the space of platform. The considerations and design procedures are described in detail.

This method was first introduced in China, and the successful use of this method to WHPJ provides valuable experience for the future redevelopment of large old oil fields.

REFERENCES

API RP 2A. Recommended Practice for Planning, Designing and Constructing Fixed Offshore Platforms -Working Stress Design. 21th Edition. Dec. 2000.

Q/HSn 3000—2002. Regulations for Offshore Ice Condition & Application in China Sea.

Applications of Virtual Reality Technology in Offshore Engineering

Ying Jiang[1], Chao Luo[1], Lina Li[1], Menglan Duan[2], Ningqiang He[1]

[1] Offshore Oil Engineering Co., Ltd., Tianjin, China
[2] Offshore Oil/Gas Research Center, China University of Petroleum, Beijing, China

Abstract Virtual simulation technology in marine engineering gradually becomes one of the main branches in offshore engineering. An introduction to simulation technology is presented at first, summarizing the applications of virtual reality technology in marine engineering and the main development trends are discussed as well. This paper also provides effective solution for some problems in marine engineering.

Key Words Virtual reality; Simulation; Offshore engineering; Operator training

INTRODUCTION

Virtual reality technology is the hotspot in recent years. It is a kind of 3D visualization technologies so that people in 3D world can obtain information and play to their creativity with ever imagined means. At present, the virtual reality technology has been widely used by some of the world's biggest companies in various industries. For instance, it's used in various aspects of construction industry, aircraft industry, car manufacturing industry, entertainment industry, petroleum industry, military and mining, etc. Virtual reality technology plays an important role for enterprises to improve development efficiency, to strengthen the data acquisition, data analysis and data processing capacity, to reduce the decision-making errors and to reduce the risks of enterprises. Virtual reality technology is widely used in the fields of engineering, construction, test, installation and maintenance in offshore engineering. By using virtual reality technology, interactive 3D virtual simulation system could be constructed; it could provide dynamic 3D real-time simulation and interactive training function such as water surface ship operating, offshore engineering installation and maintenance work. The application of virtual reality technology in offshore engineering simulates all the homework effectively, provides the reference for scientific decision-making, and trains the related operation personnel effectively. It reduces the operation risk, improves the efficiency of construction and quality.

1 VIRTUAL REALITY TECHNOLOGY

Virtual reality technology is a kind of practical technology that rises in the 1980's and early 1990's. It's a comprehensive integration technology that concerned with fields such as computer graphics technology, human-machine interaction technology, sensing technology and artificial intelligence. Virtual reality technology uses a computer to generate lifelike senses of 3D visual, hearing and smell, etc. People can naturally experience and interact with the virtual world and have more vivid feelings. Simultaneously, people can directly participate in the environment changes and interaction of things. Virtual reality technology makes the computer a system and environment that people can interact and dialog using the sensory organs, language and gestures through a more natural way. Virtual reality technology has the advantages of lower cost, improved security, vitality, repeated operation, etc, and it has get the recognitions by customers.

In recent years, the fast development of virtual reality technology has caused the attention of scholars all over the world. Virtual reality technology has already become a very active area of research. Furthermore, along with the science and technology progress, the ability of computer software and hardware has continuously developed. Virtual reality simulation technology has been gradually used to practical application from laboratory study, and it plays a more and more important role in a variety of areas (Fig. 1).

Fig. 1　Virtual reality technology

Essentially, the virtual reality is a kind of advanced computer user interface. It maximizes the user operation convenience, reduces the user burden and improves the working efficiency through providing the users all kinds of intuitive and natural real-time perception and interactive means. Virtual reality has three major characteristics: immersion, interactivity and imagination.

(1) Immersion is the most important technical characteristic of virtual reality. It refers to people can immerse to an environment created by computer system and become part of the virtual reality system, including visual immersion, tactile immersion, auditory immersion, smelling immersion, taste immersion, etc. The ideal virtual environment should make the user difficult to distinguish the true and the false, and make him immerse himself to the 3D virtual environment created by computer, just like placing himself in the real world.

(2) Interactivity refers to the human's operational level to the objects in the virtual environment and the spontaneity degree of the feedback from the environment by using special input and output devices.

(3) Imagination is the starting point of the virtual world. Designer designs virtual reality system with aid of the virtual reality technology. With the perceptual and rational knowledge from the virtual environment, designer could deepen concept, make new idea, and better use his imagination and creativity.

To sum up, the virtual reality is not only a simulation of 3D and 1D space to the real world, but also a virtualization of natural interaction mode. A complete virtual reality system could make people not only physically totally immersed, but spiritually completely immersed.

2 APPLICATIONS OF VIRTUAL REALITY TECHNOLOGY

2.1 Environment Simulation

3D virtual offshore environment contains offshore natural environment (submarine topography and geomorphology, surface cultural characteristics, etc.), climate environment (rain, fog, snow, etc.), and the entity model. The traditional expression method of offshore natural geographical environment is to use paper chart or electronic chart. No matter paper chart or electronic chart, the user can not obtain offshore geographic information directly. They can only get the offshore geographic environment understanding through the chart reading with the aid of chart pattern and standard. With the development of computer technology, a realistic offshore environment could be simulated with virtual reality technology, and satisfies people all aspects of demands intuitively.

2.2 Vessels and Equipments Simulation

With the modernization of vessel and offshore engineering equipments, the automation level and intelligence level continuously improves. Large offshore engineering equipments' complexity increases continually and increasing money is spent. In foreign, shipbuilding enterprises have already used virtual simulation technology to simulate the design results, construction process and manufacturing flow in advance before the first time construction of high added-value vessel and offshore engineering equipments. That will timely discover and correct the mistakes existing in the design, verify rationality of the plan in the actual production process and optimize the efficiency. For example, through ship simulation, the whole process (including the plate yard start, preprocessing, steel plate cutting, section production, berth block assembly, the ship launching) simulation management of ship building is realized, and the management of different stages of the simulation model and global simulation model is realized. At the same time, the key factors closely linked with the production such as shift arrangement, personnel arrangement, equipment dependence, scheduling logic, logistics mode, space limitation, berth limitation can also be considered to analyze the equipment configuration and utilization, the utilization of key logistics equipments such as tower crane, gantry crane, the key field utilization, the material output of key control point, etc. And it plays an auxiliary role of management to decisions such as scheduling and resource allocation (Fig. 2, Fig. 3).

Fig. 2　Offshore environment simulation

Fig. 3　Large offshore engineering equipment simulation

2.3 Offshore Operation Simulation

Offshore oil and gas exploration involves a great deal of offshore engineering work. Compared to land operation, offshore operation has characteristics of worse environment, more complex condition and higher safety requirements. For example, offshore pipe laying operations demand not only coordinate and efficient work among many systems such as the dynamic positioning system of deep water pipe laying and crane ship and pipe laying production line system, but also to ensure the safety and integrity of the laying pipe. For another example, offshore lifting operation especially in bad sea environment and heavy load condition is a very complicated process. It not only involves the ship stability change, ballast adjustment, collision interference and so on, but also involves multiple systems alignment operation.

Through the software environmental visual simulation of pipe laying process and lifting process, working status and pipeline state of key equipment in all systems concerned to pipe laying and lifting operation could be real-time monitored and predicted, which can provide in time reference information to all system operator, that will improve efficiency and safety of pipe laying and lifting operation.

2.4 Subsea Operation Simulation

Virtual reality system is able to conduct dynamic 3D real-time simulation for installation and recovery process, especially for the surface ship operation, underwater remote operation, etc. It allow the operator to interact with the system function, real-time control each equipment in installation and recovery process. It will stereo, intuitive, system to know the whole installation process of underwater production facilities. Thus it can further perfect the installation and recovery process. The whole operation will be more scientific and reasonable, and that will improve the efficiency of construction and quality, achieve the purpose of saving money, lowering the risk finally.

Deep underwater installation and emergency maintenance is complex. It is usually not directly visible. Operating personnel can only see a small part of the whole work through the underwater monitor, and that will affect the operator to the overall grasp of underwater operation. The application of underwater operation virtual simulation system will let operation personnel more intuitive to understand the underwater work flow. It will improve the operation personnel emergency ability, effectively reduce accident harm, reduce the difficulty of the following remedial treatment, and reduce environmental pollution and economic loss by developing training and appraisal system that integrates emergency maintenance and training system(Fig. 4, Fig. 5).

2.5 Simulator Training System

Using simulation equipments to simulate the movements and controls of aircraft, ships and submarines (including DSV) on the ground is a kind of both economical and safe method. This idea was originally used in aircraft. The risk of flight trainings threats to people's lives, forcing people to have to seek a ground-based equipment to train pilots and minimize training time in the air. The success experiences of flight simulator promote the development of marine engineering

Fig. 4 Offshore engineering operation simulation Fig. 5 Underwater operation simulation

simulator, such as submarine simulator, underwater robot simulator. The ground training research using simulator has adopted by many countries, due to the submarine movements have greater risk,

Fig. 6 Simulator training system

submarine sunk events occurrence commonly, and modern submarines technology is getting more and more complicated and the corresponding control requirement is also higher. The development and design training simulator no matter in terms of security and economy has realistic meaning, it can form effective safety training means, provide divers underwater maneuvering and operation training simulation platform and improve the training efficiency, including the system safe operation training, underwater maneuvering and work coordination training and emergency fault handling training (Fig. 6).

3 EXAMPLES OF VIRTUAL REALITY TECHNOLOGY

3.1 Operation Simulation of Subsea Production Facilities

Ormen Lange field is the second largest gas field so far discovered in Norway, and is the third in Europe. April 2004 Norwegian state council formally approved the Ormen Lange gas field development project. The Swiss ABB company provided dynamic operator simulation training system that can be used for the entire life cycle management for the project. This system can process repeated, and monitor underwater facilities and operation environment through the real time monitoring visual system in the simulation operation process. That guarantees the smooth installation of underwater facilities and the smooth laying of pipelines in extremely complex working environment. See Fig. 7.

Fig. 7 The operation interface and visual chart of operator simulation training system

3.2 Simulation of Ship Navigation

Nautis simulator is developed by NAUTIS Company. It is advanced maritime training simulator, used to provide effective simulation training system for maritime operations. Nautis has the capabilities of advanced ship dynamic simulation and realistic graphics processing, including real and accurate navigation environment, the latest hydrological geographic navigation chart, etc. Its training configuration software tool is a kind of advanced training tool, which can be configured for a wide range of training needs, such as navigation and channel planning, ship maneuvering, bridge team management, target training, emergency rescue and tug assistance and other Marine operations.

3.3 Exploration and Development of Oil and Gas Field

CNOOC (CHINA NATIONAL OFFSHORE OIL CORP.) launched the project to establish the international first-class 3D visualization research center in Bohai oil field at the end of 2005.

This system is the first set of multichannel cylindrical seamless splicing back projection system, and its image specifications are the latest in the similar system. The application of 3D visualization system to research and analysis for oil and gas field in the international industry has been quite common. The researchers can operate the data volume and attribute intuitively using 3D visualization system tools to visualization analysis the front-end collected field data, at the high immersive environments. It can greatly improve the decision efficiency, save exploration cost and time, and enhance the competitive ability. See Fig. 8 and Fig. 9.

Fig. 8 The simulation system of virtual ship development

Fig. 9 The 3D visualization center field

4 TRENDS OF VIRTUAL REALITY TECHNOLOGY IN OFFSHORE ENGINEERING

The research to virtual reality simulation starts later in China. Virtual reality simulation technology has a certain application in marine engineering field, but overall is relatively lags behind compared with advanced foreign technology. There are still many gaps, especially in the development of deepwater oil gas operation. China is in the positive research the application of virtual reality technology to various fields, due to the emergence of virtual reality technology is likely bring revolutionary effects to various fields for the future. CNOOC is carrying out the simulation of virtual reality technology in the application of offshore engineering actively, combined with the virtual reality system of installation, recovery and emergency maintenances. It will form underwater operation virtual reality system, and provide process rehearsal and training for future underwater operation.

(1) Establish virtual reality system of underwater installation operation and emergency maintenance, using virtual reality technology to establish an effective safety training means. Conventional maritime trainings need many funds, long cycle, and high risk, especially for emergency situation that the actual operation can not be carried out training in the sea. So people want to solve these problems with the help of simulation technology and virtual reality technology. It will provide underwater maneuvering and operation training simulation platform to homework personnel, including the safe operation training, underwater maneuvering and work coordination training, underwater complex environment specific operation training and emergency fault handling training, etc.

(2) Establish of the "Half physical visual simulation training system based on hydrodynamic of offshore operation of large dynamic positioning engineering ship". It is expected to be three core functions: ship driving and DP operation training, construction scheme rehearsal; offshore operation real-time tracking monitoring.

(3) Establish the 3D visualization design interface with professional software. General software is often for digital processing mode at present. Virtual reality system can transform digital processing model into 3D image place model. It combines real-time, interactivity, visualization, graphical, immersive at an organic whole, and achieves the effect of visual distinct, vivid, lifelike motion that professional software may be not available in a variety of functions.

CONCLUSIONS

The application of Virtual reality simulation technology in the field of marine engineering has very important significance. It will improve the major equipment's comprehensive performance of marine engineering, reduce system physical test times, shorten the development cycle, save development funds, improve the maintenance level, extend the life cycle, strengthen personnel training, help command personnel decisions, emergency rescue and many others. As a comprehensive scientific, virtual reality simulation system in the field of Marine engineering will play an important role.

REFERENCES

陈丽娜, 马玉杰. 2006. 虚拟现实中图像建模技术的发展[J]. 时代经贸, (08Z): 18 – 20.
程健庆, 等. 2006. 紧密结合装备发展船舶工业仿真技术[C]//中国造船工程学会电子技术学委会论文集.
贺全兵. 2008. 可视化技术的发展及应用. 中国西部科技, 7(4): 4 – 7.
胡小强. 2004. 虚拟现实技术与应用[M]. 北京: 高等教育出版社.
姜学智, 李忠华. 2004. 国内外虚拟现实技术的研究现状[J]. 辽宁工程技术大学学报, (6): 238 – 240.
赵庆国, 赵华, 湛林福, 等. 2005. 虚拟现实技术在石油勘探中的应用[J]. 石油大学学报: 自然科学版, 29(1): 31 – 33.
朱晓军, 彭飞, 朱志洁. 2003. 舰船维修虚拟训练平台研究[J]. 中国修船, (3): 34 – 37.
Javidi Giti, Sheybani Ehsan. 2008. Content-based Computer Simulation of a Networking Course: An assessment [J]. Journal of Computers, 3(3): 64 – 72.
Mitra P P, Joshi T R. 2004. Real Time Geosteering of High Tech Well in Virtual Reality and Prediction Ahead of Drill Bit for Cost Optimization and Rrisk Reduction in Mumbai High L -III Reservoir [A]. Perth, Australia: SPE Asia Pacific Oil and Gas Conference and Exhibition.
Wilson James R, Lada Emily K, Steiger Natalie M. 2006. Performance Evaluation of Recent Procedures for Steady-state Simulation Analysis[J]. IIE Transactions, 38(9): 711 – 727.

Characteristics and Adaptability of the Deepwater Oil and Gas Field Development Mode

Baoping Lu[1], Tao Zhao[2], Menglan Duan[2]

[1] Sinopec Research Institute for Petroleum Engineering, Beijing, China
[2] Offshore Oil/Gas Research Center, China University of Petroleum, Beijing, China

Abstract The deep-water oil and gas exploration and development activities are increasingly active. Several development modes in deepwater development activities are analyzed in this paper. Development modes based on floating production storage and offloading vessel, semi-submersible platform, tension leg platform and other development modes were included in the analysis. The advantages and disadvantages of various development modes were analyzed.

Key Words Development mode; FPSO; SPAR; TLP

INTRODUCTION

In recent years, the deepwater oil and gas exploration and production activities get rapid development. Large deep water oil and gas fields have been found continuously. Deepwater sea has great potential for oil and gas resources, having good prospects for exploration, and has become an important area for future oil and gas exploration. Especially in West Africa, the Gulf of Mexico, North Sea, the South China Sea, etc. exploration and development activities are very active. With the discovery of oil and gas fields, according to the geographical location and other characteristics of oil and gas fields, different development modes should be used in different oil fields.

1 DEVELOPMENT MODE BASED ON FLOATING PRODUCTION STORAGE AND OFFLOADING (FPSO)

The typical development mode based on FPSO at least needs a FPSO, the subsea production system and the shuttle tanker.

FPSO consists of production, storage and offloading parts. It played an important role in offshore oil field development activities. The FPSO has good mobility and can move to the next oil field when the old one is abandoned. So the FPSO is a cost-effective production facility in the development of offshore oil field. The FPSO can be a new construction, and it can be also converted from an old oil tanker. Constructing a new FPSO needs large investment costs and long construction time, but the new FPSO has long service life and needs low maintenance costs.

Converting a old oil tanker into a FPSO needs lower conversion costs and a shorter conversion time. But the FPSO has a short service life and needs high maintenance costs. In this development mode, mobile offshore drilling unit is rent for drilling and workover and the costs are very high. In the development of gas fields, subsea pipelines are required. FPSO must uses wet Christmas trees, thus leading to high workover costs.

The subsea production system includes subsea wellhead, manifold, various pipelines and the control lines. The subsea production system is connected to the FPSO through the risers. The production is transported to land by shuttle tanker. See Fig. 1.

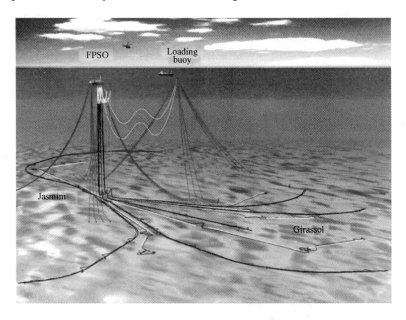

Fig. 1 Development mode based on FPSO

2 DEVELOPMENT MODE BASED ON SEMI – SUBMERSIB-LE PLATFORM

The typical development mode based on the semi-submersible platform at least needs a semi-submersible platform, the subsea production systems and pipeline.

The semi-submersible production platform is a floating system, and it can also have drilling capacity. Its main components include pontoons, columns and deck. Pontoons and columns provide buoyancy. Production equipment, living quarters and storage areas are on the main deck.

A semi-submersible drilling platform is usually converted into the semi-submersible production platform. The semi-submersible production platform and subsea production syetem form the production systems. Typical subsea production system includes manifold, Christmas trees and a number of satellite wells. The fluid production of subsea wellheads flows into the processing equipment on the semi-submersible platform through the manifold and production riser. The crude oil is processed and transported to a nearby storage tanker or oil storage system through the pump and pipeline, and then transported by shuttle tanker. Due to the semi-submersible production platform is usually converted from a semi-submersible drilling platform and modification works can be conducted on the shore shipyard, the cost and time consumed is less than marine construction and installation. When the modification works are conducted on the shore shipyard, pre-drilling can be conducted at the same time and the oil field can be put into production in advance. In addition, when the development of an oil field is over, the semi-submersible platform could be moved to other oil fields.

The semi-submersible platform can use distributed anchor, and it does not require special turret anchoring system. Advanced semi-submersible platform use dynamic position. Compared with FPSO, semi-submersible platform is more stable and the movement is smaller. The initial investment of semi-submersible platform is small. The connection to steel catenary risers is easy. But most of the semi-submersible platform have no storage capacity and the production is transported by pipeline or through other means. When the subsea trees are used, operation and maintenance work is not easy. See Fig. 2.

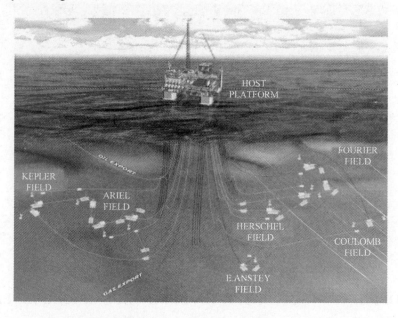

Fig. 2 Development mode based on semi-submersible platform

3 DEVELOPMENT MODE BASED ON TENSION LEG PLATFORM(TLP)

Typical development mode based on the TLP usually requires a TLP, FPSO and shuttle tanker.

The TLP is a flexible system with the vertical anchor chain. The buoyancy of the floating body or tension leg platform hull is used to maintain the tension state of the vertical anchor chain.

The TLP can be connected to rigid riser. The connection to the catenary riser is easy. The construction cost of TLP is insensitive to water depth. Water depth increases only the length of tension leg and the movement range of platform will also increase. Because its vertical swing, sway and yawing cycle is much larger than the characteristic period of the sea conditions, the vertical movement of TLP is limited. While the heave, roll and pitch cycle of TLP is much smaller than the characteristic period of the sea conditions, it avoid the resonance phenomenon in the waves. Therefore, movement of TLP is constrained, and security and stability of the platform are improved. The wellhead system is set up under the deck to facilitate data collection in the process of exploration, drilling and oil production. Field experiment can be conducted easily, and this characteristic is suitable for the exploitation of large deepwater fields. When the TLP is used in deepwater field, its overall movement performance is good and it has strong resistance to adverse environmental effect. The construction costs are low and TLP has reasonable structure. Floating body and topsides is in integration and the construction work can be conducted on land, reducing the offshore installation and maintenance costs.

Though the vertical movement of the TLP is under control, but the movement of the horizontal direction can be quite large. Its construction costs are not very low. The nonlinear dynamic response of TLP will endanger the long-period slow drift motion of the platform. The ultimate bearing capacity, fatigue fracture, reliability and maintenance of tension leg still cause certain restrictions to the usage of TLP. The tension leg platform has no storage capacity, and need subsea pipelines or FPSO to transport production. See Fig. 3.

Fig. 3　TLP

4　DEVELOPMENT MODE BASED ON SPAR PLATFORM

In the development mode based on SPAR platform, there are several choices. Dry tree can be used with SPAR and subsea production system can also be used with SPAR. The development mode usually reqires a SPAR platform, FPSO, the subsea production system and shuttle tanker.

The SPAR platform can be used in the area where the water depth is more than 300m. The main body of the SPAR platform is a large diameter floating columnar structure with irregular shape and it extend into deep water. The SPAR platform main body consists of columns and slabs, and the cylinder can be used for oil storage. Its large draft can have a good protection for the riser, while the movement response is not sensitive to the variation of water depth. Compared with TLP, the horizontal movement of SPAR platform can be controlled to some extent. The SPAR platform is considered to be another deepwater platform and has a good development prospect.

The SPAR platform is particularly suitable for deepwater operations, and it has movement stability and high security. Under the effect of the mooring system and the main buoyancy, the natural period of SPAR platform movement in six degrees is far away from the common marine energy concentrated band, and this lead to a good sport performance.

The SPAR platform has good flexibility. Because it is fixed by mooring cable system, the SPAR platform is very easy for towage and installation. After removal of the mooring system, the SPAR platform can be directly transferred to the next location to continue in service. The SPAR platform is especially suitable in the marine areas where the oil spots are more dispersed. Compared with the fixed platform, the SPAR platform is fixed by the mooring cable system, the costs of SPAR platform will not rapidly improve with the water depth increasing.

Fig. 4 Installation of the SPAR platform topside

In the mode based on the SPAR platform, the fatigue of wellhead riser and its support is quite serious. Because the platform rotation and risers rotation can be in the opposite direction, the fatigue of the riser system with the bottom of its support should be a major controlling factor. Fatigue of buoy with its support is also quit easy. The cylinder is easy to get vortex-induced vibration, leading to fatigue of buoy, risers and mooring system. In the installation of SPAR platform, large offshore lifting boat is needed to install the topside of the platform. See Fig. 4.

5 DEVELOPMENT MODE BASED ON COMPLAINT TOWER

In the development mode based on complaint tower, there are several forms. One form requires the complaint tower, the wet trees and subsea pipelines. Another form consists the complaint tower, the subsea production system and pipelines. The complaint tower can also match FPSO, and the production can be transported by shuttle tanker.

The complaint tower consists of a long narrow tower and pile foundation. Unlike fixed platforms, complaint tower is able to withstand greater lateral offset, so it can bear greater lateral force. Typically, complaint tower can be used in water depths between 305m and 915m. Similarly to the jacket platform, the lower part of complaint tower is a slender tower-like structure.

The construction of complaint tower just needs simple, conventional manufacturing methods, and installation method is flexible. Compared with conventional platform, complaint tower has tilted legs and the amount of needed steel is reduced. It has robustness to the changes of the payload.

6 TIE-BACK DEVELOPMENT MODE

The offshore field can also be tied back to existing facilities. In this development mode, the production from the subsea well is transported to infrastructure of nearby oil field by subsea pipeline. For offshore oil field development, the operators just make appropriate modifications for the infrastructure, eliminating the constuction of new production facilities, significantly saving investment costs. The tie-back development mode is a cost-effective way of offshore oil field development, especially for offshore oil field with smalle reserves. See Fig. 5.

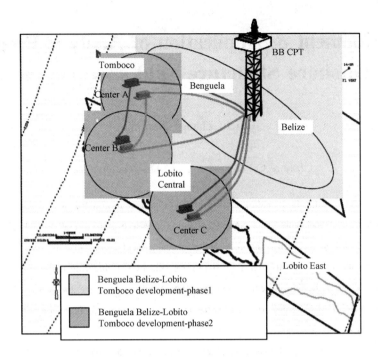

Fig. 5 Development mode based on complaint tower

CONCLUSIONS

The development mode based on FPSO need short construction period and low costs. The development mode based on semi-submersible platform saves the offshore installation cost and time, and it allow pre-drilling to put into production in advance. The development mode based on TLP has strong resistance to adverse environmental effects, and is especially suitable for deepwater field and has good security. The FPSO, semi-submersible platform, TLP and SPAR platform can be reused. Complaint tower can be used in other purposes in addition to be a production platform. In the tie-back development mode, the sea conditions has little influence on the wellhead facilities and the maintenance costs are low. The operator should choose appropriate development mode according to the specific condition of the oil field.

REFERENCES

Cees Dijkhuizen, Ton Coppens and Peter van der Graaf. 2003. Installation of the Horn Mountain SPAR Using the Enhanced DCV Balder[C]. Offshore Technology Conference.

Jean-Louis IDELOVICI and Jean-Pierre ZUNDEL. 2004. GIRASSOL: Two Years after First Oil Active Development Activity with Record Production[C]. Offshore Technology Conference.

McNeilly CC, Redwine RV, Higgs WG. 2006. Benguela-Belize Project Overview[C]. Offshore Technology Conference.

Rajasingam DT and Freckelton TP. 2004. Subsurface Development Challenges in the Ultra Deepwater Na Kika Development[C]. Offshore Technology Conference.

Richard D'Souza and Shiladitya Basu. 2011. Field Development Planning and Floating Platform Concept Selection for Globle Deepwater Development[C]. Offshore Technology Conference.

Development & Application of Analysis Program for Offshore Structure Skid shoe engineering

Huichi Zhang, Nan Wu

Offshore Oil Engineering Co., Ltd., Tianjin, China

Abstract As an important loading transmitting component in construction and loadout process of offshore structure, skidshoe bears great bending, shearing, pressure and kind of other loadings during its serving period. Based on mechanics theoretic and practical project model, in compliance with related international codes and standards, an analysis program for skidshoe engineering is developed in this paper, which utilizes Excel computation sheet's powerful data-handling capacity and Visual Basic's friendly consistent interfaces. This program considers extreme load conditions for both construction and loadout process in analysis.

Key Words Skidshoe; Manual calculation; Program development

INTRODUCTION

In the development of offshore oil industry, it is evidently that offshore structure is much more huge than before gradually, which make skidding become one of the most popular loadout methods. Engineering of skidshoe is much important to ensure a safety skidding.

1 BACKGROUND OF PROGRAM DEVELOPMENT

Skidshoe is an important loading transmitting component in construction and loadout process of offshore structure. Usually manual calculation or FME software is used for skidshoe structural strength analysis. Manual calculation analysis require a long time and the result is not obviously to get for a judgment. FME software such as ANSYS requires the designer has profound mechanics theoretic knowledge. From modeling, meshing to solution, the entire analysis process of ANSYS is complex and also a long time. In addition, abrupt stress used to occur at concentration force work point, structural geometry abnormality point, concentration stress point and boundary constraint point. Since there is not any universal rule for FME solution accuracy judgment, FME software analysis result is not acceptant for most clients.

This paper gives a new solution for above problems. Using Excel computation sheet's powerful data-handling capacity and Visual Basic's friendly consistent interfaces, a new analysis program is developed for skidshoe engineering. This program gets both maneuverability and practicability in it with pellucid interface, which makes engineering of skidshoe be simple, accurate with high efficiency. Besides those, this program adopt related international codes and standard (AISC, DNV standard etc.), which make it well accepted by clients.

2 INNOVATION OF THIS PROGRAM

(1) Parameterize: For different project, user only need to input different parameter to get the skidshoe analysis result and judgment. Program will generate calculation report automatically, which improve work efficiency and accuracy greatly.

(2) Modularize: According to the structure characteristic of skidshoe, this analysis program is divided into 4 modules. It is simple for users to detect the dangerous status and find the origin, which helps getting a correct solution.

(3) Universality: Result of this program also could be used as a reference for box girder, spreader beam strength analysis. Related modules of this program could be directly used for offshore platform weighting structure stiffener (or other similar structure) bearing and buckling strength analysis.

3 MECHANICAL SIMPLIFIED MODEL OF SKIDSHOE

This segment introduces the design and establishing process of mechanical simplified model of the skidshoe (see Fig. 1).

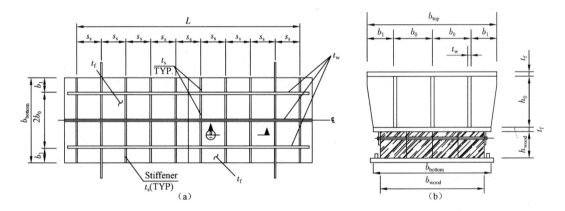

Fig. 1 Structural sketch of the skidshoe

4 STRENGTH CALCULATION, RIGIDITY CALCULATION AND STABILITY CALCULATION FOR SKIDSHOE

Skidshoe is simplified to box girder when global calculation because its length is far greater than width and height. The leg reaction acting on the skidshoe used to be simplified to concentrated force. For continuous skidshoe and large span skidshoe, the concentrated force method can affect local stress of the skidshoe but can be ignored in integral calculation. However, for small span skidshoe, this method is too conservative. Here, leg reaction is simplified to concentrated force, and corresponding calculation reduction coefficient of concentrated force for different span skidshoe is recommend for user to choose, according to project experience and related specification.

(1) Extreme load condition 1:

When structure fabricated onshore and sliding on skidway, besides the leg reaction, the skidshoe also bears the reaction force of the skidblock. For simplification, this force is simplified to distributed reaction, it is shown as following(see Fig. 2).

(2) Extreme load condition 2:

During loadout, the condition that one side of the skidshoe is on barge and the other side is on quay will be appear because of barge ballasting and loadout is not synchronic. For conservative calculation, make the skidshoe supported at both end, and reduce the leg reaction according to project experience, which make this model be closer to practical situation(see Fig. 3).

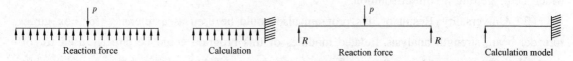

Fig. 2　Calculation model for load condition 1　　　Fig. 3　Calculation model for load condition 2

Inertia moment, section modulus, section of shear and other parameters can be calculated according to the actual sizes of skidshoe. Maximum bending moment and shear force of two extreme load condition can be calculated using the above calculation model. So it can justify if the skidshoe satisfies the requirement of integral strength, rigidity and stability.

$$y_0 = \frac{\sum A_i y_i}{\sum A_i}$$

$$I = \frac{b_{top} t_f^3}{12} + b_{top} t_f (y_{top} - y_0)^2 + n\frac{t_w h_0^3}{12} + t_w h_0 (y_{web} - y_0)^2 + \frac{b_{bottom} t_f^3}{12} + b_{bottom} t_f (y_{bottom} - y_0)^2$$

$$W_z = \frac{I_z}{y_{max}} \quad [\sigma] = 0.66\sigma_y \quad [\tau] = 0.4\sigma_y \quad \sigma = \frac{M_{max}}{W_z} \leq [\sigma] \quad \tau = \frac{Q_{max}}{A} \leq [\tau]$$

5　BEARING STRENGTH CALCULATION FOR BOTTOM PLATE OF SKIDSHOE

Skidshoes bear heavy load in construction and loadout period. Here, it is simplified to distributed force, and load magnification factor is recommended. At the same time, using the most dangerous plate from the structure formed by web and stiffener plates as the research object, carries on the analysis. As the bottom plate thickness is much larger, the former is considered to be more rigid, Taking out the bottom part as subjected to transverse distributed load and boundary as hinge point, the calculating model is as follows(see Fig. 4).

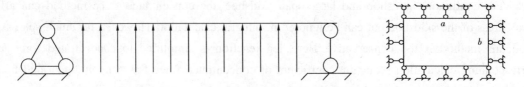

Fig. 4　Calculation model of skidshoe bottom plate

$$\sigma'_{max} = \frac{-\beta_1 q b^2}{t_f^2}$$

$$q = \frac{P}{LB}$$

a/b	1.0	1.2	1.4	1.6	1.8	2.0	∞
β_1	0.3078	0.3834	0.4356	0.4680	0.4872	0.4974	0.5000

6 BEARING STRENGTH CALCULATION FOR WEB PLATE AND STIFFENER PLATE

The reaction forces of structure will be transferred to the skidshoe through the socket. Socket contact the upper flange plate of skidshoe directly, the forces will be transferred through the upper flange plate to the stiffener plate, web plate. The contact point between socket contact and skidshoe bears great pressure. The pressures is assumed to be distributed in each point, while the load coefficients are given for users to select in specific analysis (see Fig. 5 and Fig. 6).

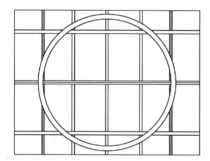

Fig. 5　Bearing mechanical model for skidshoe bottom plate

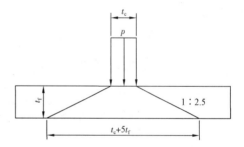

Fig. 6　Bearing stress transmitting route

Lateral buckling calculation for webs and stiffeners

According to standard DNV 30.1 1995, if the thickness of plate meet $\frac{s}{t} < \frac{2}{3}\sqrt{C\frac{E}{\sigma_k}}$, the buckling calculate is not required. First of all, the thickness of plates are judged, if meet the standard requirements, buckling calculate is not required; otherwise, calculations are carried on according to the DNV standard.

(1) Webs (see Fig. 7).

distance between stiffener plates ⩽ height of web plate: $C = 4.0$.

distance between stiffener plates ⩽ height of web plate: $C = \left(1 + \left(\frac{h_0}{s_s}\right)^2\right)^2$.

(2) Inner stiffener plate (see Fig. 8).

distance between web plates ⩽ height of web plate: $C = 4.0$.

distance between web plates ⩾ height of web plate: $C = \left(1 + \left(\frac{h_0}{s_s}\right)^2\right)^2$.

(3) Outer stiffener plate (see Fig. 9).

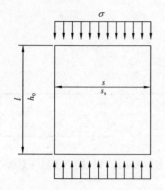
Fig. 7　Lateral bearing mechanical model of webs

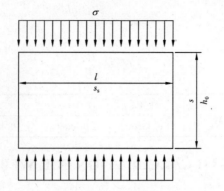
Fig. 8　Lateral bearing mechanical model of inner stiffener

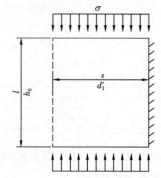
Fig. 9　Lateral bearing mechanical model of outer stiffener

distance between stiffener plates ≤ height of web plate: $C = 1.28$.

7　APPLICATION OF ANALYSIS PROGRAM

This program, based on VB language, is developed mainly for checking skidshoe in construction field (see Fig. 10).

Fig. 10　Main user interface of program

(1) Strength, Rigidity And Stability Checking Of Skidshoe (see Fig. 11).

For different projects, users only need to input relevant data for skidshoe checking, such as skidshoe size and reaction force of leg, have a choice in load condition and corresponding reduction factor. Then, the result of checking for skid-shoe is provided.

(2) Lateral Buckling Checking Of Webs And Stiffeners (see Fig. 12).

For different projects, users only need to input relevant data, such as size of compressed

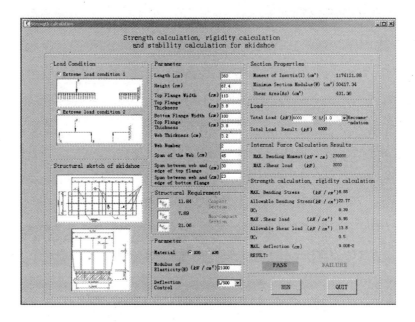

Fig. 11 User interface of checking calculation

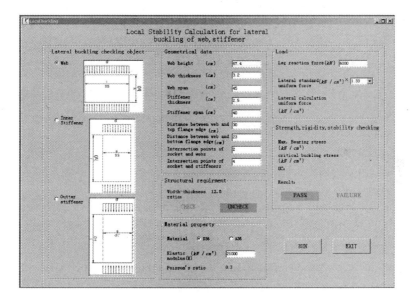

Fig. 12 User interface of lateral stability calculation

member and reaction force of leg, and have a choice in lateral compressed model and disturbing load factors. Then, the result of checking for skidshoe is provided.

(3) Bearing Strength Checking Of Bottom Plate (see Fig. 13).

For different projects, users need to input relevant data, such as span of webs and stiffeners, skidshoe parameter and reaction force of leg, have a choice in disturbing load factors. Then, the result of checking for skidshoe is provided.

(4) Bearing Strength Checking Of Webs And Stiffeners (see Fig. 14).

For different projects, users need to input relevant data, such as wall thickness of shoe, number

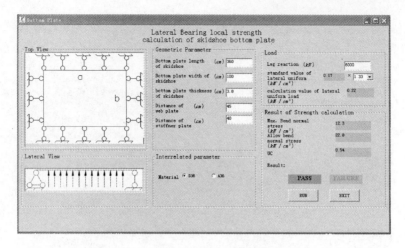

Fig. 13　User interface of local bending calculation

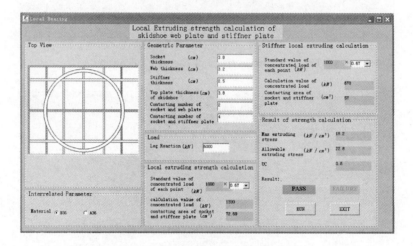

Fig. 14　User interface of bearing calculate

of points of contact of shoe and webs, skid-shoe parameters and reaction force of leg, have a choice concentrate load factors for each point. Then, the result of checking for skid-shoe is provided.

(5) Calculation Report (see Fig. 15 and Fig. 16).

This program includes various calculating sheets. Users need to input relevant data, and the calculation report of skidshoe will be generated automatically.

CONCLUSION

According to current using situation of the skidshoe in Tanggu field, the article chooses uniform section skidshoe as the research project based on engineering practice, combining with theoretically analyzing, work experience, vocational and international guide and standards, using Excel computation sheet's powerful data-handling capacity and Visual Basic's friendly consistent interfaces, developed the skidshoe designing and analyzing program which solved the problems of manual calculations and ANSYS analyzing software with the advantages of high efficiency and preciseness, bringing a breakthrough solution! The skidshoe analysis program improve the skidshoe calculation efficiency and accuracy!

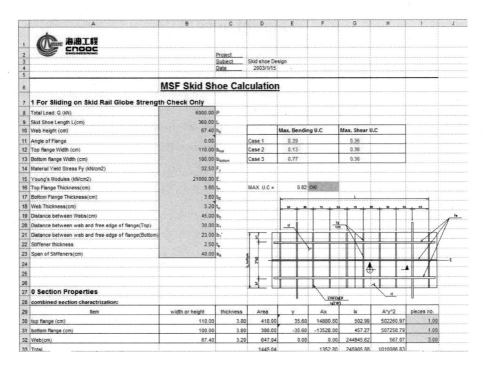

Fig. 15 Calculation report

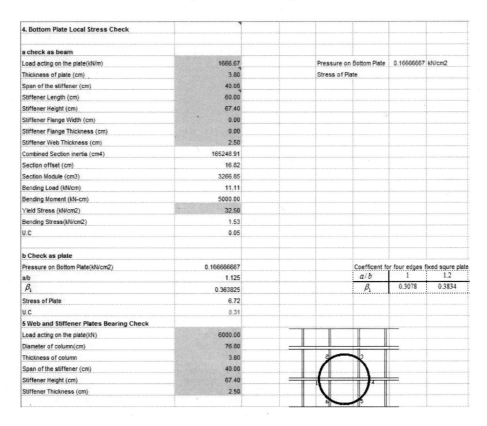

Fig. 16 Calculation report

REFERENCES

AISC 360-2005. Specification for structural steel buildings. American institute of steel construction,INC,2005.
Det Norske Veritas. 1995. Det Norske Veritas ship classification specification DNV 30. 1 1995.
Guo Qi. 2003. Visual Basic database systems development technologies (with CD-ROM)[M]. Beijing: People's Posts and Telecommunications Press.
Stephen Prokofievitch Timoshenko. 1979. Mechanics of Materials[M]. Beijing:Science Press.
Xu Zhilun. 2006. Easticity (volume 2)[M]. Beijing:Higher Education Press.

Diving Engineering —Analyses and Challenges of Air Surface-Supplied Diving on DP Vessel

Dawei Zhang, Yong Chen

COOEC, Tianjin, China

Abstract Safety risk and control methods of air Surface-Supplied diving on DP vessel are discussed in this paper. The diver's umbilical control is emphasized. In order to extend the working radius from in-water bell, a Mid-Tending-Point is set, to assist controlling the diver's umbilical. The plan would be significant to solve the analogous difficulty in offshore diving project.

Key Words DP; Surface-Supplied; Uncontrolled; Umbilical; Mid-Tending-Point; Physical hazard

INTRODUCTION

Dynamic positioning (DP) vessel can automatically maintain its position and heading by using her own propellers and thrusters. This allows operations at sea where mooring or anchoring is not feasible due to deep water, congestion on the sea bottom (pipelines, templates) or other problems. It may also position at a favorable angle towards wind, waves and current. DP vessel takes a more and more role in the offshore oil industry following the ocean development. Air Surface-Supplied diving off of a DP vessel is indispensable to accomplish the offshore project.

1 THE REQUIREMENT TO DP VESSEL WHEN AIR SURFAC-E-SUPPLIED DIVING IMPLEMENTATION

The DP vessel positions by that dynamic positioning system respond to environmental situation automatically. Losing station is the risk that must be considered. No effort should be spared to establish DP operational reliability and to avoid the possibility of that the DP vessel lose station. Any single unit that its failure could cause the vessel to move from the intended position should be 100% redundancy. And ensure that no single fault, especially the fault of the thrust units, should cause a catastrophic failure which endangers the diver in the water. But anytime, the conduct of diving operations from DP vessels, as opposed to other types, requires particular attention to be paid on the risk due to vessel uncontrolled. Not only meets the routine diving regulation and standard requirement, but also avoids the risks that are from the DP vessel itself as far as possible.

The characteristics of each vessel will affect its suitability for particular operations. Even in the short term, this may alter in the light of changes in personnel and system components. It is therefore important that not only the potential charterer but also diving contractor assess vessels suitable for their particular needs. A fundamental principle of DP diving vessel operation is that the operating requirements of the system are never allowed to exceed the vessel's capabilities in any respect.

Before a DP diving vessel undertakes DP diving operations it should undergo a full series of trials and eliminate all the faults. Otherwise all the precautions and procedures will be to no avail. The trials should include testing and tuning in harbor, followed by sea trials, during which the vessel's position keeping system should be thoroughly tested under normal and breakdown conditions, and should culminate in a DP bell dive.

The dive work must be conducted from a Class II bell (Wet Bell). Two divers are launched into water by the bell at the same time. One diver gets out carrying out the task, and the other is in the bell as the umbilical tender and the standby. The lowest people allocated for air supply diving on DP vessel are as follow: one diving supervisor, one life support technician, two divers, one standby diver, three tender.

As a minimum requirement, voice communications should be available to ensure the immediate and clear transfer of information between all responsible parties. Direct communications should be provided between DP console and dive control; dive control and bell handling control; dive control, DP console, and ship's derrick or crane. All essential voice communications systems should be provided with 100% redundancy where practicable either through duplication or provision of an alternative system. An alarm system of lights and sounds shall be provided in diving control area. And provision of a means of cancelling the audio and flashing functions of the signals from the receiving positions when they have been noted should be made.

More attention should be paid on the underwater diver's safety when DP vessel moves. A diving support vessel under stable DP control may execute changes to a previously agreed position or heading without recalling the divers to the deployment device, provided all relevant personnel have been advised, and must be execute under automatic DP Control. It is recommended that the position moves and heading changes should be executed in limited steps, e. g. maximum steps of 5metres and maximum heading changes of 5 degrees. Heading changes in all should not exceed 15 degrees or causes a 10 meters movement of the bell.

2 DIVER'S UMBILICAL CONTROL PLAN

Very great care is needed in the planning and execution of air supply diving operations to minimize the effect of thrust units on the divers. The effects of thrust unit wash or suction should be carefully considered and precautions taken to guard against them. The requirements are based on the premise that at no time should the length of umbilical from the tending point to the diver allow the diver to come into contact with the neatest thruster or propeller that is in an operating mode. The implementation way is that deselect certain thrusters or chose reasonable divers' umbilical lengths and the manner of deploying. Anytime, an umbilical control plan is necessary. The umbilical length need to be controlled as follow:

$$c^2 = a^2 + b^2 \tag{1}$$

where, a is horizontal distance from the in-water bell to the nearest physical hazard (thruster). b is vertical distance from the in-water bell to the nearest physical hazard. c is distance from the in-water bell to the nearest physical hazard (thruster).

The length of the working diver's umbilical must be restrained in such a way that it cannot reach to within six meters of any physical hazards (such as thrusters, propellers etc.). The reach of bellman/standby diver's umbilical is restrained such that three meters longer than that of the working diver's umbilical, to provide maneuverability but cannot reach to within three meters of any physical hazard. That means that the length of diver's umbilical must be less than c-6 meters, and the bellman/standby diver's umbilical must be less than c-3 meters. See Fig. 1.

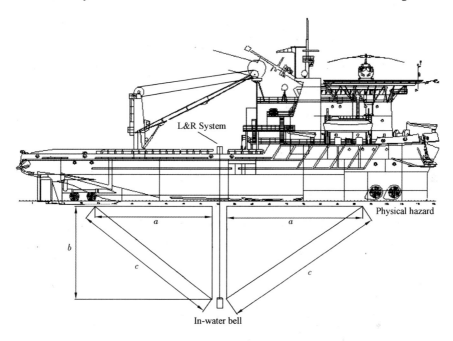

Fig. 1　Calculating safe umbilical lengths

3　DIVING WITH THE UNMANNED IN-WATER MID-TEND-ING- POINT

The job of air supply diving on DP vessel may be done close or even under the platform, such as jacket installation, pipe line and riser installation and the possible maintenances, renewal, survey, inspection project and so on. But in some case the typical restrictions such like platform overhangs, flare towers, lifeboats and bridges may stop the DP vessel close the work sites. When access to certain work sites is restricted it may be necessary to deploy divers beyond the pre-determined safe umbilical length as determined from the distance between the deployment device and the nearest hazard, e. g. thruster. This paper fuses the diving regulation and the practical experiences of the diving industry internationally, and details the ways of extending umbilical that can be safely achieved with the Unmanned In-Water Mid-tending-point (See Fig. 2). The process is as follow.

$$d \leqslant c = a^2 + b^2$$

$e <$ distance from the Unmanned In-Water Mid-Tending-Point to the nearest physical hazard
　　a—horizontal distance from the in-water bell to the nearest physical hazard (thruster); b—

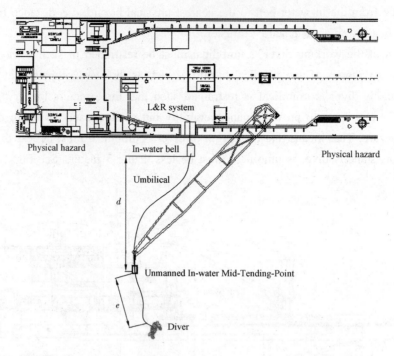

Fig. 2　The Unmanned In-Water Mid-Tending-Point

vertical distance from the in-water bell to the nearest physical hazard; c—distance from the in-water bell to the nearest physical hazard (thruster); d—distance from the in-water bell to the Unmanned In-Water Mid-Tending-Point; e—distance from the Unmanned In-Water Mid-Tending-Point to the diver

In order to the diving operation can be achieved safely, the distance d must be less than c-5 meters, and the distance e as well as the diver's umbilical length that stretched out from the Mid-Tending-Point must be 5 meters at least less than distance from the Unmanned In-Water Mid-Tending-Point to the nearest physical hazard. The bellman/standby diver's umbilical should be 2 meters length more than diver's.

The Unmanned In-Water Mid-Tending-Point can utilize a device (basket or similar). It must be suspended from crane or winch on the vessel so that hold its position relative to the vessel, especially in the case of uncontrolled movement of DP vessel.

First step the diver reaches the Unmanned In-Water Mid-Tending-Point and fix a swim line between the bell and the Mid-Tending-Point. It can hold the position of Mid-Tending-Point relative to the bell and direct for the diver and standby diver(see Fig. 3).

Then the diver secures his umbilical to the Mid-Tending-Point so that the umbilical can be held and cannot be involved to physical hazard by the current or other cause(see Fig. 4).

The diver crosses the Mid-Tending-Point and carries out his work on the work site after all the preparatory work being finished(see Fig. 5).

The Unmanned In-Water Mid-Tending-Point is a way that deploys divers beyond the pre-determined safe umbilical length as determined from the distance between the deployment device and the nearest hazard, e. g. thruster and finishes the diving project. Meanwhile it is a crapshoot. Much

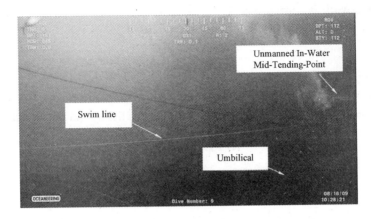

Fig. 3　Project photo-1

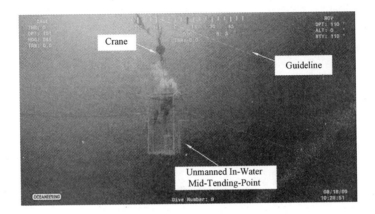

Fig. 4　Project photo-2

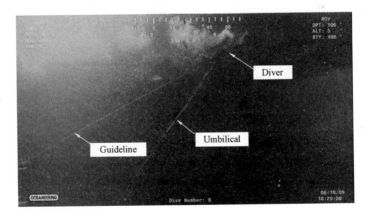

Fig. 5　Project photo-3

attention should to be paid on the risk analyses and challenges including the emergency procedure when DP vessel is uncontrolled, the minutes of diver emergency recover being extended, rescue becoming more difficult. Here needs industry insiders never stop researching and practicing to stipulate the umbilical plan that can keep diver being rescued in any emergency condition. In addition, a strong ability and skill of diver is necessary. It is important that much training and practicing to the divers before the work is carried out.

4 TAG

Now there lack the regulation and standard about diving on DP vessel in Chinese diving industry. We hope this paper would provide a little help for concerned project. Standardization of the dive on DP vessel needs to continue researching, testing, practicing and accumulating.

REFERENCES

CCC. 2004. (Underwater Engineering) Diving Operation& Safety Manual.
CCS. 2002. Dynamic Positioning System Test Guide.

Economic Analysis on Reelwell Drilling Method In Deep Water

Baojiang Sun[1], Guangai Wu[2], Linlin Yan[1], Zhankui Shang[3]

[1] School of Petroleum Engineering, China University of Petroleum, Qing dao, China;
[2] CNOOC Research Institute;
[3] Shanghai Zhenhua Heavy Industry (Group) Corporation Limited

Abstract Reelwell Drilling Method (RDM) is a reelwell drilling technology with dual concentric drill strings and supporting equipments. Based on the characteristics of deep water drilling technology, this paper presents the potential advantages of this technology and analyzes the technical feasibility and economics of this new method in deep water drilling. As a new technology, it has a series of advantages, such as wellbore structure simplified, expansion liner, lower load requirements of platform, decreases the cost of deep water drilling with less consumption of drilling fluid and casing materials, and increases the drilling efficiency. Therefore, the prospect of this technology is positive.

Key Words Deep water; Dual concentric strings; Riserless drilling; Control pressure drilling; Economic analysis

INTRODUCTION

Oil and gas exploitation in deep water facing many challenges, mainly in the following four aspects. First, the large heavy risers lead to construction difficulties and increase the use of drilling fluid. Second, the harsh environment (such as wind, waves, current, etc.) affects lifting and running risers, so the operation cost increases. Third, more accurate control of the wellbore pressure is necessary due to the narrow pore-fracture pressure window, high-pressure shallow water/air flow, seabed shallow geological disasters and other factors. Fourth, horizontal wells and extended reach drilling (ERD) are widely used of in deep water oil and gas exploitation which makes it difficult to impose weight on bit (WOB) and carry cuttings. The above challenges not only affect drilling safety and efficiency, but more likely extend operating time and increase the total costs. Therefore, it is necessary to research new equipments and technology to ensure deep water drilling efficiency and security.

Reelwell Drilling Method (RDM) Using Dual Concentric Strings is a new drilling technology. The RDM concept was first proposed by Statoil and the Norwegian Research Council in 2004. Thereafter, the feasibility study of RDM equipment and principles was carried out by Norwegian Research Council and StatoilHydra (in 2005), and the R & D and testing of Reelwell drilling key equipment was completed by JTP, StatoilHydro, Shell, and Demo2000 (in 2006). Then, the full-size model was researched on "Ullrigg" rig in Stavanger, Norway (in 2007), and the actual device was tested and performance good on "Ullrigg" rig (in 2008). In 2009, the drilling of test wells obtained success on land and at sea, which proved RDM technology to be an effective tool for managed pressure drilling, extended reach drilling, deep water drilling and challenging formation drilling. Soon after this new technology won the OTC Spotlight on New Technology Award nominations (in 2009), it was promoted and applied in the field (in Calgary, Canada, 2010) and received recognition of RC Energy and Nabors Canada. So far, Reelwell technology has not been used in the practical production.

RDM using dual concentric strings is mainly based on the flow control of drilling fluid. The key equipments include dual concentric strings, top injection joint, sliding piston, flow converter, dual concentric shut-off valve, etc.

Dual concentric string is the crucial tool for reverse circulation, and the drilling fluid was injected into and back out from its annulus and inner pipe. The top injection joint, installed between the top drive and dual concentric strings, deals with the drilling fluid injecting and cuttings discharging when drilling. Fixed on the outside wall of the dual concentric string, the piston can move with drill pipe (the pipe will be rotate when drilling) and seal upper wellbore annulus together with the underwater blowout preventer (BOP) stack, the flow converter, the rotating control head and the intermediate casing. If high pressure fluid is injected into the sealed annulus, the pressure can be converted into weight by the seal and be imposed on the bit. The drilling fluid carrying cuttings goes through the flow converter near the bit and flow back through the inner pipe of the dual concentric strings. While stopping the pump, the shut-off valve shuts down the annulus and the inner pipe flow channel, so that bottom hole pressure is isolated.

During drilling, the mud pump injects drilling fluid through the top injection joint bypass into the dual concentric string annulus. Ejected from drilling bit, the drilling fluid carrying cuttings goes through the flow converter into the inner pipe of dual concentric string, then discharged from the wellhead. After flowing through choke manifold, the drilling fluid is treated by disposal system (vibrosieve, knockout drum, etc.) and returns to the mud pit, finally achieve the closed circulation.

Compared with conventional drilling methods, RDM using dual concentric strings has the following advantages. Riserless drilling can reduces the requirements of deck space and its carrying capacity. In conventional offshore drilling, drilling fluid is pumped through drill pipe, carrying cuttings and flowing along the annulus and riser annulus. While in dual concentric strings drilling, drilling fluid is pumped through the annulus of dual concentric string into the well, and back to the wellhead along the inner pipe of dual concentric string after carrying cuttings. The high return speed also solves the problem of cuttings carrying in large diameter risers. Riserless drilling can save space and reduce the weight of platform. For extended reach wells and horizontal wells, it is hard to impose WOB effectively due to the friction between the drill pipe and the borehole wall, which lead to low drilling speed and low efficiency. This problem can be properly solved in the dual concentric strings drilling. The sliding piston is fixed on the outer drill pipe, and form the sealed annulus together with underwater BOP, RCD and wellbore. The underwater injection system injects heavy weight drilling fluid to the sealed annulus, and gives a thrust to the piston. The pressure transmits to the drilling bit through the lower part of the drill pipe, thus effectively imposed

WOB. On account of the closed circulation path of drilling fluid, the dual concentric strings drilling can control pressure in the wellbore while drilling. The drilling cost is reduced with riserless drilling, less drilling fluid, simplified wellbore structure, lower load requirements of drilling platform.

1 TECHNICAL FEASIBILITY

Dual Concentric Strings drilling system mainly consists of the following parts:

(1) Top Injection Joint. Installed between the dual concentric string and top drive, the top injection joint is the essential equipment in RDM. Drilling fluids are injected from top injection joint by-pass into the dual concentric string annulus, and then discharge through the inner pipe with cuttings.

(2) Dual concentric string, also called concentric dual string. The outer pipe consists of API standard pipe ϕ127 mm or ϕ168 mm, and the inner pipe is aluminum pipe designed on the basis of RDM drilling theory. When making a connection, both the joint inner pipe and the joint outer pipe are placed on the drill rack, then lifted and connected like traditional drilling.

(3) The upper BHA. Installed on the drill string about 400 meters above the lower BHA, it is composed by sliding piston and other devices, It can be used to control bottom hole pressure and WOB. The sliding piston is fixed on the outside wall of dual concentric string, moving with the string, allowing string to rotate in the hole that has been casing. The piston can seal the annulus, and the high pressure fluid which is sealed in annulus act on the bit indirectly, giving the bit hydraulic drilling pressure. Additionally, the sliding piston can also be used to liner expansion because of enough traction it provided.

(4) The lower BHA, including some conventional modules, dual concentric shutoff valve and flow convertor, etc. The shutoff valve is fixed under the dual concentric string, it can close or open dual channels simultaneously or individually. When the bottom overflow occurs, we can prevent fluid into the inter pipe and the annulus via opening the valve. If meeting with leakage formation, it can isolate bottom by stopping drilling and opening the valve, thus, a series of complicated underground accidents can be avoided. In short, the valve is mainly used to assure drilling safety. Furthermore, if the annulus shut-off valve is closed individually and appropriate drilling pump is chosen, the hydraulic fracturing test can be easily achieved in the hole directly.

(5) Intermediate liner, a liner of any length located between the upper BHA and the lower BHA. It will be expanded in the hole after drilling out new section.

(6) Surface data acquisition system and sensing device, containing pressure sensor and flowmeter. The drilling fluid inflow flowmeter is installed on riser, and the outflow flowmeter is installed in the place of return pipeline choke. The control system can accomplish some basic automatic diagnosis by computers.

(7) Wellhead pressure control equipment, including swivel control head and the conventional BOP.

(8) The upper annulus control system.

Flow control unit. There is a special valve used to control the amount of drilling fluid inflow and outflow, for the purpose of precise pressure control when drilling and making connection. The pipelines where drilling fluid flow in and flow out all reduced the drilling cost. Furthermore, efficiency and recovery ratio improve significantly by the usage of cross practices (such as drilling with expansion liner and fracturing without lifting) and MPD method.

(9) Exterior are equipped with pressure sensor and flowmeter, meanwhile, computers are used to monitoring the hole conditions at remote.

(10) Solid control system and other auxiliary equipments

The basic process of dual concentric strings drilling can be explained as follows: ① To hang the liner on wellhead; ② To lay down lower BHA and dual concentric strings, and connecting with liner bottom; ③ To lay down upper BHA and connecting with liner top; ④ To run drill string and start drilling; ⑤ To boost pressure on annulus fluid which is above the sliding piston, in order to control bottom hole pressure, to impose WOB and to advance drill pipe; ⑥ To break the liner away from BHA through the hydraulic or mechanical action when the new drilling section reaches the liner length, and then finish the expansion operation; ⑦ To trip out; ⑧ To repeat the process until drilling to designed depth.

The dual concentric strings drilling system provides a closed-circuit circulation system, and can use computer to control downhole pressure and the amount of drilling fluid accurately. If bottom hole pressure is higher than formation fracture pressure, we can open the throttle to reduce back pressure. However, if bottom hole pressure is below the formation pressure, we can close the throttle to increase back pressure. This technique can ensure a constantly steady pressure when stopping the pump or making a connection, so as to avoid complicated accidents caused by pressure fluctuation. The property of closed-circuit circulation makes RDM system break the limits of the conventional managed pressure drilling (MPD), the underbalanced drilling (UBD) and external reach wells drilling (ERD). Particularly, this system can also be applied in deep water drilling.

In the aspect of drilling equipments, the dual concentric strings drilling system has many differences with conventional drilling technology in the top drive adapter, sliding piston, dual float valve, flow converter, intermediate liner and surface hydraulic system, but other equipments including the rig, pipes, BHA, the bit, throttle device, well control equipment, drilling pump, solid control system and auxiliary equipment can use the conventional drilling equipments. The cost is not expensive even if improving parts of the equipments. Therefore, as long as the mentioned key equipment technology solved, dual concentric strings drilling can be easily achieved.

2 ECONOMIC ANALYSIS

A series of new technologies are introduced to the dual concentric strings drilling system, and this reduces the cost of deep water drilling and enhances the oil field development economical efficiency, which can be illustrated in the following aspects:

(1) Cost saving. Dual concentric strings provide the in-flow and out-flow channel for drilling fluid, so the riserless drilling can be achieved. With the tremendous-heavy riser replaced by drill pipe, the expense on riser can be saved. Besides, the platform deck load decreases and the operation water depth increases, making it possible that the platform which lacks the capacity before can also operate in deepwater areas, and thus the platform's daily rent reduces.

The reduced cost is:

$$P_1 = (\Delta P_{pd1} - \Delta P_{pd2}) \times d \tag{1}$$

where ΔP_{pd1} is platform's daily rent of conventional drilling method, ΔP_{pd2} is platform's daily rent of riseless drilling method, d is the total drilling time.

(2) Work efficiency improved for the parallel operations. The liner and BHA can be directly jointed together and then lay down into the hole. When the new drilling section reaches the length of the liner, the liner can be released. Certain quantitative cement is then injected to the open hole section and annulus pressure increases to provide the piston thrust which promotes the expansion head carried by BHA to expand the liner and finish the cementing operation. Compared to the conventional cementing operation, this reduces a trip's time of lifting and a trip's time of casing down. Simultaneously, operation without risers saves relevant running and pulling time and cost.

The reduced cost is:

$$P_2 = P_{td} \times (t_1 + t_2 + t_3) \tag{2}$$

where P_{td} is the total daily cost, t_1 is the lifting time, t_2 is the casing down time, t_3 is the running and pulling time of risers.

(3) Wellbore structure simplified and the cost of materials reduced. Because liner drilling technology can be used, the hole diameter will remains the same along a longer interval and wellbore structure is simplified for the expansion liner drilling is used after the small hole drilled in the surface section. And this also improves ROP, reduces the corresponding drilling time. At the same time, simplified wellbore structure also reduces the use of casing steel and cementation materials:

The reduced cost is:

$$P_3 = P_{td} \times t_4 + P_{cs} + P_{cmt} \tag{3}$$

where t_4 is the reduced drilling time, P_{cs} is the reduced casing cost, P_{cmt} is the reduced cost of cementation materials.

(4) Less consumption of drilling fluid and other materials reduces total cost. The required amount of drilling fluid decreases greatly owing to the introduction of riserless drilling technology and the single diameter wellbore structure. In addition, the sliding piston separates the fluid in the hole annulus from the circling drilling fluid, which saves approximately another 50% drilling fluid. If well killing is needed, the kill fluid consumption is reduced as well.

The reduced cost is:

$$P_4 = P_{dpu} \times \Delta V_d + P_{kpu} \times \Delta V_k \tag{4}$$

where P_{dpu} is the cost of drilling fluid per unit, ΔV_d is the volume of reduced drilling fluid, P_{kpu} is the cost of kill fluid per unit, ΔV_k is the volume of reduced kill fluid.

(5) In spite of total drilling cost saving, some expenses also increase for adopting many unconventional drilling equipments, including dual concentric strings, top drive adapter, sliding piston, dual float value, flow converter, etc.

The added cost is:

$$P_{add} = \sum P_i \tag{5}$$

in which P_i represents the added equipment cost on each aspect.

Through the above analysis and comparison to the conventional deepwater drilling technology, the total saved cost reduced by dual concentric strings drilling technology equals:

$$P_{sum} = P_1 + P_2 + P_3 + P_4 - P_{add} \tag{6}$$

3 EXAMPLES

To calculate the drilling cost of a well (water depth 1,500 m, well depth 4,760 m) by both conventional method and RDM using dual concentric strings.

(1) Conventional drilling method.

Table 1 shows the design of wellbore structure.

Table 1 Wellbore structure of conventional drilling

Casing diameter (in)	Grade	Casing depth (well depth) (m)
36	X56	1,600
20	X56	2,237
13 3/8	N80	3,498
9 5/8	P110	4,755

The drilling time is 34.5 days, and the detailed process is shown in Table 2.

Table 2 Operating procedures and duration of conventional drilling

Program	Time(d)	Accumulated time(d)
Moving ships	1.5	1.5
Injecting and installing the 36in conductor	1.0	2.5
Drilling the 26in hole	1.5	4.0
Running 20in casing, cementing	1.7	5.7
Running BOP and riser	3.6	9.3
Drilling 17 1/2 in hole	3.0	12.3
Running 13 3/8 in casing, cementing	5.2	17.5
Drilling 12 1/4 in hole	8	25.5
Electrical logging	1.9	27.4
Running 9 5/8 in casing, cementing	3.0	30.4
Wiper trip, cementing qualify test	1.5	31.9
running bridge plug	1.0	32.9
Tripping out BOP	1.6	34.5

Using the 6th generation drilling platform, the rent is 515,000 $/d, the total cost of a single well drilling is 36,920,000 $. Detailed expenses are shown in Table 3.

Table 3 Operation cost of conventional drilling

Item	Cost(thousand dollars)
Rig	20,404.0
Logistics organization	3,663.0
Consumable material	2,874.0
Service	4,967.0
Formation evaluation	3,332.0
Supervision	645.0
Public administration	1,035.0
Single well development	36,920.0

(2) RDM using dual concentric strings.

Using riserless drilling technology and single diameter wellbore structure design, the result is shown in Table 4.

Table 4 The wellbore structure of the dual concentric strings drilling

Casing outside diameter (in)	Grade	Casing depth(well depth)(m)
20	X56	1,600
10¾	Expansion liner(original size 8⅝ in)	4,755

On the foundation of the research on RDM above and the actual formation property of local field, the drilling operation procedures and the corresponding construction time are formulated. The total time is 28.5 days and Table 5 shows the details.

Table 5 Operating procedures and duration of dual concentric strings drilling

No.	Operating procedure	Time (d)	Accumulated time (d)
1	Shifting ship	1.5	1.5
2	Injecting and installing the 20in conductor	1.0	2.5
3	Drilling the 12¼in hole	1.5	4.0
4	Running 10¾in casing, cementing	1.5	5.5
5	Running BOP and RCD	3.0	8.5
6	Running dual concentric strings system, drilling 12¼in hole (400 m)	0.8	9.3
7	Expanding liner from 8⅝in to 10¾in	0.2	9.5
8	Cementing	1.5	11.0
9	Repeating step 6–8 for 7 times to the designed well depth	17.5	28.5

With riserless drilling technology, requirements for platform's ability reduce. If the platform rent is 400,000 $ per day, the single well drilling cost will be 29,885,000 $. Details are shown in Table 6.

Table 6 Operation cost of dual concentric strings drilling

Item	Total cost (thousand dollars)
Rig	15,369.0
Logistics organization	3,423.0
Consumable material	2,249.0
Service	4,000.0
Formation evaluation	3,210.0
Supervision	608.0
Public administration	1,027.0
Single well development	29,885.0

According to calculation, the drilling time of RDM decreases by 6 days (17.4%) and drilling cost decreases by 7,035,000 $ (19%), compared with the conventional deepwater drilling technology.

4 CONCLUSIONS

At present, the research on the dual concentric strings drilling is basically a blank in our country. The dual concentric strings drilling technology has many advantages. With crucial dual concentric strings, drilling fluid is pumped into the hole through the dual concentric string annulus and returns to the surface through the inner drill pipe carrying cuttings, so as to realize the closed-circuit circulation.

Through comparison, we know that dual concentric strings drilling can shorten the construction period by about 17.4% and reduce the cost by about 19%, which indicates a great superiority over conventional deepwater drilling technology. Moreover, it shows great advantages in precise bottom hole pressure control, WOB control, extended reach well drilling, hole cleaning, liner drilling, deepwater drilling, the single diameter wellbore structure drilling, safety improving etc. Therefore, this new technology has great potential in deepwater drilling application.

Although this technology has obtained many achievements, it still has many deficiencies as a new drilling technology. Numerous tests and essential experiments are still necessary to find out problems and solve them, making dual concentric strings drilling technology more perfect. With the proven reserves in deepwater field increases, it will be of great significance to strengthen the research, test and application on this technology.

ACKNOWLEDGEMENTS

The above work is supported by National Nature Science Foundation Project (No. 51034007, No. 51104172, No. 51004113), State Oil and Gas Major Projects (2011ZX

05026-001-02) and Innovative Research Team in University of the Ministry of Education of China (No. IRT1086).

REFERENCES

Chen Yingjie, Ma Tiantao, Zeng Xin, et al. 2010. Foreign ReelWell Drilling Technology and Application [J]. China Petroleum Machinery, 38(8):87 – 92 (in Chinese).

MirRajabi M, Nergaard A I. 2009. Riserless Reelwell Drilling Method to Address Many Deepwater Drilling Chanllenges. IADC/SPE 126148.

Ola Vestavik, ScottKerr, StuartBrown. 2009. Reelwell Drilling Method. SPE/IADC 119491.

Sun Baojiang, Cao Shijing, Li Hao, et al. 2011. Status and Development Trend of Deepwater Drilling Technology and Equipment[J]. China Petroleum Drilling Techniques, vol. 39(2):8 – 15 (in Chinese).

Vestavik O M, Kerr S, Brown S, et al. 2009. Reelwell Drilling Method—A Unique Combination of MPD and Liner Drilling. SPE 124891.

Key Technologies and Future Trend for Deepwater Platform

Dagang Zhang

DMAR Engineering, Inc., Houston, Texas, USA
Deepwater Engineering Research Center, Harbin Engineering University, Harbin, China

Abstract The development of deepwater platform has reached a new stage in recent years with the substantial increase in water depth and various types of new structure concepts. The previous research-dominated design approach has shifted into designed-oriented engineering approach. But the researches of various new issues still play an important role in deepwater platform due to the lack of industry experience on the complexity of some new systems. The shifting of the industry approaches reflects some of the new development of the industry: (1) maturation of the industry in recent years; (2) the increasing need of quickly developing the discovered oil/gas fields; (3) substantial increased number of deepwater projects; (4) the increased complexity and the demanding of more efficient installation; and (5) worldwide participation of deepwater field exploration and production.

Technology has always been the driving force in deepwater projects. With the recent advancements in deepwater exploration, there are some changes in key technologies in deepwater. Some old challenging problems have been solved, and some new technologies have emerged. These cover the areas of design, analysis, construction, and installation. The importance of the floating system in offshore projects can be demonstrated in the following several areas: the substantial dynamic structure responses due to wave loading and current loading; the limited motion requirements of risers in deepwater; and the increasing difficulty of installation for different components of the system.

With those changes, the industry has now focused more on the integration technologies which produce the break-through for the entire system and results significant advantages or savings of project; than the used-to-be concentrating on single key technology. Some examples are: coupled time domain analysis, integrated equipment module, stress joint for risers, subsea integrated system, integrated platform for offshore installation, just name a few.

China is on the verge of quickly moving into deepwater exploration now. With the demanding of developing offshore oil filed in South China Sea, there are many new technologies to be encountered. Some of these are the current industry is facing; and more will be new challenges. To understand and to get prepared for the future development are becoming our major tasks now and in near future.

This review will target the current key technologies of deepwater platforms and future developing trend, evaluate the main tasks during the design, construction, installation, and associated major technical requirements, and address the major technical challenges to be encountered during the deepwater platform project.

Key Words Deepwater; Platform; Technology; Floating system; Installation

INTRODUCTION

The industry of oil and gas exploration and production worldwide has gone through tremendous developments for the last one and half decades. During this period, the allowable water depth for production platforms has increased from several hundred meters to several thousand meters. The active fields have also spread from a couple of concentrated areas, such as North Sea and Gulf of Mexico, to sites worldwide. This rapid expansion of the industry not only requires the advancement of the hardware in use, but also puts a high demand on the technical capability of the engineering society. Fig. 1 shows the depth of deepwater development has increased rapidly in recent years.

Floating systems are now becoming the leading tools for expanding the production of oil and gas in offshore energy fields. Most future increase of production will come from floating production systems. These floating systems range from water depths of several hundred meters to several thousand meters. Different types of floating systems have to fit into this wide spectrum of water depth. Currently, there are four major types of floating production systems: tension leg platforms (TLP), SPAR platforms, semi-submersible platforms, and FPSO. Depending on the purpose of application, these platforms can be divided into two major categories: dry tree and wet tree. Dry tree platform supports BOP on the top of deck, and can access the well same way as fixed platform does. Wet tree platform has BOP at the bottom of sea floor, and need much more complicate operation for well access. Only two deepwater dry tree platform concepts are field proven: TLP and SPAR. There are quite a few studies comparing the weight and cost efficiencies with regard to variations in payload and water depth for TLP and SPAR.

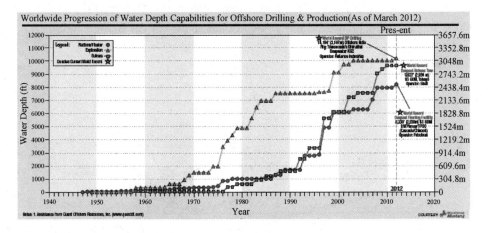

Fig. 1 Depth of deepwater development has increased rapidly in recent years

Offshore activity is being driven by growth in demand for energy, the need for new sources of hydrocarbons and advances in technology that have dramatically lowered the cost of finding and developing offshore oil and gas. Radical technology changes in the offshore sector have increased the success rate and reduced the cost of finding and developing oil by 40 percent since the early 1990s. Driving these improvements are advances in seismic technology, use of directional drilling and multiple completions, improvements in subsea systems, improved production techniques, better coordination among field operators, etc. Chevron estimates that, using Shell's Auger TLP installed in 1995 as a base, new technology and the learning curve have cut both capital intensity and construction/ installation time in half. The most challenge facing an offshore floating structure engineer now is how to produce an efficient structure in an almost always-fast track schedule project. Most of projects do not have schedule for engineering recycle, while at the mean time an efficient structure is always demanded in order to keep project cost down.

The oil field development in China offshore areas has a history of more than 40 years, over which time a vigorous oil industry has been established. Entering the twenty-first century, oil field development in China offshore areas is rapidly expanding into deepwater sites. The South China Sea is very rich in oil and gas distribution, and has been called "the second Gulf of Mexico". The conventional fixed platform models cannot satisfy the requirement of new field developments. Floating systems, such as the TLP or semi-submersible, need to be introduced into the oil field development (Zhang D, 2007).

1 TECHNOLOGY OF DEEPWATER DEVELOPMENT

With the advancement of technologies in floating system and subsea, field development techniques have made major improvement in recent years. Using different types of platforms as combined system to develop the field has gained popularity due to its economics, operability and flexibility. Such recent examples include Kizomba field development in West Africa for Exxon, West Seno field development in South East Asia for Chevron and Okume Complex in Equatorial Guinea for Hess. Fig. 2 illustrated the field layout of such a development trend: the field uses a combined system of two TLPs, one FPSO and 4 fixed platforms. The central process platform is located on one of the fixed platform, and FPSO functions as an oil output system for the processed oil.

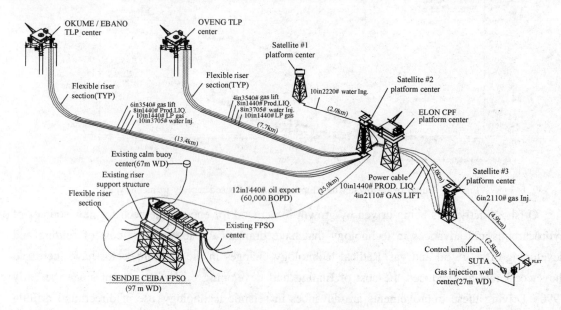

Fig. 2   Okume complex field layout

Dynamic global action is a very important aspect of floating structures. Platform is anchored to the sea bottom and needs to resist various external environmental loads, while the structure itself also takes on other complicated forces such as global split forces (or called pry/squeeze loads), global torsional forces, etc. These forces act together on the structure and constitute the challenge of the global structure design. In the design of floating system, global load plays a very important role. Almost all of the major connection areas are dominated by the global action loads, which could be generated from global pry/squeeze load, global acceleration load, mooring attachments, etc. As the column gets taller and the environmental loads get larger, the global action on the structure gets bigger and stresses get higher, thus constitutes more challenge on the structure design from both strength and fatigue aspects. Stability of the overall plate/shell structure is also affected due to the increased global stress from the larger global loads and the increased design head from higher draft. The design characteristics of some major members will also behavior differently. It is becoming more challenging for the designer to balance these design requirements and to get an optimized structure.

Offshore directional drilling is a new technology used in deepwater oil and gas exploration in recent years. Drilling normally started from sea bottom, and can be remotely controlled in different directions: vertically, horizontally, angularly, or in any desired direction. The current drilling technology can go up to 10 km. Directional drilling is an important breakthrough in offshore industry. Directional wells can increase the exposed section length through the reservoir by drilling through the reservoir at an angle. It can drill into the reservoir where vertical access is difficult or impossible. It allows more wellheads to be grouped together on one surface location which results in fewer rig moves, less surface area disturbance, and thus make it easier and cheaper to complete and produce the wells as well as reducing environmental impact. It allows drilling a relief well to relieve the pressure of a well producing without restraint (a "blow out"). In this scenario, another well could be drilled starting at a safe distance away from the blow out, but intersecting the troubled wellbore. Then, heavy fluid (kill fluid) is pumped into the relief wellbore to suppress the high pressure in the original wellbore causing the blowout. During the rescue of the Deepwater Horizon, BP PLC drilled two relief wells to prepare the intersecting of the troubled well. Fig. 3 is the illustration of the relief wells used in the operation. The wells on two ends are the relief wells.

Advanced remote control system is an important integral part of deepwater technology. Any underwater operation is always under the assistance of the remote operated vehicle (ROV), which ranges from well operation and foundation installation to the installation of risers and umbilical. The design and operation of the remote control system is significantly complicated due to the presence of harsh environment, such as strong current and turbulent waves. The precise operation requirement also imposes technical challenges. Taking a GOM accident for example, as shown in Fig. 4, ROV is used to guide delicate work such as the precise cutting and installing and tightening screws underwater 1,500 m. All the operations are controlled on the ships at water level, which are subjected to continued environmental loads.

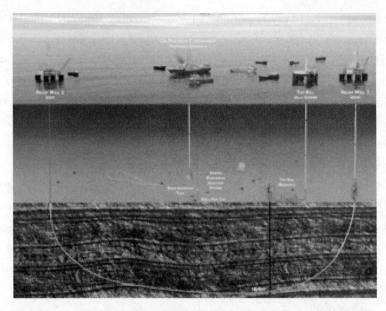

Fig. 3　Directional drilling technology used under sea bottom can reach out 10km and can be precise to a couple of feet

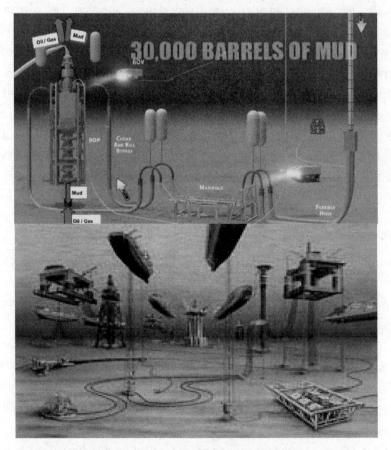

Fig. 4　ROV technology is a necessary and important part of the deepwater exploration

Advanced remote control system is an important integral part of deepwater technology. Any underwater operation is always under the assistance of the remote operated vehicle, which ranges from well operation and foundation installation to the installation of risers and umbilical. The design and operation of the remote control system is significantly complicated due to the presence of harsh environment, such as strong current and turbulent waves. The precise operation requirement also imposes technical challenges. Taking a GOM accident for example, as shown in Fig. 4, ROV is used to guide delicate work such as the precise cutting and installing and tightening screws underwater 1,500 m. All the operations are controlled on the ships at water level, which are subjected to continued environmental loads.

Tender assisted drilling (TAD) is another development trend in floating system. Floater with TAD system has significantly reduced the payload of the platform and decreased the investment on floater system substantially. This opens the door for many new deepwater field developments to use TLP and SPAR platform. The advantage of the floater combined with TAD system is more significant when several floaters are used for the continuous development of the field. One of the applications for a floater with TAD system can be at the development of offshore marginal field. Due to the increase of water depth, the conventional fixed platform model for exploration of these fields becomes uneconomical. It also would be too expensive to use large scale of floater structure for these marginal fields due to the large amount of initial investment. Floater with TAD system can be used to economically develop these fields. There are many marginal fields in China offshore, especially in shelf areas. The application of this field developing model, combining with the existing field developing experience in China, will open the door for many marginal field developments.

2 TECHNICAL CHALLENGES AND DEVELOPMENTS

Most of deepwater projects are at Frontier Areas. Lack of data and information to the new frontier areas has created major challenges to the development. From the safety consideration, conservative approach has been used most of time. In most cases, deepwater exploration reaches out to untouched areas, and has always been associated with harsh environment. In these underdeveloped areas, there is no preset facility available, and even the basic environmental data, such as wave, current, and geotechnical data, etc. are usually scarce. It is a big challenge to engineering and operation to put a drilling and production facility in these areas. Even for the well developed area GOM, the understanding of the environmental condition has also under major changes in the past decade. Here are just a couple of examples. Current is a major dominant factor in the areas of riser and mooring design for floating structure. The design current has increased from 1.5 knots to >5 knots from the middle of 1990s. 100 year hurricane has always been the extreme design condition for floating system in GOM. But the 3 hurricanes happened in 2004 and 2005 within one year span has put the design condition in question. Each of these three conditions exceeded the normal design condition for 100 year hurricane case. How to closely describe the environmental condition is definitely the most important step in correctly designing deepwater platforms.

There are always potential risks associated with deepwater oil field development. One of the major risks is the uncertainty of the reservoir size. Although the technology has been advanced very significantly in recent years in reservoir determination, it still is a major challenge when it comes to determining the quantity and the size of the reserve. There are still plenty of disappointing stories of project explorations. Improving the technology of geotechnical exploration and reservoir sizing will continue to be a challenge to geoscientist. Due to the high capital investment for deepwater projects, there are high cost penalties for mistakes associated with deepwater projects. The investment for deepwater project is typically at $ 500 millions plus range.

High Drilling Cost is the first challenge for deepwater. Current cost of drilling a deepwater well ranges from 30 millions to 60 millions US dollars per well. It takes anywhere from 3 months to 5 months depending on water depth and field conditions.

Another challenge is that deepwater projects are likely to be far away from shoreline and there is no pipeline laid yet. Lack of infrastructure is very common to deepwater. This will require either major pipeline work or oil storage tanker as part of the development. As a result, the cost and complexity of the projects will increase substantially. The long flow distance of crude oil for deepwater also poses the flow assurance challenge. The study of flow assurance has been one of the major tasks for any deepwater projects.

The challenge that offshore development project is facing nowadays is how to produce an efficient system in an almost always-fast track schedule project. Most of projects do not have schedule for engineering recycle, while at the mean time an efficient system is always demanded in order to keep project cost down. One of the ways to counter this is using multiple phases of field development. The field development can be broken down into several phases, with much less budget for each phase. Once the first phase comes on production and the reservoir size and flow rate are determined, the second phase then could be planned accordingly.

Field architecture is very important to the successful drilling operation. The key challenge for a multiple platforms combined system and TAD system is mooring the tender in deepwater and keeping a safe working distance between the tender and the floater and/or FPSO/FPU under all the weather conditions. The locations of floaters, FPSO/FPU mooring lines, offloading buoys, pipeline and flowline can strongly influence the design of the tender mooring system. An overly cluttered system arrangement can preclude the use of TAD since there may be no suitable places to set mooring lines of tender vessel. With many flowlines and mooring lines from different units, the field is usually very crowded.

Floating structure concept plays a very important role in deepwater projects, and the design of the floating structure is one of the most important tasks in the project. The importance of the floating structure in offshore projects can be demonstrated in the following several areas: the substantial dynamic structure responses due to wave loading and current loading, the limited motion requirements of risers in deepwater and the increasing difficulty of installation for different components of the system.

Three major technical aspects have to be considered, that is the strength of the structure, the fatigue resistance capacity of the system and local and global stability of the structure. Strength design has to consider both the characteristics of structure geometry and wave environment. Depending on the type of the structures, the governing forces can be the maximum acceleration loading, the maximum pry/squeeze loading or the maximum mooring induced loading. To resist the ever-changing dynamic environment, fatigue design plays a very important role in the design of floating structures. This is the most complicated design requirement and often causes major difference between floating production structures and other types of structures.

There are also other challenges for deepwater projects, such as complex topography, harsh environment, long mooring legs-big & heavy, high installation cost, engineering capabilities limitation, and high quality requirement for fabrication.

3 FUTURE DEEPWATER DEVELOPMENT

Both majors and large independents are increasingly focusing their finding and development budgets on new deepwater fields. Advances in technology and compressed development cycles, both a function of floating production and subsea development, make this segment the most attractive on a life cycle basis.

The deepest water around, with the highest number of imminent prospects, is still the deepwater Gulf of Mexico. This is followed by Brazil and offshore West Africa. All three segments are in what we refer to as the South Atlantic region. The resultant drilling and construction activity in all three areas helps explain the current migration of deepwater drilling and construction equipment into the region, primarily from the North Atlantic. Shell, BP, Exxon and Chevron are particularly active in the U. S. Gulf. Total is very active off West Africa. Petrobras is aggressively pursuing deepwater activities off Brazil. All of the majors are now concentrating their exploration and development activity on deepwater (exceeding 350 meters) and ultra deepwater (exceeding 1,200 meters).

In a recent industry survey, respondents said that the most attractive investment at this time is deepwater exploration and development. This interest in deepwater is evident in Gulf of Mexico leasing activity. Although there were fewer bids vs. the prior years, the total funds exposed were greater. Significantly, the auction was held in an oil price environment that was less than supportive. Basically, fields developed with floating production systems are a subset of a larger set of fields utilizing subsea completions. With the exception of certain deepwater fields in the U. S. Gulf using TLPs and SPARs, both of which allow the use of dry trees, the bulk of deepwater development worldwide utilizes wet trees tied back to floating production monohulls, semi-submersibles and to a lesser degree, to converted jackups. The choice between monohull and semi is a function of onboard storage. In general, FPSOs are favored where there is a lack of pipeline and fixed platform infrastructure, while semi-submersibles are preferred where infrastructure connections are convenient, or at least possible. Jackups are limited by water depth.

In a significant number of cases, the tieback is not to a mobile floating unit, but rather to an existing fixed structure in shallower water. These latter units are concentrated on gas and oil production in regions where paraffin and hydrate problems can be minimized. Techniques include subsea separation, multi-phase pumping, pre-treatment, injection, pigging and various approaches to pipeline insulation, including deepwater burial. Significant technical effort is going into extending the tieback range.

What will the new systems look like? Because of the opportunities in ultra deepwater, the Gulf of Mexico will remain the domain of TLPs, both mini and full size, SPARs, and Semi-submersibles. Storage will be a secondary issue and the challenge will be to develop fields, as well as their tiebacks to existing infrastructure, in a capital and time-efficient manner. Significant funds will be spent on new installation technologies, particularly for new ultra deepwater pipelay capability, in order to make this possible.

Deepwater production offshore West Africa and Brazil will be largely limited to TLP, monohull and barge-based floating systems, reflecting the kinder weather environment and the absence of developed pipeline infrastructure in these regions.

Offshore Southeast Asia is a big place, with virtually every potential option for field development. For the foreseeable future, most development will be in relatively shallow water with benign climatic conditions, in areas lacking infrastructure. The result will be a number of fields developed with platforms and monohulls that incorporate onboard storage capability. Shell is pursuing deepwater development of the Malikai field in the Malaysia; and various groups are developing deepwater reservoirs, offshore New Zealand and Australia, with platforms monohulls.

In the North Atlantic, with its harsher weather and relatively "shallower" deepwater, we expect to see semi-submersible and monohull production units, perhaps an occasional platform. Our definition of North Atlantic includes fields off Canada's eastern provinces and north of Norway. Terra Nova, offshore Eastern Canada, is utilizing a floater. While water depths are not onerous, iceberg avoidance and subsea scouring are definite issues.

The oil field development in China offshore has more than 40 years history, and has established its vigorous oil industry. Entering the twenty-first century, oil field development in offshore China is rapidly expanding into deepwater. The SouthChina Sea is very rich in oil and gas, and has been called "The Second Gulf of Mexico". The conventional fixed platform models cannot satisfy the requirement of new field development. Floating system, such as TLP or Semi-submersible, will be needed to introduce into the oil field development.

There are many good examples of field developments and project histories around the world, which can be used in the development of China deepwater oil fields. Most of the field development represented the current advanced industry technologies, and have a proven history of both technical viability and economic benefits. TLP/SPAR combined with other floating system(s) provides good option for exploring both deepwater and offshore marginal fields. This approach can have several advantages in the following areas. Floater can provide a host for both drilling and production of dry trees. The drilling option is accomplished through tender rig. It will help to solve the current situation of lacking of drilling vessels in Nanhai area. The current international market for drilling vessel is very tight, and the cost of each deepwater well drilling can range up to 50 million US dollars. TLP/SPAR is also the effective way to provide the dry tree production and avoid the expensive subsea development cost.

There is a lack of pipeline infrastructure in offshore China. FPSO/FPU is always the best alternative to this kind of situation. It can provide an effective way to store and offloading the produced oil. There is also a lot of experience of using FPSO in China oil industry. The combined concept with a floater structure and tender (semi-submersible or barge) unit permits a drilling system between the two platforms. Provided that tender is converted drilling semi-submersible or a barge with acceptable deck carrying capacity, the tender can maintain in its mud and cementing systems and support the wellhead platform via flexible hoses. Compared to using a self-contained platform drilling rig, using a TAD rig has significant advantages in terms of payload reduction, and a significantly reduced construction cost and project cost. The option of combined floater with FPSO/FPU and TAD provides the opportunity for new field development expanding into deepwater drilling. It allows the use of dry wellheads, and compared to a subsea solution it produces significantly lower drilling cost, lower initial capital investment, cheaper well intervention and workover, and much less reduced field development risk. The main technical challenges to use such a system include the followings: field architecture of the overall system, fully coupled time domain analysis for this two- or three-body system, hawser strength and fatigue design and analysis, floater lateral mooring and TAD vessel mooring design and analysis including mooring buoys and preset system, and field installation. Introducing floater with TAD system to the development of shelf fields will give more options for China offshore industry. It can provide an effective way to develop these fields and produce significant economic benefit for the field development.

CONCLUSIONS

Deepwater development will continue to face technical challenges in various areas as the exploration gets deeper. Technology will play a very important role in overcoming these challenges and help accelerating the development. This paper studied the current status of the industry, reviewed the recent developed technologies driving the industry, and discussed some challenges of the deepwater development. Attention has also been concentrated on some technical aspects of field development and floating system design, which are two important areas in deepwater projects.

Offshore structure design is a complex process and needs to consider many factors, such as strength, fatigue, buckling and fabrication. Global response of the structure is always an important aspect of the design and analysis. Introducing the major members into the structure, such as lateral braces and knee braces, can change the internal load path of the structure and have major impact on the structure performance. The dynamic response from pry/squeeze load is a key factor in the design of platform structure. Most of the critical connections are governed by the maximum stress and fatigue created from this dynamic load.

REFERENCES

Beato C, et al. 2002. Coupled Tender Mooring for Spar and TLP Designs [C]//Proceedings of the 11th Offshore Symposium.

De Kat JO. 1990. Installation of Floating Production System in Timor Sea[J]. Schip En Werf, 57, 231 – 237.

Sandstr R, Huang J, Danaczko M. 2006. Development of Close-Moored Deepwater Systems: Kizomba A and B projects[C]. Offshore Technology Conference, Houston, Paper 17919.

Simmon J. 2005. The Okume Complex Project. Presentation on Society of Naval Architect and Marine Engineers, Houston, 1 – 25.

Van Hoorn F, Devoy SD. 1990. The Dry Transport of the Green Canyon Tension Leg Wellhead Platform by a Semi-submersible Heavy-lift Ship. Offshore Technology Conference, Houston, Paper 6471.

Wetch S, Wybro P. 2004. West Seno: Facilities Approach, Innovations and Benchmarking. Offshore Technology Conference, Houston, Paper 16521.

Zhang D, Deng Z, Yan F. 2009. An Introduction to Hull Design Practices for Deepwater Floating Structures. Journal of Marine Science and Application, 8(2): 123 – 131.

Zhang D, Deng Z, Yan F. 2010. The Application of Lateral Braces in the Design of Floating Structures [C]. Proceedings of the Twentieth (2010) International Offshore and Polar Engineering Conference, Beijing, 614 – 623.

Zhang D. 2007. Application of Tension Leg Platform Combined with Other Systems in the Development of China Deepwater Oil Fields. China Ocean Engineering, 20(4): 517 – 528.

Zimmer R, Figgers R, Geesling J, Kim H, Yeung D, Banatwala Z, Sidwell G, Stout J, Harris B, Van Vugt M. 1999. Design and Fabrication of the URSA TLP Deck Modules and TLP Hull Upper Column Frame [C]. Offshore Technology Conference, Houston, Paper 10755.

Full Mission Bridge Simulations-
a Must for (Complex) Offshore Projects

Jakob Pinkster

Head R&D, STC Group, Rotterdam, The Netherlands

Abstract A Full Mission Bridge Simulator (FMB), DNV Class A (or equivalent), is a ship simulator with a ship's bridge where a bridge management team (e.g. captain, pilot, mate, helmsman) can carry out all possible missions (e.g. entering/leaving a harbor, navigation through a busy channel, etc.) that should be carried out on board of that ship. Such FMB simulators are generally used to train seafarers and pilots in the art of sailing a ship under all kinds of different conditions (including emergencies) and they are also used for nautical (research) studies such as accessibility (new) harbors and sometimes installation of offshore units.

In this paper, it is shown that the use of FMB simulations is a must for (complex) offshore projects and that it is, amongst others, "penny wise and pound foolish" not to do so.

Key Words Full mission bridge simulations; (complex) Offshore projects; Successful; Quality; Payback; Necessity

NOMENCLATURE

CNOOEC	China National Offshore Oil Engineering Company, Ltd
COOEC	China Offshore Oil Engineering Company, Ltd
ECDIS	Electronic Chart Display and Information System
FMB	Full Mission Bridge
FPSO	Floating, Production, Storage and Offloading
IOS	Instructor Operator Station
RIRO	Rubbish In ···Rubbish Out
ROV	Remotely Operated Vehicles
UAV	Unmanned Aerial Vehicle, (UAV also known as a drone)
VLCC	Very Large Crude Carrier
VTS	Vessel Traffic Services

INTRODUCTION

A Full Mission Bridge Simulation (FMB) is a ship simulator with a ship's bridge where a bridge management team (e.g. captain, pilot, mate, helmsman) can carry out all possible missions (e.g. entering/leaving a harbor, navigation through a busy channel, etc.). In short, FMB simulators, DNV Maritime Simulator Class A (or equivalent), are generally used to train seafarers (both navy and merchant navy) and pilots in the art of sailing a ship under all kinds of different conditions (including emergencies).

FMB Simulations are now well known in the fields of maritime education of seafarers, (merchant) navies for further training of key personnel, marine pilots. They are also used for nautical studies such as, research questions involving harbor accessibility for certain ship types and size (e. g. can the largest of all container ships The triple E Class of Maersk be accommodated in the harbor of the port of Colombo? can a 120,000 tons DWT tanker be berthed at a new liquid tanker jetty planned within the existing harbor configuration of Sohar?). Such key questions are often asked by harbor engineering contractors & harbor authorities and these require answering in an early design stage. This way engineering concepts can be checked and, if necessary, be corrected on time (a stitch in time save nine!) and save much wasted effort, money, time, etc.

The purpose of this paper is to show the offshore industry that full mission bridge simulations (FMB) are a must for (complex) offshore projects and that it is, amongst others, "penny wise and pound foolish" not to do so. Of course, such simulations may contain a high degree of complexity and take a great deal of effort and time to set up, execute and evaluate, but, in all fairness, they can (and often do) deliver enormous results. Results which may mean the difference between failure and success of the offshore project at hand.

The foregoing goal will be reached by means of the following steps in the paper:

(1) A description of a single FMB simulator;
(2) Use of more FMB simulator in a virtual world;
(3) Tug boats (interactive and vector type);
(4) Advantages/disadvantages of a FMB simulator;
(5) Examples of the use of FMB simulators;
(6) Examples of the use of FMB simulators for offshore project;
(7) Example of Offshore Installation simulation;
(8) Possible offshore projects in the future.

1 A DESCRIPTION OF A SINGLE FMB SIMULATOR

In essence, generally a single FMB simulator consists of

(1) A ship's navigation bridge equipped with control equipment for steering, engine settings, bow/stern thruster settings, radars, ECDIS, communication, binoculars, (engine) sound generators, etc.

(2) One or more computers with simulation program software.

(3) One or more computers for projection of visuals on the screen(s).

(4) A projection screen area upon which the outer world of the ship's bridge is projected.

When is a simulator used? Simulators are used when (human) experience is sought with certain equipment in a given environment without using the actual equipment and environment itself. In order to be able to get the required experience, simulations use models of reality along with the human element in carefully chosen scenario's. Scenario's used in simulations, depict the situations that are to be met in reality and are generally run in real time (i. e. in real time simulations, everything happens in real time therefore one second of simulation time is exactly equal to one second real time).

Each simulator has an own ship (i. e. the ship that is virtually sailed by the bridge team on the ship's bridge), possibly some traffic vessels (i. e. other vessels that are present in the area sailed in) and an environment (i. e. sailing area with water depth and environmental conditions (wind, waves and current).

Advantage of real time simulations is that human factors in the simulations is not effected by a difference in time scale. Therefore the simulation results regarding human behavior are directly applicable to the same situation in real life (e. g. situational awareness, distraction, concentration span, confusion, breakdown in communication, improper Conn/lookout, departure from original (project) plan, violation of rules/regulations, reaction time, fatigue, etc.)

If the ship's bridge is fully equipped with all navigation equipment and aids such that all missions can be carried out on the bridge and the model of the own ship reacts realistically then one speaks of a full mission bridge.

Depending on the demands placed on the underlying simulation program, a FMB simulator may require one or more computers to accommodate the necessary software.

The projection screen area upon which the outer world of the ship's bridge is projected determines the type of simulator; i. e. a 360°simulator has an all-round view from the bridge of 360 degrees (see Fig. 1), a 240° simulator has a view from the bridge of 240 degrees. Often the latter also has a large screen behind which shows a view from the bridge of the aft part of the vessel and its outer surroundings.

A special type of simulator is a blind simulator. A blind simulator is the term used for a simulator with no view projectors and no projection screen area, i. e. view from the bridge is 0 degree and therefore no view screens are present.

In general, the more panoramic view (i. e. degrees view) one has from the bridge of a simulator, the more projectors that are used in the hardware setup and the more expensive the system is to build and to operate. Each view projector requires a separate computer with separate (high definition) fast graphic card, etc.

Fig. 1　STC's 360°FMB simulator (1) and IOS with classroom (2)

There are times when a blind simulator can be very useful and emulates the real world very well (e. g. when the person in the simulation has no view of the outer world, think of a VTS operator, a UAV (DRONE) operator in a container control unit setup, etc.) or an operator of a remotely controlled underwater vehicle. In such cases, necessary information for the VTS/UAV operator within the simulation is shown via single or multiple computer screens on a desk or in consoles.

2 USE OF MORE FMB SIMULATORS IN A VIRTUAL WORLD

There are simulations thinkable when more than one own ship is the subject of interest in one and the same virtual world. The following two examples illustrate such usage. Firstly, two vessels simultaneously sailing the same fairway together as two separate entities in the same or opposite direction will require two separate FMB simulators each running in one common virtual environment. Secondly, a total of no less than six separate simulators would be required to work together in one virtual world for the installation of one FPSO (own ship on a blind simulator) and five tug boats (each an own ship in a FMB simulator). In the latter example, the FPSO could be represented by a blind simulator whereas and five tug boats may each be running on a FMB simulator.

3 TUG BOATS (INTERACTIVE AND VECTOR TYPE)

If the effect of the human factor on board of the tug boat is subject of investigation during simulations then it is imperative that use be made of an interactive tug boat. An interactive tug boat is a tug boat which is during simulations is sailed by a bridge team consisting of one or more persons. Fig. 2 shows the bridge of a FMB simulator where the own ship is a tug boat (interactive) which is pushing a tanker sideways in an attempt to assist the larger vessel to berth alongside another tanker at a busy oil terminal.

If the effect of the human factor on board of the tug boat is not a subject of investigation during simulations then use may be made of a vector tug. A vector tug is a tug boat which is operated by the simulator operator/instructor at the IOS; the operator thereby adjusts the tug force vector (both in magnitude and direction) and thereby the actions of the tug boat with the exception of the human factor as would be when active on the FMB. Note that there is no FMB bridge and no bridge team at all for the tug boat in question. Because of this, when regarding simulation costs, vector tug boats are cheaper than interactive tug boats.

   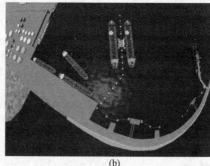

(a)   (b)

Fig. 2 An interactive tug boat bridge on a FMB simulator (a), busy oil terminal (b)

4 ADVANTAGES/DISADVANTAGES OF A FMB SIMULATOR

There are a number of distinctive advantages and disadvantages regarding FMB simulators. These will now be named and followed by a short qualification if deemed necessary:

Advantages FMB simulators:

(1) Does not require the real object and environment.

(2) Excellent quality when simulations are performed correctly.

(3) Cheaper than hiring/buying the real thing/going to the location where the action really should be undertaken.

(4) May lead to reduced insurance premiums (that may well far exceed the simulation costs involved) as "no stone was left unturned" in the preparations for the actual project execution.

(5) All aspects of a given task may be part of a simulation setup (i. e. task related skill (seamanship, ship handling, communication (internal/external))), behavior in case of stressful conditions (such as heavy mental workloads, emergencies/contingencies, crisis, etc.).

(6) There is always a non-destructive outcome of simulation (ship may ground but never any bottom damage, etc. whatever happened).

(7) Can be carried out in a controlled environment (all condition may be properly set and not spoilt by sudden bad weather conditions, etc. ···unless so required - then this can be introduced into the simulation by the simulator operator).

(8) Can be started and stopped at any time.

(9) Can be repeated any time (partly or wholly) (if you don't succeed at first, try, try again; if you need more data from simulation, run the simulation again).

(10) Everything that happens during simulation can be recorded (lots of (technical) data collected!). This includes the human factor (visual and audio recording throughout the whole simulation).

(11) Results can be analyzed in-depth producing "hard" objective results & findings.

(12) If results are not what they should be then alter the scenario until it is (i. e. change the goal posts and/or modus operandi; find out that what can be (better) done/how it can be done instead of maybe only finding out that it cannot be done).

(13) Find and define the role of the human factor in the task(s) to be done (find out possible operational limitations, stress factors, etc.).

(14) Assess the suitability of (key) project personnel involved for a given task (i. e. regarding situational awareness, specific knowledge necessary, motivation, ability to work well together, etc.).

(15) Can be used for teambuilding (i. e. by simulating a given project with all (key) team members present, a form of team building is also being carried out simultaneously).

Disadvantages:

(1) Simulation requires a lot of input data in order to make the visuals, mathematical models, scenarios, etc..

(2) Rubbish in = rubbish out (i. e. if the models are not correct then the results will not be correct either; this holds true for the human factors part of the simulations just as well).

5 EXAMPLES OF THE USE OF FMB SIMULATORS

FMB simulators can be used for the following tasks (and more):

(1) Nautical Safety Analyses;

(2) Traffic Analyses;
(3) Port Optimization Studies;
(4) Human Factor Analysis;
(5) Ship handling courses;
(6) Bridge resource management courses;
(7) ECDIS courses;
(8) Tug Operation Studies;
(9) New building Simulations;
(10) Accident Analysis.

6 EXAMPLES OF THE USE OF FMB SIMULATORS IN OFFSHORE

FMB simulators can be used in offshore projects for example for the following tasks (each described more in detail later below):

(1) Offshore exploration simulation.

Ship handling, bridge resource management (including communication) and training in sailing certain pathways (tracks) for underwater surveys for oil/gas fields. Such exploration vessels may be outfitted with large (expensive and complex) systems of acoustic equipment aft of own ship.

(2) Offshore ship-ship transfer of cargo.

Ship handling and bridge resource management (including communication mooring master with assisting tug(s) and "mother ship").

(3) Offshore float outs.

Simulations of float-out of offshore construction from building dock with the aid of shore side winches and tug boats (including communication tow master with assisting tug(s) and shore facilities).

(4) Offshore tow outs.

Simulations of tow outs of offshore construction from protected waters near shipyard out to sea/further outfitting area with the aid of tug boats.

(5) Offshore tow legs.

Simulations of tow of offshore construction along different (difficult e. g. narrow straights, etc.) areas of a long tow from building location to destination location (including communication tow master with assisting tug(s), other vessels and shore facilities).

(6) Offshore change of towing configuration.

Simulations of change of towing configuration (e. g. change of offshore tow configuration of an offshore drilling rig from 1 tug boat configuration to 5 tug boats configuration in the vicinity of drilling location area in order to go into the next installation phase of the rig), (including communication tow master with assisting tug(s), other vessels and shore facilities).

(7) Offshore Installation Simulation.

Simulation of installation of offshore unit(s). Example given later in this paper concerning the installation of a VLCC class FPSO in Penglai 19-3 field in Bohai Bay, China Sea.

(8) Offshore float overs.

Simulation of installation of offshore unit(s) topsides via float over system (including communication tow master with assisting tug(s), other vessels and shore facilities).

(9) Offshore diving support simulations.

Simulation of Offshore diving support operation (including communication diving master with diver(s), other vessels and shore facilities).

(10) Dynamic positioning.

Simulation of offshore dynamic positioning operation (including communication positioning master with other vessels and shore facilities).

(11) ROV simulations.

Simulation of offshore ROV operation (including communication ROV operation master with mother ship, other vessels and shore facilities).

7 EXAMPLE OF OFFSHORE INSTALLATION SIMULATION CARRIED OUT AT STC's SIMULATOR FACITITIES

From 1991 onwards STC's simulator facilities (ex MSR, Rotterdam) have been utilized for many different simulations. Some of these simulations have been related to offshore projects. In June 2008, STC B. V. conducted a complex Offshore Installation Simulation, Assessment and Training program at STC B. V.'s simulation center, Wilhelminakade 701, 1072AP, The Netherlands. The whole simulation center (consisting in those days of four FMB simulators) was then booked out for one full week and around 30 different project participants (coming from CNOOEC, COOEC and other companies involved) were actively present in our facilities. The different parties present from these companies were:

(1) Positioning Manager;
(2) Positioning Masters;
(3) Tug Masters;
(4) Installation Manager;
(5) Superintendents;
(6) Engineers;
(7) Positioning Engineers;
(8) Positioning surveyors.

From the side of the simulation center, parties present included:

(1) Project manager;
(2) Assistant project manager;
(3) Offshore operations expert (exercise coordinator);
(4) Simulator operators.

The subject of the simulations was the installation of the largest FPSO ever to be installed in Chinese waters, Hai Yang Shi You 117, or HYSY117, which was to be moored to a special single point mooring system (i.e. a rotational Soft Yoke Mooring System (SYMS)) at the Penglai 19-3 field in the Bohai Bay, the inner most gulf of the Yellow Sea on the coast of North Eastern

China, in March 2009. The main particulars of this VLCC class FPSO are shown in Table 1 and a models of the vessel and the SYMS along with a picture of one of the assisting tug boats during a simulation exercise is shown in Fig. 3.

Table 1 Main Particulars of HYSY117 FPSO

Parameters	Value
Water Depth (m)	27.6
Length over all (m)	323.000
Length between perpendiculars (m)	313.000
Breadth Moulded (m)	63.000
Depth Moulded (m)	32.500
Design Draft Moulded (m)	20.000
Mating Draft even trim & even heel (m)	14.000
Displacement at mating draft = 14m (t)	272365

Fig. 3 Simulation models of Hai Yang Shi You 117, China's largest FPSO (a) and the SYMS with jacket (b) and the captain's view from the starboard side forward tug boat assisting the FPSO during one of the simulations to approach the SYMS (c)

For the simulations to be undertaken, mathematical (ship) models were made of the FPSO, SYMS, tug boats and visual models were made of the FPSO, SYMS, tug boats and the relevant environment of Penglai 19-3 field in Bohai bay. Fig. 4 shows a screenshot of the area along with the main players and Fig. 5 the five tug mooring spread used to assist the FPSO when approaching (and later mating with) the SYMS.

Fig. 4 Simulation screenshot of Hai Yang Shi You 117 FPSO, SYMS, tug boats and the relevant environment of Penglai 19-3 field in Bohai Bay, China

All models (both ship and visuals) were tested on correct (maneuver) behavior and correct visualization before simulations took place. Test criteria. It is of the utmost importance that all modeling is correct in simulations, otherwise RIRO and that truly leads us nowhere!

When can one speak of a good simulation? A good simulation (irrespective of the outcome) is based on the combined effort of good simulator models, environment and proper scenarios. A simulator scenario is a combination of simulator models and environmental conditions (location, wind, waves, current, water depth) and a given task or mission. A simulator is nothing if there are no suitable scenario for it, then it is like a computer without any electrical power source, i. e. useless! Simulation scenarios are usually defined by the client and STC together.

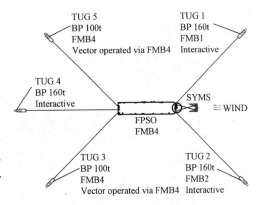

Fig. 5 The five tug mooring spread used to assist the FPSO when approaching the SYMS. Later also used for the mating thereof

For the Hai Yang Shi You 117 FPSO, simulations, a number of scenarios were defined and used in the simulations which depicted all the main stages of the actual FPSO installation as planned within the project procedures. These scenarios included:

(1) a number of positioning trials (for familiarization with simulators, the 5 tug mooring spread and to try out system workability, etc.).

(2) station keeping (familiarization with the 5 tug mooring spread, try out system workability).

(3) temporary mooring to the pre-installed SYMS mooring tower.

(4) pendulum mating (method used mate the 2 systems).

(5) ballasting of the SYMS yoke to a storm-safe situation.

As one can well imagine, life is not always as peaceful as it could be and, to model this effect, the above scenarios were not only conducted under normal conditions but also under a number of stressful conditions. The latter included contingency situations such as black out of primary tugboat, breaking of mooring lines, deadly accidents and, of course, one of mother nature's oldest tricks …the occurrence of sudden bad weather conditions and then just when you needed it the most (i. e. at the critical moment of hook up between FPSO and SYMS)!

All told, it took a five full days to run all the simulation exercises (including familiarization runs, project management, etc.). Resulting from the execution of all these scenarios a lot of objective and subjective data was collected together. Objective data came from digital files taken from each run and subjective data came from questionnaires directly filled in after each run by key personnel involved in the simulation run (i. e. position master and tug boat masters). This data was organized and analyzed and, together with the opinion of the offshore expert (exercise coordinator), formed a sound basis which helped to check the original execution plan, and/or operation procedures and to identify any shortcomings therein. The same held true for the assessment question of key installation personnel present regarding their performance, communication skills, suitability, etc. For this very important and demanding assessment task, STC has their own team of well qualified and experienced assessors under leadership of a psychologist. Needless to say, such assessment results are confidential and will only be shared with those directly involved (see Fig. 6).

(a)

(b)

Fig. 6 Scenario screenshot showing the view from the rear of the port side forward tug as the FPSO approaches the SYMS (a) and a part of key personnel assessment team in action (b)

8 BRIEFING/DEBRIEFING

Before each scenario (i.e. simulation run) was executed, all relevant personnel were briefed during a briefing session on the task to be undertaken. Directly thereafter the run was executed.

Generally, directly after each run was executed, a debriefing session was held whereby the exercises were thoroughly evaluated. During such debriefing sessions, the whole exercise can be viewed by all on a projector screen in real time steps via STC's powerful debriefing tool (see Fig. 7). Besides visualization of the exercise (possible both in 2D and 3D views), data is also available such as position, speeds, headings, rate of turns, line forces, etc. and this can be displayed at any time for all objects simulated (in this case the FPSO, tug boats and the SYMS). Other data available are engine settings, bow thruster setting, rudder settings of the tug boats, and, of course, environmental data such as wind, waves, currents and water depth.

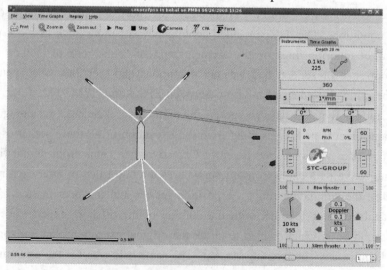

Fig. 7 Screenshot showing 2D views and relevant vessel data produced by STC's powerful debriefing tool

The comprehensive training and assessment program, as described here, consisted of all aspects related to the challenges presented by the FPSO/SYMS installation, i. e. exercises in positioning, station keeping, FPSO approach to the SYMS, connecting of the mooring lines, station keeping whilst the pendulums are being hooked up to the SYMS, bringing the installation barge alongside of the FPSO, contingency measures, etc.

For such simulation programs, it is important to:

(1) model (positions and visuals, there where necessary) the offshore field (including all relevant fixed and floating objects and no-go areas such as (underwater) pipe lines, mooring wires, etc.).

(2) make the numerical models of all of the FPSO, SYMS, tug boats (these are to be controlled by the positioning master(s) (FPSO) and by the bridge teams (tug boats), thus introducing the (all-important) human factors into the simulations).

(3) model the meteorology and oceanology environment.

(4) The actual performance of the key installation personnel when executing the installation work (via observations there of made during some of the simulations and analysis thereof).

Basically speaking, the complete real time simulation program in essence provided a virtual platform where all the key personnel involved were familiarized with all operational procedures, all the main FPSO installation topics were addressed (i. e. recognized, performed and assessed). Moreover (and certainly also of prime interest) these "true to life" simulations at the STC facility, also helped to find any (possible) errors or (possible) critical points regarding operation procedures, execution plan, and performance of key installation personnel (i. e. communication skills, position/role suitability, etc.).

One final departing statement from one of the key personnel participants upon leaving the last simulation meeting in June 2008 by STC in Rotterdam rounded the whole simulation week up in a nut shell.

"It wasn't cheap, but it was worth every single penny".

That this was not an understatement was proved after the actual installation had taken place and Wang A. M, et al. (2010) stated (regarding the above simulations) that:

"Their various findings have been successfully used to guide the different installation phases of the mating operation between the FPSO and the SYMS in shallow water. The mating operation went quite smoothly. The methodology of the virtual simulation addressed here can make an excellent complement to numerical simulations, model tests and field tests. It is certain that virtual simulations will play more and more important roles in different offshore installation applications".

9 POSSIBLE OFFSHORE PROJECTS IN THE FUTURE

As far as FMB simulator use is concerned, possible future offshore projects will most probably remain focused in and around unit (e. g. FPSO, etc.) installation operations.

Also, the tendencies of such operations is to go to greater water depths, and, in some cases, much lower ambient temperatures and harsh adverse weather conditions and/or even more desolate regions.

These tendencies will definitely have an impact on the contents of FMB simulations. Amongst others, it may well be advantageous to simultaneously include actual ROV operations within such future simulations as well. ROV operations can fail to perform properly when, for example, it's umbilical's become entangled around an offshore underwater structure (e. g. legs of a jacket, mooring stem, etc.). Such entanglement problems may lead to very detrimental effects for the project at hand (even loss of life!) and cause large downtimes.

The above tendencies will result in even more pressure being exerted on getting things right the first time and thereby lead to more checking of the working of the project execution plan, training and assessment of the relevant key members of the project team. It may be expected that the results from such FMB simulations will be very valuable to all concerned and that the benefits thereof will well supersede the costs involved.

FMB simulations carried out on time within the project's time line, clearly give a large boost to the probability for the operation to succeed. However, in case of the appearance of any issues at all during the actual operation, the same FMB simulations will certainly help to find the answers to the awkward questions (i. e. regarding performance, training and experience of the key personnel, etc.) that are usually ask to be fully and unambiguously answered after an offshore operation has been unsuccessfully executed resulting in loss of life, damages, time loss, etc. be they large or small.

CONCLUSIONS

Based on the description of a single FMB simulator as well as the use of more FMB simulator in a single virtual world and the advantages/disadvantages of a FMB simulator and the different types of offshore projects, the following conclusions can be stated that FMB simulations :

(1) can be highly successful to assist offshore (installation) projects to be successful.

(2) can give key personnel the required experiences that they (probably) did not already possess.

(3) can leads to a happier and more confident installation crew (a much better team!).

(4) can lead to reductions in insurance premiums which go beyond the costs of the actual simulations.

(5) may well be beneficial for future offshore projects.

RECOMMENDATIONS

Looking at the growing number of oil fields with FPSO's, drilling rigs, etc., it would be advised to:

(1) train and assess a specially dedicated local contingency group by means of FMB simulators to be able to act accordingly in times of sudden issues such as abandoning of mooring, due to bad weather conditions, fire, etc..

(2) develop, together with the offshore industry, extensions of existing simulation methods in order to be able to facilitate the (complex) offshore projects of the future.

ACKNOWLEDGEMENTS

Acknowledgements are hereby made to Dr. Alan Wang (CNOOEC), and Capt. Dave Betts of POSH SEMCO, for their assistance in making the Hai Yang Shi You 117 FPSO virtual simulation project such a success and for their continuous efforts to utilize such simulations in (complex) offshore projects. They have hereby helped to set the industry standards for such projects for the coming years.

REFERENCES

Wang A M, et al. 2010. Virtual Simulations of VLCC Class FPSO-SYMS Mating Operation[C]. Proc 20th (2010) International Offshore and Polar Engineering Conference, Beijing, China, June 20 – 25, 2010, ISO PE 2010, Volume I: 478 – 485.

Motion Analysis and Weather Window Selection for Towing of "Nan Hai Tiao Zhan" Semi-submersible Platform

Hui Yang, Bo Liu, Guanglei Zhang, Nan Zhou

Floater Department, Offshore Oil Engineering Co., Ltd., Tian Jin, China

Abstract Towing motion performance of "Nan Hai Tiao Zhan" semi-submersible platform is analyzed by using SESAM program, which is based on Morison's equation and 3D potential theory. According to the design maximum allowable roll or pitch amplitude of 6 degrees, the maximum allowable significant wave heights are determined for a range of wave period with a cyclic iterative method. And the weather widow suitable for towing is recommended on consideration the limited sea states, climate characteristic of South China Sea and the whole update modification schedule.

Key Words LH11-1 FPS; Upgrade modification; Motion analysis; Towing; Weather window

INTRODUCTION

The Nan Hai Tiao Zhan FPS is a major component of the Liuhua 11 – 1 Development System, which is located in approximately 310m of water in the South China Sea. The FPS is a converted SEDCO 700 series drilling rig built in 1975 and converted to a floating production unit in 1995 in Singapore, and then deployed at the Liuhua 11 – 1 oil field about 130 miles southeast of Hong Kong, permanently moored by 11 mooring legs. It was originally designed for a service life of 10 years (from 1995) and has been continuously operating for more than 15 years. Also, a new oil field Liuhua 4 – 1, about 11km away from Liuhua 11 – 1, was planned to be produced by the Liuhua 11 – 1 FPS facilities via subsea tieback, thus requiring additional topsides capacities and life extension of the aged semi-submersible FPS. For service life extension and upgrade purpose, the FPS was disconnected from field in November of 2011 and transported to a shipyard in Guang Zhou for dry-dock upgrade modification.

As the environmental condition of South China Sea is complexity and capricious, motion analysis and appropriate weather window selection for the disconnection, voyage and field installation are the key points for making the whole project schedule. In this paper, the motion performance of the FPS during towing is analyzed by using the 3D potential theory. SESAM program developed by DNV is used in the analysis. By using a cyclic iterative method of frequency analysis to do the short term statics, the limited sea condition satisfied the maximum allowable roll and pitch amplitude not exceed 6 degrees are obtained. Then the whole update modification project schedule is recommended according to the weather window selection of towing.

1 METHODOLOGY

Normally the motion response prediction of a floating structure is based on the specified environment condition, potential theory and the spectrum analysis method. The relationship of wave energy and motion response is usually assumed as a linear system, and the response spectrum can be described as:

$$S_R(\omega) = S_w(\omega) |H(\omega)|^2 \quad (1)$$

Where, $S_R(\omega)$ and $S_w(\omega)$ are the response spectrum and wave spectrum respectively, $H(\omega)$ is the transfer function.

After getting the platform response spectrum, the platform motion response with a certain exceeding probability can be determined by the extreme value distribution statistics.

While the objective of this paper is to determine the environment condition and to select whether window appropriate to operation at field and towing, it's an inverse question of response prediction.

According to the Operation Manual of the FPS, the motion amplitude of roll and pitch shouldn't exceed 6 degrees during towing. when roll or pitch motions are in excess of this 6 degrees limit, the FPS is to be ballasted down to its survival draft of 70 ft. Based on the design maximum allowable roll or pitch amplitude of 6 degrees, the maximum allowable significant wave heights, H_s, are calculated for a range of mean wave periods, which gives the limited sea state for towing. The allowable H_s for a given wave period is determined by assuming an initial value for H_s and iteratively changing this value, while checking each of the motion amplitude requirements, until the allowable H_s was determined. Then the weather window for towing can be determined according to the environmental characteristic of South China Sea.

The transfer function $H(\omega)$ is obtained by using the SESAM program package developed by DNV. For large bodies, the wave force, additional mass and damping are solved by the 3D diffraction potential theory. For slender cylinders (cylinder diameter D should be much less than the wave length L and D/L is about 0.1 to 0.2), the wave force and viscous damping are solved by the Morison equation. The RAOs are obtained by solving the motion equations (2):

$$\sum_{j=1}^{6} m_{ij}\ddot{x}_{ij} + \sum_{j=1}^{6} c_{ij}\dot{x}_{ij} + \sum_{j=1}^{6} k_{ij}x_{ij} = \sum_{j=1}^{6} F_{ij} + F_m \quad (i = 1,2,\cdots,6) \quad (2)$$

2 MOTION ANALYSIS OF TOWING

2.1 Hydrodynamic Model

The FPS consists of two lower pontoons (295ft × 50ft × 21ft) which are connected by one cross pontoon (21ft × 12ft) at each end. The centerline transverse spacing of pontoons is 195ft, and the centerline longitudinal spacing of cross pontoons is 225ft. From the lower hulls four corner column + sponsons (18ft × 30ft), and four intermediate columns rise up to support the rectangular main deck. The diameter of four corner columns is 30ft and 18ft for intermediate columns. Transverse trusses and braces are arranged between the four pairs of columns and the main deck transverse girders. The main deck height is 130ft from the baseline. Fig. 1 gives the lateral view of the FPS.

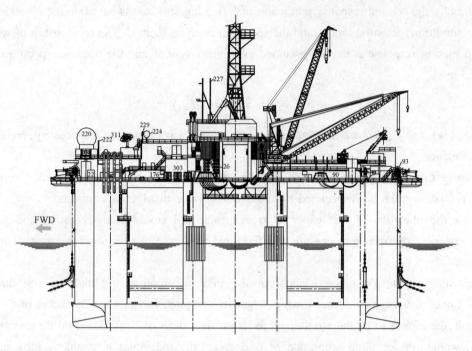

Fig. 1　"Nan Hai Tiao Zhan" FPS

3D hydrodynamic model of the FPS includes panel model and Morison model, both which of them are built by using the Genie module of SESAM program. According to the symmetry characteristic of the FPS, only 1/4 of wet surface model is built and the whole panel model is obtained by mirror, see Fig. 2. Transverse trusses and braces are built as Morison model with actual size, see Fig. 3. In order to consider the viscous effect, pontoons and columns are also simulated as the virtual Morison cylinders with 0.01m diameter. The towing velocity of 1.2m/s is expected, and is omitted in the analysis.

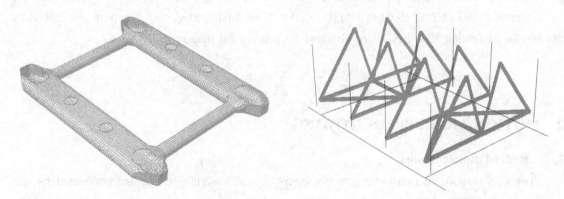

Fig. 2　Panel model (draft = 7m)　　　　　　Fig. 3　Morison model

According to the towing loading condition, the draft of voyage is 7m and the displacement is 17715 tons. Property of the mass distribute is defined by the radius of gyration. As the draft is near the depth of the pontoons, nonlinear effect of pontoon-wave interaction is considered by adding 7% and 4% critical damping for roll and pitch respectively.

2.2 Transfer Function of Motion Response

By using the HydroD module of SESAM for motion performance analysis, transfer function of motion response for roll and pitch are obtained, see Fig. 4 and Fig. 5. For reference, heave transfer functions are also plotted in Fig. 6. It's noted that the peak response period of the roll is 9s in transverse wave direction and 10s for pitch in heading direction.

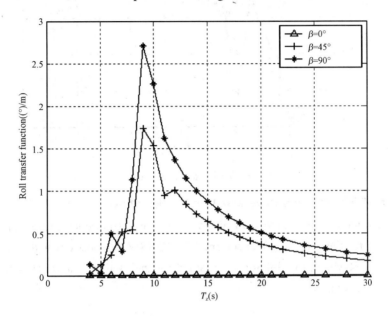

Fig. 4　Roll transfer function

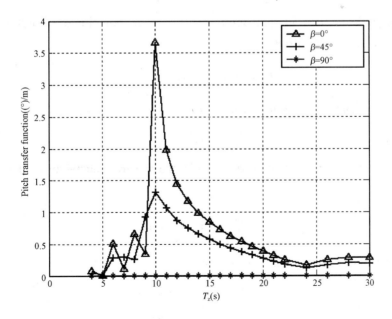

Fig. 5　Pitch transfer function

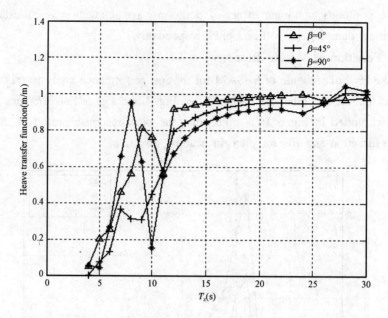

Fig. 6 Heave transfer function

2.3 Prediction of Motion Amplitude and Limited Sea States

After obtained the transfer function of motion response, response spectrum can be obtained according to Eq. (1) for the given wave spectrum. The JONSWAP spectrum with $\gamma = 3.3$ is used for the motion analysis. Rayleigh distribution is assumed for the motion amplitude statistics. This Rayleigh distribution is given by:

$$f(x) = \frac{x}{\sigma^2} \exp\left(-\frac{x^2}{2\sigma^2}\right) \tag{3}$$

With this Rayleigh distribution, the probability that the motion amplitude exceeds a chosen threshold value, a, can be calculated using:

$$F(x > a) = \exp\left(-\frac{a^2}{2\sigma^2}\right) \tag{4}$$

Then response value which probability of exceeding is $1/N$ can be given by:

$$a = \sqrt{2\sigma^2 \ln N} = \sqrt{2m'_0 \ln N} \tag{5}$$

Where m'_0 is the zero order moment corrected with the spectral bandwidth of the response spectrum:

$$m'_0 = (1 - \frac{\varepsilon^2}{2}) m_0 \tag{6}$$

The spectral bandwidth parameter, ε, is given by:

$$\varepsilon = \sqrt{1 - \frac{m_2^2}{m_0 m_4}} \tag{7}$$

m_n denotes the nth order moment given in this case by:

$$m_n = \int_0^\infty \omega^n S_R(\omega) \, d\omega \tag{8}$$

A range of wave periods from 5s to 17s are analyzed for calculating the maximum allowable significant wave heights for roll and pitch respectively. The allowable H_s for a given wave period is determined by assuming an initial value for H_s and iteratively changing this value for the 3 - hour short term response predictions, while checking each of the motion amplitude requirements of not exceeding 6 degrees, until the allowable H_s was determined.

These limited sea states are given in Fig. 7. It's noted that the minimum limited wave height H_s is 3.4m at $T_z = 8$s, which limited sea state is governed by pitch.

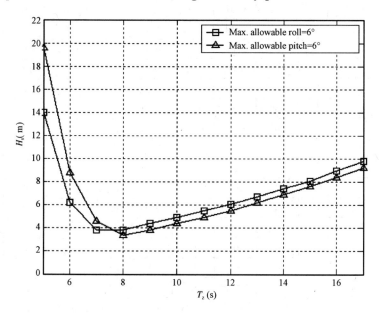

Fig. 7 Limited sea states for wet tow of the FPS

3 WEATHER WINDOW SELECTION FOR TOWING

Liuhua 11 - 1 oil field locates in the low-latitude area of South China Sea that belong tropical marine monsoon climate. From October to March of next year, northeast monsoon prevails. By the cold wave hitting, the weather is dry, cold and sunnier in the early stages, while cold and rainy in the later stages. From May to September, southwest monsoon prevails and tropical cyclone is frequently. The temperature is high and weather is wet and has much storms. Affected by monsoon climate, the South China Sea storms have obvious seasonal variation. South direction wave prevails in winter and northeast waves prevails in summer.

The joint distribution of H_s versus T_s for platform area in one year is listed in Table 1, where T_s is the significant wave period. The relationship between T_s and T_z is: $T_s = 1.20 T_z$, where T_z is zero up-crossing wave period. It can be seen from Table 1 that normally in this sea area, the wave period T_s is between 4s and 10s and the H_s less than 4m. Table 2 ~ Table 4 give the probability statistic of the sea states that satisfy the requirements of towing and different working conditions respectively.

Table 1　Combining Probability Distribution of the Sea Area of LH11−1 oil field

Significant wave height(m)	Significant wave period(s)										Total
	<2	2~3	3~4	4~6	6~8	8~10	10~12	12~16	16~20	>20	
>0.0≤0.5	0	0	0.03	3.11	0.62	0.03	0	0	0	0.04	3.82
0.5≤1.0	0	0	0.34	13.81	4.59	0.18	0	0	0	0	18.91
1.0≤1.5	0	0	0.04	10.39	7.37	0.36	0.01	0	0	0	18.17
1.5≤2.0	0	0	0	6.33	10.93	0.43	0.03	0	0	0	17.71
2.0≤2.5	0	0	0	2.36	11.84	0.41	0.08	0	0	0	14.69
2.5≤3.0	0	0	0	0.86	7.35	0.43	0.04	0	0	0	8.68
3.0≤3.5	0	0	0	0.31	5.37	1.19	0.04	0	0	0	6.91
3.5≤4.0	0	0	0	0.12	2.14	1.73	0.03	0	0	0	4.03
4.0≤4.5	0	0	0	0.09	1.04	0.82	0	0	0	0.01	1.96
4.5≤5.0	0	0	0	0.06	0.82	0.59	0.05	0	0.01	0	1.53
5.0≤5.5	0	0	0	0.01	0.47	0.50	0.10	0	0	0	1.08
5.5≤6.0	0	0	0	0	0.24	0.43	0.12	0	0	0	0.79
6.0≤6.5	0	0	0	0	0.12	0.39	0.04	0	0	0	0.55
6.5≤7.0	0	0	0	0	0.04	0.45	0.15	0	0	0	0.64
7.0≤7.5	0	0	0	0	0.02	0.24	0.16	0.01	0	0	0.42
7.5≤8.0	0	0	0	0	0	0.03	0.04	0	0	0	0.06
8.0≤8.5	0	0	0	0	0	0.03	0.02	0	0	0	0.04
8.5≤9.0	0	0	0	0	0	0	0.01	0	0	0	0.01
9.0≤9.5	0	0	0	0	0	0	0	0	0	0	0
9.5≤10.0	0	0	0	0	0	0	0	0	0	0	0
>10.0	0	0	0	0	0	0	0	0	0	0	0
Total	0	0	0.41	37.44	52.98	8.22	0.89	0.01	0.01	0.04	100.00
Mean(m)	0.0	0.0	0.7	1.2	2.2	3.9	5.4	7.4	4.9	0.9	
Max(m)	0.0	0.0	1.1	5.1	7.4	8.4	8.9	7.4	4.9	4.3	

Table 2　Probability Statistic of Meeting the Limited Sea States of Towing ($H_s < 3.4$m)

Month	1	2	3	4	5	6
Probability(%)	86.42	86.24	100	96.91	100	98.28
Month	7	8	9	10	11	12
Probability(%)	99.7	94.45	88.07	88.13	58.71	86.96

Table 3　Probability Statistic of Meeting the Limited Sea States of Installation ($H_s < 2$m)

Month	1	2	3	4	5	6
Days(d)	13	9	20	20	27	25
Probability(%)	43.3	30	66.7	66.7	90	83.3
Month	7	8	9	10	11	12
Days(d)	21	19	22	24	6	6
Probability(%)	70	63.3	73.3	80	20	20

Table 4 Probability Statistic of Meeting the Limited Sea States of ROV Working ($H_s < 2.5$m)

Month	1	2	3	4	5	6
Days(d)	19	13	25	24	30	28
Probability(%)	63.3	43.3	83.3	80	100	93.3
Month	7	8	9	10	11	12
Days(d)	27	25	24	26	10	10
Probability(%)	90	83.3	80	86.7	33.3	33.3

It can be seen from the above tables that environment condition from November to March of the next year is relatively poor and not suitable for disconnecting, installing or producing. As to complete the update modification needs about 4~5 months for the FPS and a single voyage from field to shipyard needs about 3~4 days, it's recommended that to conduct disconnection at field and complete the towing to shipyard in October, which is the end of the favorable weather season from March to October. Then the winter storm season can be used for the update modification. And the favorable weather season can be expected for towing back and re-connecting installation at field after the dry-dock upgrade modification. This schedule arrangement can reduce impact on the production of the FPS to a minimum, and limited sea states requirement for towing and operating at sea can be easily meet at the same time.

CONCLUSION

By analysis of the motion performance of the "Nan Hai Tiao Zhan" semi-submersible platform, limited sea states satisfied the amplitude of roll and pitch not exceed 6 degrees have been obtained under the condition of 7m draft of towing. By statistic analysis for environmental condition of South China Sea and general consideration on the term of production, towing and drydock modification, the weather window suitable for towing is recommended from the perspective of economy and security. The FPS has been successfully towed to Guang Zhou Huangpu Shipbuilding Co. for modification in the early November in 2011. And the upgraded Liuhua 11 - 1 FPS was towed beck to oilfield in May of 2012. Now, the field has been producing again successfully. The motion analysis and weather window selection for towing played a very important role on making schedule of this project.

REFERENCES

Bryant J H, Methvin J R, Dague E E, Zhu M C, Wang H K. 1996. Liuhua 11 - 1 Development-Field Development Overview[C]. Offshore Technology Conference, Paper 8172. Houston, Texas, May 6 - 9.

DNV. 2011. SESAM User Manual.

Sorrel D, Green J D, Florida M J, Chen M, Sheridan J. 1996. Liuhua 11 - 1 development-Phase I Installation [C]. Presented at the 1996 Offshore Technology Conference, Paper 8186. Houston, Texas, May 6 - 9.

Numerical Simulation on SPAR platform's Cable Tension in Deep Sea under Wave and Current

Song Sang, Changdong Li, Lei Li, Xueliang Jiang

College of Engineering, Ocean University of China, Qingdao, China

Abstract The mooring system of SPAR platform is critical, while the cable tension plays a very important role at the design of mooring system. It takes some classic SPAR for example which worked in Nanhai. The SPAR motion equation is set up and the cable tension is studied by using AQWA software. The numerical simulation of platform motions and cable tension is processed from the following three aspects: different wave-current angles, different layout of mooring lines and when one cable is broken. Some useful conclusions are obtained by the analysis of the results, and it may provide some reference for the design of mooring system.

Key Words SPAR platform; Mooring system; Wave and current; Cable tension

INTRODUCTION

Deep-sea drilling and production platform suffers external loads mainly including wind, waves and currents. Because of these loads, the mooring positioning system is needed to keep the platform in the work area. So the study, design and production of mooring system may face with new challenges (Li, Xu and Wang, 2010). The design of mooring system is very complex, the cable tension and the displacement of SPAR platform under constraints of mooring system are strictly limited according to all related specifications (Peng, 2009).

Analysis of the cable tension is very critical in the study of mooring system. Now, there are three methods to calculate the cable tension with respect to time domain method: lumped mass method, finite difference method and finite element method. Yang Mindong et al. (2009) analyzed the static equilibrium configuration and tension distribution of the mooring lines affected by different current force and concluded that when current force is small, loads increase from the bottom up; when current force increases, loads decrease and then increase; when the flow rate is particularly large, the cable tension is getting smaller and smaller. Furthermore, as shown in the configuration graph, elastic elongation of the polyester mooring lines is large under the external forces. Zhang Huoming et al. (2010) studied the static properties of multi-component mooring ropes, proposed a quick and efficient computation method for the static characteristics of the complex mooring system in the optimal design of equivalent water-depth truncation system. Gao Xifeng et al. (2010) performed the coupled dynamic analysis of mooring systems of SPAR platform and verified the correctness of it.

Previous researches did many studies on the shape and tension of the mooring line, but did few on the cable tension in the interaction of waves and current. And no literatures about cable tension analysis based on AQWA software are available. This paper studied cable tension based on AQWA software and demonstrated good results.

1 MOTION EQUATION OF MOORING FLOATING STRUCTURES (XIAO AND WANG, 2005)

Motion equation of mooring floating structures subjected to waves and currents may be expressed as

$$M\{\ddot{x}\} + C\{\dot{x}\} + K\{x\} = F_W + F_{Current} + F_M \quad (1)$$

Where M Mis the added mass coefficient matrix; C is the damping coefficient matrix; K is the restoring force coefficient matrix; $\{\ddot{x}\}$, $\{\dot{x}\}$ and $\{x\}$ correspond to generalized acceleration array, generalized velocity array and generalized displacement array, respectively; F_W is the wave force; $F_{Current}$ is the current force; F_M is the mooring force.

The added mass coefficient and the damping coefficient may be computed using frequency domain function, respectively and the restoring force coefficient may be computed using the ship statics theory. In the flow field fluid motion is irrotational based on the assumption of the inviscid fluid. Because the heave motion cycle and the pitch motion cycle are almost equal to wave period, this paper merely studied the heave motion and the pitch motion and the effects of first-order wave forces (diffraction force and radiation force) on platform. The formula for calculating first-order wave forces has been derived by Zhang Haiyan and Zhao Wenbin(Zhao, 2007; Zhang, 2008). Here is the brief introduction:

F, representing the total wave forces applied on the cylinder, may be expressed as:

$$F = \iint_s p \cdot n \mathrm{d}s \quad (2)$$

M, representing the total wave moment, may be expressed as:

$$M = \iint_s p \cdot (r \times n) \mathrm{d}s \quad (3)$$

The p in both expressions refer to the pressure of arbitrary point in the flow field, expressed as $p = -\rho \frac{\partial \Phi}{\partial t}$; The Φ corresponds to velocity potential of the flow field; Vector $n(n_x, n_y, n_z)$ is the unit outward normal of the cylinder; Vector r is radius vector from an arbitrary point of the cylinder surface to the point of moment; The s is the cylinder wetted surface.

Because of the geometric symmetry, linear wave forces loaded on the cylinder are just in the x direction (surge) and z direction (heave) and wave moment is not zero merely in the y direction (pitching). Force in the heaving direction and moment in the pitching direction may be expressed as:

$$F_z = \iint_s p \cdot n_z \mathrm{d}s = -\int_0^{2\pi}\int_0^a \rho \cdot \frac{\partial \Phi_1}{\partial t}\bigg|_{z=-h} r \mathrm{d}r \mathrm{d}\theta \quad (4)$$

$$M_y = \iint_s p \cdot (zn_x - xn_z)\,ds = -\int_0^{2\pi}\int_{-h}^{0} \rho \cdot z \cdot \frac{\partial \Phi_2}{\partial t}\bigg|_{r=a} a \cdot \cos\theta\,dz\,d\theta$$

$$-\int_0^{2\pi}\int_0^{a} \rho \cdot \frac{\partial \Phi_1}{\partial t}\bigg|_{z=-h} r^2 \cdot \cos\theta\,dr\,d\theta \qquad (5)$$

where h is the platform draft, a is the platform radius, Φ_1 is the internal velocity potential, Φ_2 is the external velocity potential.

Current force may be computed using expression as follows (Zhu, 1991):

$$F_{\text{Current}} = \frac{1}{2} C_D \rho A v^2 \qquad (6)$$

Here, C_D refers to drag coefficient vertical to the axis and may be desirable for 0.6~1.0 for circular structures, ρ is the density of sea water. v is the design current velocity, A is the unit projected area of the structures vertical to the current direction.

Restriction effect of the mooring system on deep-draft platform are great. And the nonlinear mooring forces produced by a single mooring line may be expressed, based on classical simulative mooring line formula, as (Nie, Liu, 2002; Sarkar A, Eatock Taylor R, 2001; Ellermann, Kreuzer and Markiewicz, 2002)

$$F_M = ks_m + \gamma k s_m^3 \qquad (7)$$

Where F_M is the cable tension, S_m is the displacement of the platform in the axial direction, k is the total linear rigidities, γ is nonlinear stiffness (expressed by stiffness) with the unit of $1/m^2$.

2 ANALYSIS OF EXAMPLES

This paper used the survival condition in which wave angle is 180°(in AQWA, wave or current angle refer to angle between propagation direction and x-axis counterclockwise direction) and wave cycle is 29.3s just as the natural period of heave in the case of internal resonance. Coordinate system is shown in Fig. 1.

Current velocity is 0.93m/s (Shi, 2011) as characteristic velocity of the South China Sea. The SPAR platform example (Wang, 2008) come from "Deep Star Project", and the water depth is 1018m. The other main parameters are given in Table 1.

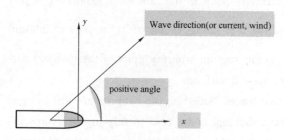

Fig. 1 Explanation of wave direction (or current, wind) in coordinate system

Table 1 The main parameters of platform

Designation	Unit	Quantity
Mass (except mooring system)	t	212,200
Diameter	m	37.2
Design draft	m	198.1
Height of centre of gravity	m	89
Density of sea water	kg/m³	1,025
Heaving damp ratio		0.005144
Pitching damp ratio		0.004308
Surging damp ratio		0.002759
Initial metacentric height	m	10.08
Heaving natural period	s	29.3
Pitching natural period	s	55.4

Table 2 Parameters of mooring lines

Item	Bottom chain	Middle wire	Top chain
Material	K4 unstudded cable	Steel wire	K4 unstudded cable
Length (m)	400	1200	230
Mass in air (kg/m)	191.4	48.7	191.4
Mass in water (kg/m)	166.5	38.6	166.5
Cross-sectional area (m²)	0.007	0.007	0.007
EA(N)	7.0445×10^8	7.8464×10^8	7.0445×10^8
Breaking strength (N)	9.515×10^6	7.465×10^6	9.515×10^6

In the case every line is made up of three parts: top chain, middle wire, and bottom chain, and they are arranged in four groups with each including three lines, evenly distributed within the 360° direction. The mooring parameters (Shi and Yang, 2010) are shown in Table 2, and their layout is shown in Fig. 2.

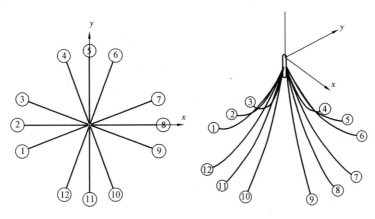

Fig. 2 Arrangement of mooring lines

2.1 Impact of Different Wave-current Angle on the Cable Tension

During the analysis, the height of wave is 8m and the incident angle are 180°, 90°, 0°. Since the layout of mooring lines is symmetrical, the study just selected No. 2, No. 8 and No. 11 from all 12 lines. Point A, B and C, which point A is the anchor point (starting point) belonging to anchor chain, point B belongs to steel wire, point C is fairlead belonging to anchor chain, can reflect the tension conditions of the whole line. The distance between A and B along the mooring line is 1575.5m (In AQWA, there are 51 points in one line, the first one is the anchor point and the last is in the fairlead. The distance from every point to the anchor point is equal to the length of catenary, calculated by the software. The point connected steel wire and mooring chain is not available for the shortcomings of the software, so the point close to cable reel, point B, is picked to be analyzed, which is 1575.5m away from the anchor point.).

The shape of mooring line in the AQWA is shown in Fig. 3.

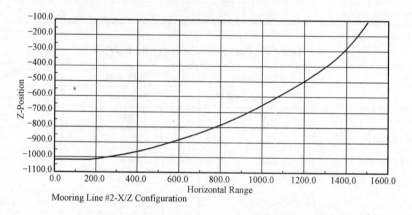

Fig. 3 Shape of mooring line

When current incident angle is 180°, that is, wave and current are in the same direction, the tensions of No. 8, No. 11 and No. 2 under the influence of wave and current are plotted in Fig. 4, Fig. 5 and Fig. 6.

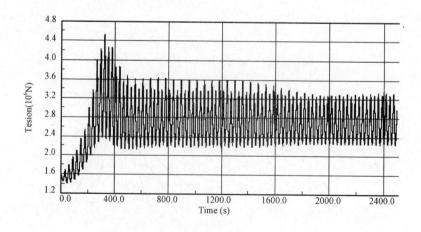

Fig. 4 Tension at point C of No. 8

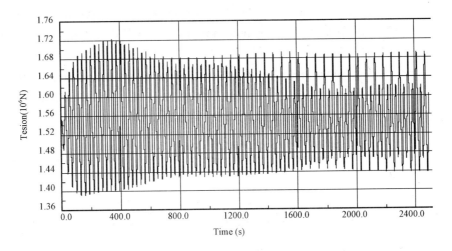

Fig. 5 Tension at point C of No. 11

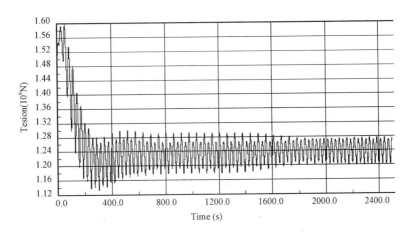

Fig. 6 Tension at point C of No. 2

The tension at point C of No. 8 is about 4500kN, No. 11 is about 1720kN, and No. 2 is about 1600kN. So the tension produced in No. 8 is the biggest. Safety factors of point C (safety factor is the ratio of breaking tension and maximum tension) are listed in Table 3.

Table 3 Safety factors of point C

	No. 8	No. 11	No. 2
Tension(kN)	4500	1720	1600
Safety factor	2.114	5.532	5.946

The tension at point A of No. 8 is 3700kN, the tension at point B is 4300kN, the tension at point C is 4500kN shown in Fig. 4 as the biggest. For the fact that on the No. 8 safety factor at point A is 2.571, at point B is 1.734, at point C is 2.114, and safety factor required by specification is 1.67, so the No. 8 is safe.

In summary, when current incident angle is 180°, the tension at point C of No. 8 is the biggest and safety factors at point A, B, C satisfy the specification. That is to say the mooring line No. 8 is safe when the height of wave reaches 8m. Conclusion may be drawn from the above figures that tension increases gradually from the anchor point to the fair-lead along the line, and reaches the maximum at the fair-lead.

When current incident angle is 90°, under the influence of wave and current, the tension at point C of No. 8 is about 1950kN, No. 11 is about 3250kN, and No. 2 is about 1880kN. So the tension produced in No. 11 is the biggest. Safety factors of point C are listed in Table 4.

Table 4 Safety factors of point C (current incident angle 90°)

	No. 8	No. 11	No. 2
Tension (kN)	1950	3250	1880
Safety factor	4.879	2.927	5.061

In summary, when current incident angle is 90°, the tension at point C of No. 11 is the biggest and safety factor satisfies the specification. The tension at point B of No. 11 is about 2950kN and the safety factor at point B is 2.530 meeting the requirements of specification. So the No. 11 is safe.

When current incident angle is 0°, that is, wave and current are in the opposite directions, under the influence of wave and current, the tension at point C of No. 8 is about 1520kN, No. 11 is about 1740kN, and No. 2 is about 3200kN. So the tension produced in No. 2 is the biggest. Safety factors of point C are listed in Table 5.

Table 5 Safety factors of point C (current incident angle 0°)

	No. 8	No. 11	No. 2
Tension(kN)	1520	1740	3200
Safety factor	6.259	5.468	2.973

In summary, when current incident angle is 0°, the tension at point C of No. 2 is the biggest and safety factor satisfies the specification. The tension at point B of No. 2 is about 2880kN and the safety factor at point B is 2.592 meeting the requirements of specification. So the No. 2 is safe. Therefor the entire mooring system is operating safely under such conditions.

Changing the combined angle of wave and current to study the cable tension and the conclusion is that the tension at the fair-lead is the greatest, while the tension at the connection points of the different materials of compound mooring lines should be checked in order to ensure the safety of the entire mooring system, the change of included angle of wave and current may have a certain impact on the cable tension, the cable tension on the meeting-wave side reaches the maximum when wave and current are in the same direction (180°).

2.2 Impact of Mooring Way on the Cable Tension

This section compares the cable tension in two different mooring ways: three-groups way and four-groups way. Values of parameters in both mooring ways are exactly equal. All mooring lines, divided into three groups, are uniformly distributed in the range of 360° with each group composed of four lines. During the analysis, the height of wave is 8m and the incident angle is 180°, and tension at point A, B, C of No. 2 is to be studied. The layout of mooring lines arranged in three-groups way is shown in Fig. 7.

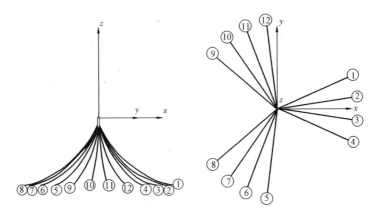

Fig. 7 Layout of mooring lines in three-groups way

When the height of wave is 8m, the tension at point A, B, C of No. 2 arranged in three-groups way is shown in Fig. 8, Fig. 9 and Fig. 10.

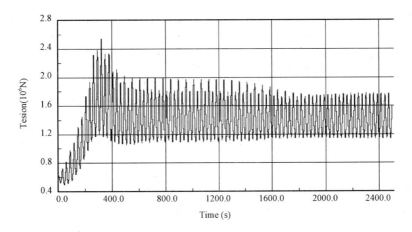

Fig. 8 Tension at point A of No. 2

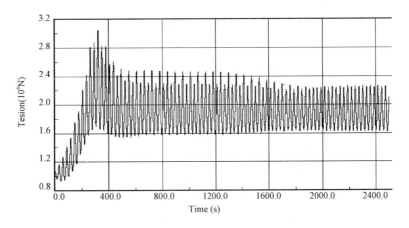

Fig. 9 Tension at point B of No. 2

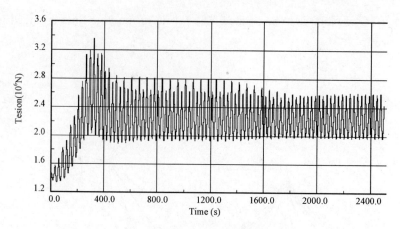

Fig. 10　Tension at point C of No. 2

As is shown in Fig. 8, the tension at point A is about 2600kN, point B is about 3050kN and point C is 3400kN.

Previous section has given the tension of No. 8 arranged in four-groups mooring way. Safety factors and tensions of No. 2 arranged in three-groups way are listed in Table 6 combined with those of No. 8 arranged in four-groups way.

Table 6　Safety factors and tensions of mooring line

		A	B	C
Three-groups (No. 2)	Tension (kN)	2,600	3,050	3,400
	Safety factor	3.659	2.444	2.789
Four-groups (No. 8)	Tension (kN)	3,700	4,300	4,500
	Safety factor	2.571	1.734	2.114

In summary, it can be concluded from Table 6 that the tension at fair-lead of mooring line arranged in four-groups way is larger than that arranged in three-groups way when wave and current are in the same direction due to the consideration of cable tension on the meeting-wave side and the ignoration of the others.

2.3　Impact on Cable Tension and Surging, Heaving and Pitching Motion with One Mooring Line Broken

During the analysis the mooring lines are arranged into four groups (see Fig. 2). In order to study the cable tension with one mooring line broken, condition parameters are set as that wave and current incident angle is 180°, wave period is 29.3s and the height of wave is 8m. The mooring line No. 8 on the meeting-wave side is assumed to be broken. The line No. 7 adjacent to the broken No. 8 is selected to do analysis considering the tension of mooring line on the meeting-wave side is the largest, calculating the tension at point A, B and C. The number and layout of mooring line are shown in Fig. 11 with one broken line.

The tensions at point A, B, C of No. 7 are shown in Fig. 12, Fig. 13, Fig. 14 respectively.

The tension at point A is about 5700kN shown in Fig. 12, the tension at point B is about 6100kN shown in Fig. 13, the tension at point C is about 4000kN shown in Fig. 14.

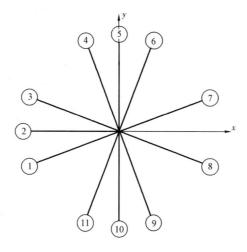

Fig. 11　Layout of mooring line

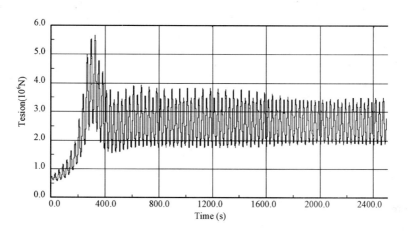

Fig. 12　Tension at point A of No. 7

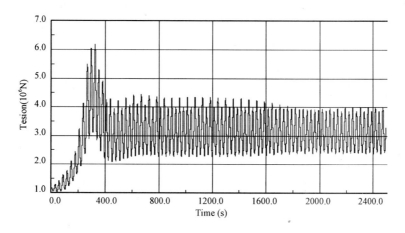

Fig. 13　Tension at point B of No. 7

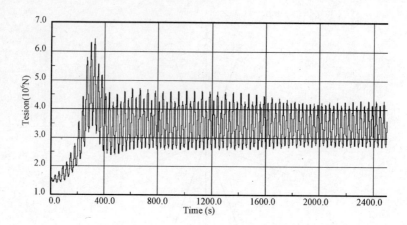

Fig. 14 Tension at point C of No. 7

Before the line No. 8 is broken, the tensions at point A, B, C of No. 7 are 3200kN, 3650kN, 4000kN respectively.

Tensions and safety factors at point A, B, C of No. 7 before and after the mooring breaks are listed in Table 7 for comparison.

Table 7 Safety factors and tensions of No. 7 before and after the mooring breaks

		A	B	C
No. 7 (Unbroken)	Tension(kN)	3,200	3,650	4,000
	Safety factor	2.973	2.045	2.378
No. 7 (Broken)	Tension(kN)	5,700	6,100	6,500
	Safety factor	1.669	1.222	1.463

From Table 7, it can be concluded that before the mooring line breaks tensions at point A, B, C of No. 7 are small relatively with safety factors conforming to the specification, tensions of No. 7 increase greatly with safety factors decreasing sharply after the mooring line breaks. Under the broken condition, safety factors at point A, C of No. 7 meet the requirement of the specification which requires the safety factors of mooring lines reaching 1.25 with one line broken. Safety factor at point B is slightly less than the required value leading to some potential danger, but that does not prevent the platform from working under certain sea conditions.

Before and after the mooring line breaks, time domain responses of the three type of motion as surge, heave and pitch are shown in Fig. 15, Fig. 16, Fig. 17.

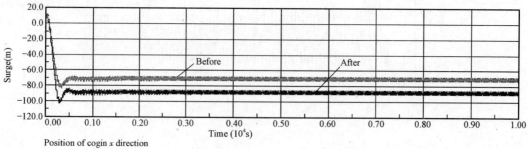

Fig. 15 Comparison of surging motion before and after the mooring breaks

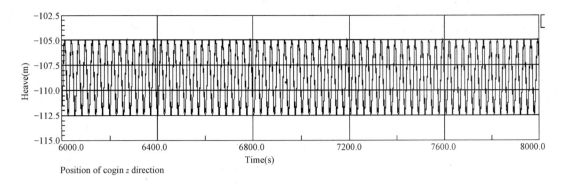

Fig. 16 Comparison of heaving motion before and after the mooring breaks

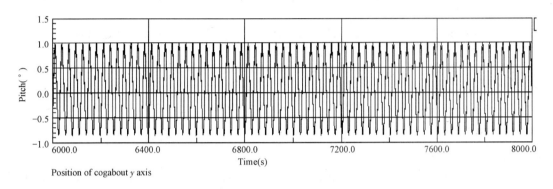

Fig. 17 Comparison of pitching motion before and after the mooring breaks

The surging motion performance has a marked increment after the fracture of one mooring line while the motion performance of heave and pitch change insignificantly as displayed in Fig. 17. This suggests that the fracture of one mooring line has a significant impact on surging motion and a smaller impact on heave and pitch. It can, therefore, be said that the platform can still run normally with only one line broken under certain conditions.

To sum up, when wave-current incident angle is 180° and one mooring line on the wave-meeting side is broken, cable tensions of the two mooring lines nearest the broken one increase greatly with the underlying danger of fracture of mooring lines. Furthermore the fracture of one mooring line may cause a larger surging motion performance and has a smaller impact on heave and pitch compared with the unbroken mooring system. As long as the rest mooring lines are functioning, the platform can still run normally.

CONCLUSION

From the calculation of example in this paper, conclusion can be drawn as follow:

(1) The cable tension at the fair-lead is the largest along the entire mooring line and it reaches the maximum when wave and current are in the same direction. While the safety factor at the fair-lead then is within the specification meaning the line will not break at this point. Compared to the point at fair-lead, the connection point (point B) of steel wire and anchor chain is easier to break with lower breaking strength.

(2) When wave and current are in the same direction, the maximum tension at the fair-lead of mooring line arranged in four-groups way is larger compared with the three-groups mooring way under the same conditions as the result of analysis on impact of mooring way on the cable tension.

(3) In a certain sea condition the breaking of one mooring line can affect the adjacent lines and even cause the fracture of the adjacent lines and thereby the instability or capsizing of the platform.

REFERENCES

Ellermann K, Kreuzer E, Markiewicz M. 2002. Nonlinear Dynamic of Floating Cranes. Nonlinear Dynamics, 27(2): 107 – 183.

Gao X F, Zhang Z Y, Yin B B. 2010. Coupled Dynamic Analysis of Mooring Systems of SPAR Platform [J]. Ocean Technology, 29(1): 70 – 73.

Katrin E, Edwin K, and Marian M. 2002. Nonlinear Dynamics of Floating Cranes [J]. Nonlinear Dynamics, 27: 107 – 183.

Li Z Z, Xu X P, and Wang H L. 2010. Development of Offshore Pplatform Mooring Systems [J]. Oil Field Equipment, 39(5): 75 – 78.

Nie W, Liu Y Q. 2002. Structural Dynamic Analysis of Ocean Engineering [M]. Harbin: Harbin Engineering University Press.

Peng C, et al. 2009. Study On Relationship Between the Response of Spar and the Mooring Lines [J]. China Offshore Platform, 24 (4): 9 – 20.

Sarkar A, Eatock Taylor R. 2001. Low-frequency Responses of Nonlinear Moored Vessels in Random Waves: Coupled Surge, Pitch and Heave Motions [J]. Journal of Fluids and Structures, 15: 133 – 150.

Shi Q Q. 2011. Research on Kinetic and Dynamic Characteristics of A Deepwater Drilling Semi-submersible Platform [M]. Shanghai: Shanghai Jiaotong University Press.

Shi Q Q, Yang J M. 2010. Research on Hydrodynamic Characteristics of A Semi-submersible Platform and Its Mooring System [J]. The Ocean Engineering, 28(4): 1 – 8.

Wang W W. 2008. Numerical Simulation of Heave-pitch Coupled Nonlinear Motions for Spar Platform [M]. Tianjin: Tianjin University Press.

Xiao Y, Wang Y Y. 2005. The Calculation and Analysis for the Mooring System [J]. China Offshore Platform, 20 (6): 37 – 42.

Yang M D, et al. 2009. Static Analysis of Mooring Lines Using Nonlinear Finite Element Method [J]. The Ocean Engineering, 27(2): 14 – 20.

Zhang H M, Zhang X F, Yang J M. 2010. Static Characteristic Analysis of Multi-component Mooring Line Based on Optimization Thinking [J]. Ship Science and Technology, 32(10): 114 – 121.

Zhang H Y. 2008. Study on Nonlinear Motion Behavior of Coupled Heave-pitch for the Spar Platform [M]. Tianjin: Tianjin University Press.

Zhao W B. 2007. Analysis of Hydrodynamic Loads and Coupled Heave-pitch Motion for Spar Platform [M]. Tianjin: Tianjin University Press.

Zhu Y R. 2002. Wave Mechanics in Ocean Engineering [M]. Tianjin: Tianjin University Press.

Research on Hydrodynamic Characteristics of Heave Damping Plate on Deepwater SPAR Platform

Weiwu Wu[1], Miaomiao Huang[1], Xiaofeng Kuang[1], Quanming Miao[1]

[1] China ship scientific research center, Wuxi, China

Abstract Firstly, Hydrodynamic forces of circular heave damping plate with different porosity are studied and simulated. Numerical results of the drag coefficient and added-mass coefficient are given in this paper, and the influence of the porosity on the drag and added mass coefficient is also investigated. The hydrodynamic forces of heave damping plate with various numbers of holes and various shapes are compared at the same porosity. Secondly, different distances between two heave damping plates are studied. Finally, numerical simulation was established to simulate complex three-dimensional heave damping plate on SPAR platform, The numerical results indicate that the small corner steel structure play an important role in flow field. The hydrodynamic coefficients are quite different from each other; It is shown that practical plate with complex structure can't be simplified as smooth plate for this reason.

These results will give some useful instruction for the optimization design and improvement of the heave damping plate of deepwater SPAR platform.

Key Words Heave damping plate; SPAR; Hydrodynamic

NOMENCLATURE

Keulegan-Carpenter Number $\quad KC = \dfrac{2\pi a}{D}$

D is the diameter of circular heave damping plate; a is the amplitude of displacement; f is the frequency of heave movement; C_d is the drag coefficient; C_a is the added mass coefficient.

INTRODUCTION

As the increasing of the crude oil requirements, many countries invest in exploitation of ocean oil and gas. A lot of different kinds of new platform are designed, SPAR is one of them. The heave performance of SPAR is expected to be very good because of the rigid risers and other production equipments on the platform. In order to avoid the sympathetic vibration of platform and wave and make sure that the platform has a good movement performance, generally, this can be done by making the natural period of heave much bigger than the period of wave. Heave damping plate can produce big heave damping and added mass to SPAR platform, and it can prolong the heave natural period of SPAR. The appearance of heave plate changed the situation that SPAR must have big draft to make sure the big heave natural period, which makes that the construction cost reduce obviously. So, many academicians focus on the research field of hydrodynamic performance of heave damping plate.

Research on hydrodynamic performance of heave damping plate included model test and numerical simulation. There are a lot of research works on these two aspects, for example, in model test: Thiagarajan and Troesch (1998) test the heave damping of a vertical circular cylinder with a heave damping plate in the bottom, it is proved that the heave damping plate can added the heave damping of the circular cylinder. Prislin (1998) have done a decay test of a single square plate in still water, and the relationship among the drag efficient and KC number and Re number is presented. Johnson (1995) had done a model test research on hydrodynamic performance of several square plates. Tao (2008) accomplished the forced oscillation test of heave damping plate by Planar Motion Mechanism (PMM), the influence of the porosity on the drag and added mass coefficients are investigated. Ji (2003) has done a forced oscillation test of a triangle heave damping plate. Zhang (2008) evaluated the influence of heave damping plate to the heave performance of a deepwater platform. In theoretic analysis and numerical simulation: Tao (2003) simulated the wake flow of a forced oscillation cylinder with a circular heave damping plate, the hydrodynamic coefficients of heave damping plate and circular cylinder are presented.

The researches above are based on the object of simplified heave damping plate. This kind of researches are very important, for example, saving computation resource and shortening research period, however, in the practical design of SPAR platform, in order to satisfy the security requirement the structure of heave damping plate are generally very complex. This complex structure should be considered because it can affect the hydrodynamic coefficients and wake flow a lot. This paper also presents a hydrodynamic performance numerical simulation of a complex structure heave damping plate at the first time.

1 NUMERICAL METHOD

1.1 Mathematics Model

The heave damping of solid heave damping plate is due to the vortex shedding on the brim of the plate. Using the potential theory can't simulate this phenomenon, so the viscous theory method is adopted to simulate the wake flow in this paper, and the inertial frame and dynamic mesh technology are used to simulate the heave movement of the heave damping plate. The numerical methods of dynamic mesh as follow.

When the amplitude in vertical direction of heave damping plate is small, the control equations of flow field are:

Continuity equation:

$$\nabla \cdot v = 0 \qquad (1)$$

While v is the vector of velocity.

Momentum equation:

$$\frac{\partial}{\partial t}(\rho v) + \nabla \cdot (\rho v\, v) = -\nabla p + \nabla \cdot [\mu(\nabla v + \nabla v^T)] \tag{2}$$

While ρ is density of fluid, p is pressure, μ is the viscosity of fluid.

Dynamic mesh should satisfy the geometry conservation, and the differential coefficient of control volume is calculated by the following equation:

$$\frac{dV}{dt} = \oint_{\partial\Omega} \boldsymbol{u}_g \cdot d\boldsymbol{S} = \sum_{n_f} \boldsymbol{u}_{g,j} \cdot \boldsymbol{S}_j \tag{3}$$

n_f is the number of control surface, $\boldsymbol{S}_j$ is the vector area of the surface j.
$\boldsymbol{u}_{g,j} \cdot \boldsymbol{S}_j$ is calculate by Eq. (4):

$$\boldsymbol{u}_{g,j} \cdot \boldsymbol{S}_j = \frac{\delta V_j}{\Delta t} \tag{4}$$

δV_j is the volume variation in one time step of surface j.

The RNG $k-\varepsilon$ turbulence model is adopted in this numerical simulation, it can simulate the flow with big strain ratio and curving streamline very well.

1.2 Computational Domain and Grid

In order to mitigate the influence of boundary to the flow field of heave damping plate, the computational domain is needed to be big enough. The computational domain in this paper is showed in Fig. 1. The distance on top and under the heave damping plate is about 4.0L, the distance to the side of heave damping plate is also about 4.0L.

Structured mesh is adopted in all calculation cases in this paper, near the wall of heave damping plate the mesh is dense and this can also be showed in Fig. 1.

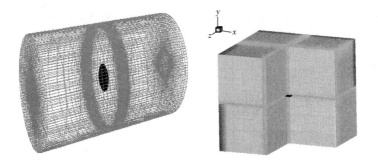

Fig. 1 Computational domain and mesh

1.3 Computational Objects Surface Grid

Geometry model of computational objects are showed in Fig. 2 and Fig. 3. The surface grid of circular heave damping plates are showed in Fig. 2 and the surface grid of complex structure heave damping plates with different shape of corner steer are showed in Fig. 3.

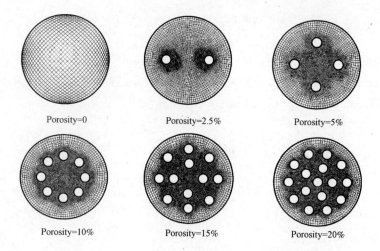

Fig. 2 Surface grid of circular heave damping plates

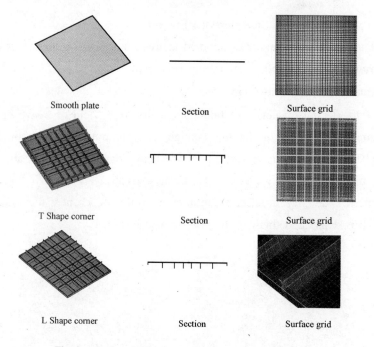

Fig. 3 Surface grid of complex heave damping plates

2 HYDRODYNAMIC CHARACTERISTICS ANALYSIS

2.1 Hydrodynamic Coefficients Analysis

The heave damping and added mass coefficient can be derived from the force and displacement time history by the method of Fourier average. The heave hydrodynamic force is equal to the pressure integral on the surface of the heave damping plate. The viscous drag can be expressed as the Morison equation as following.

$$F_d(t) = \frac{1}{2}\rho S C_d u(t) |u(t)| \tag{5}$$

$u(t) = -a\omega\cos(\omega t)$ is the velocity of heave damping plate, it shows that the heave damping plate forced to oscillate in sinusoid track. The drag and added mass coefficient can be derived from the time history of hydrodynamic force by Fourier method, which was presented by Sarpkaya & Isaacson (1981).

$$C_d = \frac{3\omega}{4\rho S v_m^2} \int_0^T F_t \cos(\omega t) \, dt \tag{6}$$

$$C_a = -\frac{1}{\pi \rho \forall v_m} \int_0^T F_t \sin(\omega t) \, dt \tag{7}$$

$v_m = a\omega$ is velocity amplitude of heave, T is the heave period, S is the area of the heave damping plate, $\forall$ is the volume of the heave damping plate.

Fig. 4 is the typical time history of displacement and hydrodynamic force, it shows that there is a obvious phase between the displacement and force, which is equal to 28.8°. Time histories in one period of different porosity are showed in Fig. 5.

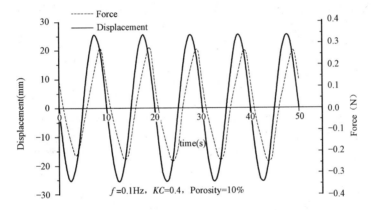

Fig. 4 Time history of displacement and force

Fig. 5 shows that the phase between displacement and force increasing as the porosity increasing. When the porosity is equal to 20%, $f = 0.1$Hz and $KC = 0.2$, the phase is the maximum equal to 37.5°. When the frequency is the same, the phase between displacement and force reducing as the KC increasing. When KC is the same, frequency affect the phase evidently, as the frequency increasing from 0.1Hz to 0.5Hz, the phase reduce rapidly.

The hydrodynamic force of heave damping plate with various numbers of holes at the same porosity is presented in Fig. 6. It indicated that the hole number has little influence to the hydrodynamic coefficient, so in this numerical simulation, big hole can be chosen instead of very small holes, which can make the meshing more simply and save the computation resource greatly.

Fig. 7 presented the hydrodynamic coefficients at different porosity when $KC = 0.2, 0.4$ and $f = 0.1$Hz, 0.5Hz.

For circular heave damping plate, Fig. 7 shows that the drag coefficient is increasing and then decreasing as the porosity increasing, when the porosity is about 10%, the drag coefficient has a maximum value, while the added mass coefficient is decreasing observably as the porosity increasing. Porosity plate can make the drag coefficient bigger at low KC number than at high KC number.

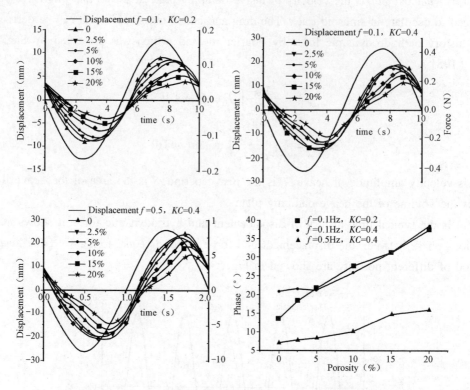

Fig. 5　Phase between force and displacement of heave damping plates

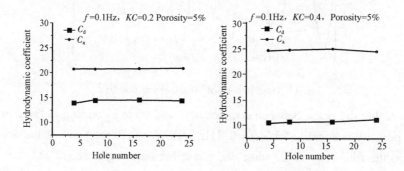

Fig. 6　Influence of hole number to Hydrodynamic

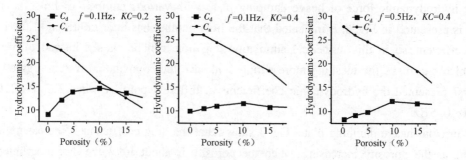

Fig. 7　Hydrodynamic coefficient vs. porosity

Fig. 8 shows the hydrodynamic coefficients results of heave damping plate with different shape at the same situation, $KC = 0.2, f = 0.1 \text{Hz}$.

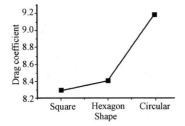
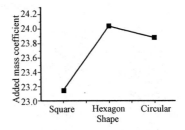

Fig. 8 Hydrodynamic coefficient of plates with different shape

The distance between two heave damping plates is very important parameter which can affect the heave hydrodynamic coefficients evidently. Fig. 9 shows the results of different distance of circular plate at the same porosity equal to 10%.

Fig. 9 shows that when the distance is bigger than $1.0D$, it can affect the added mass coefficient little. The drag coefficient is almost affected little by the distance when the distance is bigger than $1.5D$. The maximum hydrodynamic coefficients are close to the situation of single plate.

In fact, in the practical applications, the heave damping plate generally has different shape other than circular plate, and the plate often has a very complex structure with different shape of corner steer other than flat.

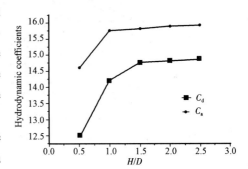

Fig. 9 Results of distance optimize computation between two plate

Using the numerical method above, the hydrodynamic coefficients of heave damping plates with complex structure on a certain SPAR platform have been evaluated in this paper. Different shape of corner steel are simulated and compared with the smooth plate, the outline and surface grid showed in Fig. 3. The results of hydrodynamic coefficients listed in the following Table 1 and showed that the influence of the complex structure of different shape corner steel is obvious. It indicated that the complex structure of corner steel should be considered in the practical engineering of SPAR design.

Table1 Hydrodynamic coefficients of complex and smooth plates

item	Unit	L shape corner	T shape corner	Smooth
Maximum drag	N	1.85×10^7	1.68×10^7	1.80×10^7
phase	(°)	20.16	13.68	20.16
Drag coefficient	—	8.78	5.86	8.58
Added mass	kg	2.03×10^7	1.93×10^7	2.06×10^7

2.2 Flow Field Analysis

The flow fields of two complex structure heave damping plates are compared with the flow field of smooth plate. Fig. 10 ~ Fig. 12 show that the vortex of the T shape corner steel plate can't develop completely and shed fully, which lead to smaller drag coefficient than the smooth plate and L shape plate. Fig. 13 ~ Fig. 15 is the instantaneous velocity fields of the three plates.

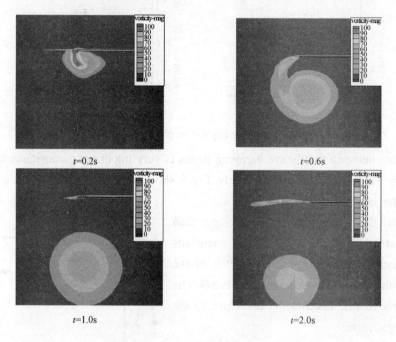

Fig. 10　Vortex contours of smooth plate

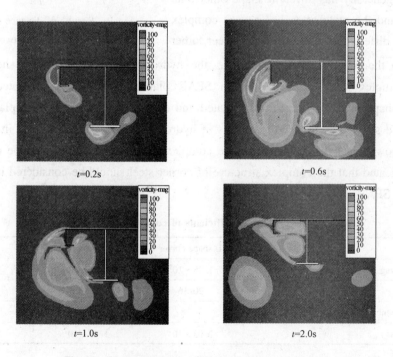

Fig. 11　Vortex contours of T shape corner steel plate

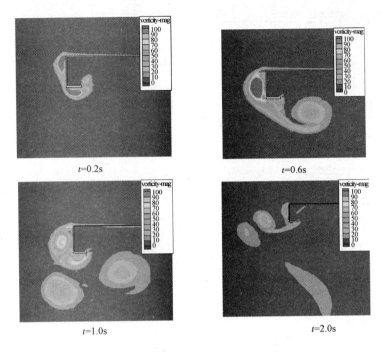

Fig. 12　Vortex contours of L shape corner steel plate

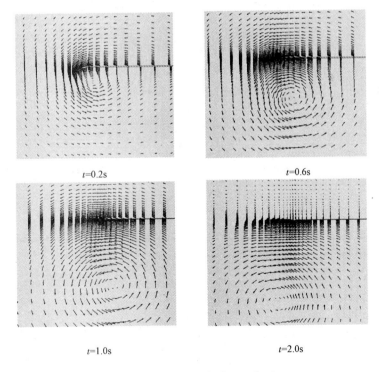

Fig. 13　Velocity field of smooth plate

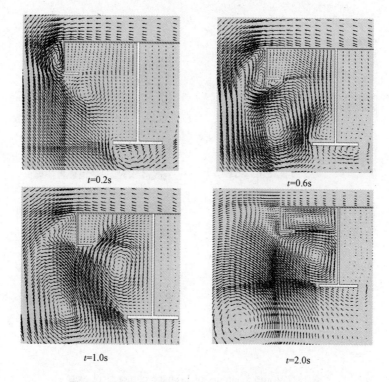

Fig. 14 Velocity field of T shape corner steel plate

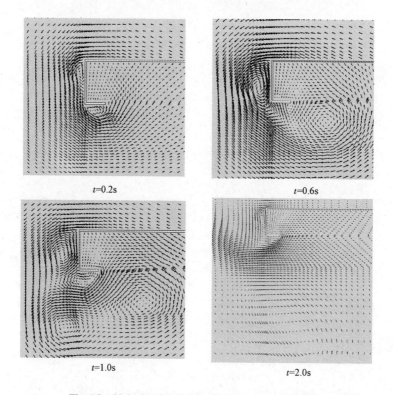

Fig. 15 Velocity field of L shape corner steel plate

DISCUSSION

According to this research results, some discussions can be made as experiential guidelines to optimize the hydrodynamic characteristics of heave damping plate on SPAR platform.

First, at low KC number for SPAR platform, porosity plate can add the heave damping much better than the solid plate, when the porosity equal to 10% the drag coefficient has the maximum value.

Second, circular plate can produce bigger drag coefficient than square plate and hexagon plate with the same area in the vertical direction.

Third, when the distance is bigger than $1.0D$, it can affect the added mass coefficient little. The drag coefficient is almost affected little by the distance when the distance is bigger than $1.5D$. The maximum hydrodynamic coefficients are close to the situation of single plate.

CONCLUSIONS

CFD dynamic mesh technology is used to simulate the flow field of the heave damping plate in this paper. The results can be usable for establish the guidelines for the design of heave damping plate on SPAR platform.

From the numerical simulation results of the complex structure plate, it indicated that the complex structure of heave damping plate can't be ignored in the particle engineering of SPAR design.

ACKNOWLEDGEMENTS

The authors are grateful to the financing of the ministry of industry and information department. Also thanks the readers for their suggestions to this paper.

REFERENCES

Ji Hengbin, Huang Guoliang, Fan Ju. 2003. Forced Oscillation Model Test of Heave Damping Plate[J]. journal of Shang Hai Jiao Tong university, 37(7): 977 – 980.

Johnson E. 1995. Added Mass and Damping of Truss Spar[R]. Deep oil technology, International report, California, USA. .

Prislin I, Blevins R D, Halkyard J E. 1998. Viscous Damping and Added Mass of Solid Square Plates [C]. Proceedings of the international conference on offshore mechanics and arctic engineering, OMAE, Lisbon, Portugal.

Tao L, Thiagarajan K. 2003. Low KC Flow Regimes of Oscillating Sharp edges Ⅰ. Vortex shedding observation [J]. Applied ocean research, 25: 21 – 35.

Tao L, Thiagarajan K. 2003. Low KC Flow Regimes of Oscillating Sharp edges Ⅱ. Hydrodynamic forces [J]. Applied ocean research, 25: 53 – 62.

Tao L, Dray D. 2008. Hydrodynamic Performance of Solid and Porous Heave Plates[J]. Ocean engineering, 35: 1006 – 1014.

Thiagarajan K, Troesch A. 1998. Effects of Appendages and Small Currents on the Hydrodynamic Heave Damping of TLP Columns[J]. Journal on offshore mechanics and arctic engineering, 120: 37 – 42.

Zhang Fan. 2008. Conception Design and Hydrodynamic Research of Deep Water SPAR Platform [D]. Shang Hai Jiao Tong University.

Technique Design of Model Tests for a Typical Wave Energy Converter

Xinyun Ni, Quanming Miao, Xiaofeng Kuang, Bo Lan

China Ship Scientific Research Center, Wuxi, China

Abstract As a kind of powerful renewable energy, wave energy has been paid much attention by researchers and engineers. Nowadays, the wave energy converters (WECs) can be mainly classified into three types (Oscillating Water Column, Oscillating Float and Overtopping). The Oscillating Float Converter is a popular one over the world for its higher efficiency and simplicity. In this paper, authors will focus on designing model test schemes and critical techniques, finally, establishing fundamental experiment procedures to evaluate output power generated by Oscillating Float Converter. The essential part of the test model is a heaving buoy generating electricity with a low frequency generator. It can explain the basic converting mechanism of the WEC, for instance, the motion of the buoy, energy transformation and electricity generation. Referring to model test of WEC, two important respects, including the buoy's motion in waves and the wave energy conversion efficiency, should be considered. For this purpose, the basic steps of WEC model test had been established and model tests with different wave parameters had been completed in CSSRC's wave basin. According to the research above, the test steps and measurement technologies are feasible.

Key Words Wave energy; Buoy; Efficiency; Model test

INTRODUCTION

Energy is an important basic support for the development of society. Every field, for instance, industry, agriculture, manufacturing, needs a large amount of electricity, which is one kind of energy. As the developing of economy, much more energy has been required. At present, energy is mainly converted from oil, natural gas and coal. The usage of these fuels will cause pollution due to the emission of CO_2. Meanwhile, oil, natural gas and coal are limited and will decrease as being consumed.

Impressed by large amount of wave energy, more than 1,500 patents had been invented for utilizing wave power for individuals' requirements (Stahl, 1892). The earliest user of wave power was Gerald in 1799. As the oil crisis in 1973, the research on WEC had developed dramatically in universities and institutions. During 1980s, when the oil price declined, the budget paid on wave energy research was reduced, leading to the dramatic decrease of research in WEC area (Ross, 1995). Recently, with the issue on the CO_2 pollution to the atmosphere and less and less fossil fuel reserve, people once more paid attention to wave energy research and development. By now, many types of WECs had been designed and tested in Japan, UK, Netherlands, and USA and other countries.

For one kind of ocean energy, the storage is abundant for its exploitation. Many countries which have put their efforts on wave energy exploitation have invented multiform WECs, some of which have been applied to generate electricity for commercial use. For example, The LIMPET OWC was one of the OWC converts developed by WaveGen Ltd. in Scotland (Heath, Whittaker and Boake, 2000). Another typical OWC WEC is Mighty Whale, which is now the most successful one developed by Japan and a 120kW prototype had been installed 1.5km off shoreline at 40m water depth (Washio, Osawa and Nagata, 2000). Wave Dragon is one of Wave Overtopping type devices, which was developed by a group of companies. Both 1/50 scale and 1/3.5 scale models had been tested in laboratory (Sørensen, Hansen, Friis-Madsen, et al., 2000). Pelamis is an Oscillating Buoy Converter, which had been developed successfully in UK (Yemm, 1999). It had been applied in commerce for generating electricity. The oscillating type device Salter Duck, which could covert both the kinetic and potential energies to electricity (Thorpe, 2000), was introduced by S. Salter in 1974. These devices showed above are part of WECs. There are many more wave energy devices being developed in coastal countries.

Different typical WECs have been developed preliminarily. However, the efficiency, reliability and survivability of wave converters must be concerned and considered before they are applied in real ocean conditions. Model test is the most direct way to evaluate the performance of WEC. Many published papers had reported the methods and results for their model tests (Budal, Falnes, 1980; Henderson, 2006; Ferreira, Hadden, 2009; Stalberg, 2005; Bhinder, Mingham, et al., 2009). The model test of WEC has played a pivotal role on WECs' design. In ITTC 2011, the WEC model test was also an essential part of the documents.

In this paper, the main purpose of author is to establish a model test method in basin for oscillating buoy WEC. To achieve this purpose, a schematic graph of oscillating buoy converter had been shown which will be used in test. After that, a technique design including buoy's dimension, generator selection, speed-up gear design, balance weight design, and similarity criterion had been expressed. In the following section, the model tests and results have been displayed. The test is performed in regular waves and irregular waves with different parameters.

1 A TYPICAL WAVE ENERGY CONVERTER

Many types of WEC have been invented by now. At the preliminary stage of developing WEC, researchers mainly focused on oscillating water column and received some successes. However, the requirement of enhancing the conversion efficiency is always an important element in the process of inventing WEC. Then, researchers have gradually focused on one kind of WECs, the oscillating buoy, for its simpler structure and higher efficiency. The means of oscillating buoy WECs absorbing energy from wave work mainly based on the motion of heave and/or pitch. A typical oscillating buoy WEC had been described by Eriksson, Isberg and Leijon (2005). It absorbs wave energy through the up and down movement of buoy, namely the heave motion. The mechanical energy is converted into electricity through the connection between buoy and linear generator. The connection between buoy and generator may use different kinds of accessories which depend on the type of generator including permanent linear, hydraulic turbine and so on. The generator of WEC described former was installed on the sea bottom. If the installment of generator in model test was similar to the method mention above, which is hard for us to do in lab circumstance, the cost of model test would increase significantly. In this model test, there are two purposes, which are establishing a model test method for oscillating buoy WEC and analyzing the effect of wave parameters on the WEC efficiency. Considering these purposes, it is unnecessary to use the device mentioned in the reference. In this model test, a much simpler WEC was received and the sketch is shown in Fig. 1. Obviously, it can explain the principle how this typical WEC actually works.

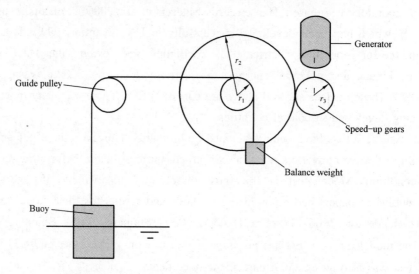

Fig. 1 The sketch of wave energy conversion model test device

The model test device includes cylinder buoy, steel cable, synchronous belt and wheel, speed-up gears, balance weights, generator and data requiring facility. The working process of the model test device is shown in following steps:

(1) The buoy will have a heave motion by the excitation of incident wave. When the buoy is raised up, the synchronous wheel with a radius r_1 will rotate at the clockwise direction driven by the balance weight. The wheel with radius r_2 will have synchronal rotation angular velocity with wheel r_1. Based on the linear velocity equality between wheel r_2 and r_3, the wheel r_3 will have higher rotation angular velocity at the counter-clockwise direction. Then, the generator will be driven by wheel r_3.

(2) When the buoy is falling, the generator's shaft will rotate at the clockwise direction. The generator will not work at this direction owing to the function of ratchet wheel installed in the generator.

Based on this process, only the raised up motion will produce electricity in an incident wave period. If up and down motion all play a role on producing electricity, the energy output can be double.

2 TECHNIQUE DESIGN OF MODEL TESTS

2.1 Buoy's Dimension

The calculation of buoy's characteristic had been done in another article (Ni, Kuang, Miao,

2011). The results showed that cylinder with smaller diameter and larger draft will get larger response wave energy, namely input wave energy. The dimension of the prototype cylinder will be set as height 10m, diameter 10m and draft 7m. In this condition, the buoy will get a better wave energy input.

The pitch motion also has some effects on the energy transmission. Therefore, the pitch amplitude should be controlled at a small range. The calculation results had shown that the effect of height of COG on the pitch amplitude can be ignored. To ensure the stability of buoy and easily operate ballast, the height of COG will be set as 3.0m and radius of inertia as 2m. In the model test, the scaling factor is 1:15 considering the output power and rated power of generator.

2.2 Generator Selection

In model test, the power evaluation in theory should be done firstly in order to select an appropriate generator. For a fluid mechanics system, the power P can be expressed as (Mccormick, 1983):

$$P = C\rho A v^3 \tag{1}$$

Where, C is a constant, ρ is water density, A is cross section area and v is velocity. The relation of power between model and prototype can be obtained by:

$$\frac{P_m}{P_p} = \frac{A_m}{A_p}\left(\frac{v_m}{v_p}\right)^3 = n^{\frac{7}{2}} \tag{2}$$

Where, m denotes the model parameters and P denotes the prototype values. Based on this formula, the output power of buoy in model test can be deduced from the power generated through prototype which is calculated in theory. Then, the power generated in model test can be decided based on above formula. Then, the generator will be chosen correctly. In the wave condition, which is period 6.7s, wave height 2m, the average power produced by the model is 16W.

Comparing to the ordinary generator working frequency, wave motion is low frequency one. In order to decrease the frequency difference between wave and generator, a type of low frequency generator was selected to adapt to the low frequency motion. Eventually, a generator with rated power 30W, voltage 15V, current 2A and average rotation frequency 3Hz was selected.

2.3 Balance Weight Design

The selection of balance weight exactly plays an important role on the successful execution of test model. The balance weight should have enough mass to overcome the resistance produced by generator and wheel friction. The resistance produced by generator will vary with the input wave parameters. The balance weight can be adjusted according to the generator resistance and wheel friction at different wave parameters conditions. The cable tied between balance weight and buoy will bear tension force because of the role of balance weight. The cable need to be strengthened enough to bear the tension force and no distortion. The balance should satisfy the following formula:

$$G_p > F_{rm} \tag{3}$$

Where, G_p is balance weight and F_{rm} is maximal value of resistance. At the initial time, the buoy and balance weight should be at a balance state, the equations of forces are:

$$G_p = F_T \qquad (4)$$

$$G_c = F_T + F_b \qquad (5)$$

Where, F_T is tension force of rope, F_b is buoyancy of buoy and G_c is weight of buoy. When the buoy goes upward, the tension force in cable will be released to offer force to overcome the generator resistance and wheel friction. Then, the balance weight will drive the generator to work. When the buoy goes downward, the additional gravity of buoy because of buoyancy loss will overcome the wheel friction and raise the balance weight.

2.4 Similarity Criterion

Model test should satisfy geometry similarity, kinematic similarity and gravity similarity.

First, the Fr number of model should be equal to the corresponding value of prototype. The equation is:

$$\frac{v_m}{\sqrt{gL_m}} = \frac{v_p}{\sqrt{gL_p}} \qquad (6)$$

Where, v is velocity, g is gravity acceleration and L is characteristic length. The subscript m denotes model values and P denotes that of prototype. Then, the ratio of v_m to v_p is deduced. The formula is:

$$\frac{v_m}{v_p} = \sqrt{\frac{L_m}{L_p}} = \sqrt{n} \qquad (7)$$

Where n is scale. Another number St should also be satisfied. The equation is:

$$\frac{f_m L_m}{v_m} = \frac{f_p L_p}{v_p} \qquad (8)$$

Where, f is the frequency. Then, the ratio of T_m to T_p is:

$$\frac{T_m}{T_p} = \frac{L_m}{L_p}\left(\frac{v_p}{v_m}\right) = \sqrt{n} \qquad (9)$$

Where, T is the period. Based on the equality of Fr and St between model and prototype, the wave parameters in model test can be confirmed deduce from wave parameters in real wave conditions.

3 MODEL TEST ITEMS

3.1 Wave Parameters

In this model test, the scale is 1:15. In prototype, the natural period of adopted buoy in this paper is about 7.0s. In order to enhance the heave amplitude, the wave period should be set approximately 7s. The wave periods were varied from 5s to 9s in prototype conditions, while the corresponding values were from 1.291s to 2.324s in model test based on similarity criterion. The wave height also has influence on energy output efficiency. Therefore, three kinds of wave heights were set, with 0.1m, 0.133m, 0.167m in model dimension. In model test, both regular wave and irregular wave were considered. Then, PM wave spectrum with a certain characteristic period 1.73s, and significant wave height 0.133m were used to simulate irregular wave. In model test, the test contents contained the time history of buoy's displacements, output power and incident wave heights.

3.2 Model Test

The model test had been executed in China Ship Scientific Research Center (CSSRC) basin. The dimension of the basin is length 69m, width 46m, depth 4m. It can generate regular wave, irregular wave and short-crested wave. The maximal wave height of regular wave can reach 0.5m. The period range is 0.5 ~ 5s. The significant wave height of irregular wave can reach 0.5m.

In this paper, the project mainly focuses on the feasibility of experimental test technique on oscillating buoy WEC. The optimization of WEC had not been considered completely in this model test. The wave parameters included regular and irregular waves in this test. The required data contains three items: wave height of incident wave (collecting data by wave-height gauge), response amplitude of buoy (collecting data by 6 degree of freedom no-touching device) and voltages output by generator (collecting data by measuring voltages system).

Through observation, data collecting and processing, estimation for WEC could be made in terms of motion performance and efficiency. In the model test, 4 ropes with springs were tied to buoy at waterline position, forbidding buoy's drift in waves. The effect of spring on buoy is controlled to ensure the natural characteristic in heave motion.

Fig. 2 shows the combination system of generator, data acquisition equipment for voltages and wave heights, and electric circuit with resistance and light bulb. Fig. 3 shows the model test in regular wave with wave height 0.133m, period 2.066s. Fig. 4 shows the test in irregular wave with significant wave height 0.133m, characteristic period 1.73s.

Fig. 2 Generator and data acquisition device

Fig. 3 Model test in regular wave
($H = 0.133\text{m}, T_m = 2.066\text{s}$)

3.3 Results

Fig. 4　Model test in irregular wave
($H_s = 0.133\text{m}, T_{01} = 1.73\text{s}$)

In this paper, authors are mainly focusing on the R & D for WEC model test rather than the optimization of WEC for higher efficiency. The generator's damping coefficient was not selected specially, only considering its rated power, voltage and current. The damping coefficient of generator plays an essential role on output power. In this model test, the heave response amplitude of buoy was controlled by the generator with a large damping coefficient. Although it has no effect on establishment of model test method in basin, the damping coefficient of generator should be optimized when WEC will be designed in researches in the future.

Fig. 5 ~ Fig. 11 display a few model test results revealing some helpful information. Fig. 5 shows the output power in regular wave with wave height 0.1m and period 1.291s. In this condition, the output power is close to zero and the heave motion of the buoy is completely limited by the damping function of generator. In Fig. 6, wave height is identical to the height in Fig. 5 and the period is chosen as 1.807s. It shows that the output power oscillates around 0.6W. Comparing with wave parameters in Fig. 5, the wave height has been changed to 0.133m in Fig. 7, considering the effect of wave height on output power. The result shows that the output power fluctuates around 0.5W. Fig. 8 shows the relationship between averaged output power and wave height with a given period 1.807s. It demonstrates that the power increases gradually with the wave height with a non-linear relationship. Fig. 9 shows the relationship between averaged output power and period with a given wave height 0.133m.

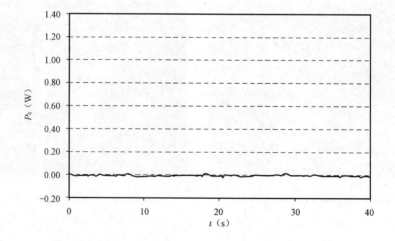

Fig. 5　Time history of output power ($H = 0.1\text{m}, T = 1.291\text{s}$)

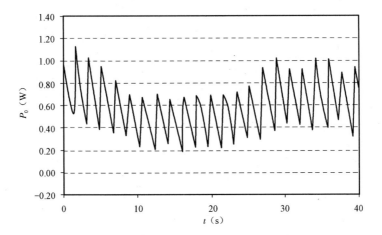

Fig. 6 Time history of output power ($H = 0.1\text{m}, T = 1.807\text{s}$)

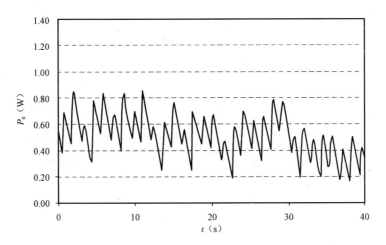

Fig. 7 Time history of output power ($H = 0.133\text{m}, T = 1.291\text{s}$)

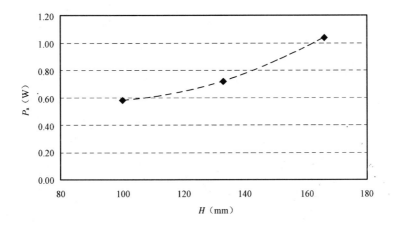

Fig. 8 Output power varied with wave height at period 1.807s

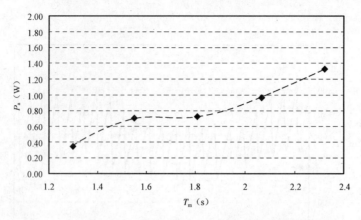

Fig. 9 Output power varied with wave period at height 0.133m

The test in irregular wave had also been performed. The results have been displayed in Fig. 10 and Fig. 11. In Fig. 10, it is the result about response amplitude of buoy varied with time, while Fig. 11 shows the time history of output power. The maximum power is 3.77W and averaged power is 0.66W.

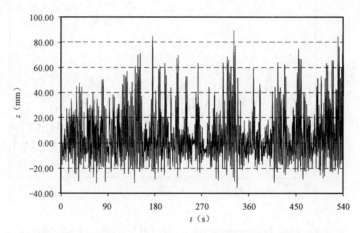

Fig. 10 Time history of response amplitude of buoy in irregular wave
($H_s = 0.133\text{m}, T_{01} = 1.73\text{s}$)

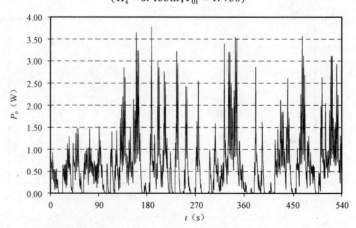

Fig. 11 Time history of output power generated by generator in irregular wave ($H_s = 0.133\text{m}, T_{01} = 1.73\text{s}$)

CONCLUSIONS

In this paper, an oscillating buoy WEC is taken into account to establish its model test method in basin. This type WEC model can be operated easily. It is a proper facility for us at the initial stage of doing model research on wave energy conversion. The optimizing of the WEC adopted in this test is not considered deeply.

The detailed model test methods described in this paper includes generator selection, speed-up gear design, and balance weight design and so on. Some suggestions on wave parameters in model test and required data have been shown in this paper. Through this project, the model test method had been established successfully. In test, the incident wave heights, the heave motions of buoy and output voltages were measured. Although only primary results had been shown in test, some useful information had been received.

In the following work, researchers will focus on the optimization of WECs. The optimization items are comprised of spring coefficient, damping coefficient, amplitude control, and phase control and so on. Model test and numerical simulation will play an important role in the optimization.

ACKNOWLEDGMENTS

The work reported here is supported by CSSRC innovation fund. The authors thank Jianliang Zhu for doing the drawing of model device and also other colleagues for their advices.

REFERENCES

Bhinder M A, Mingham C G, et al. 2009. Numerical and Experimental Study of a Surging Point Absorber Wave Energy Converter. Proceedings of the 8th European Wave and Tidal Energy Conference, Uppsala, Sweden.

Budal K, Falnes J. 1980. Interacting Point Absorbers with Controlled Motion, In: Count, B. (Ed.). Power from Sea Waves. Academic Press, London: 381 – 399.

Eriksson M, Isberg J, Leijon M. 2005. Hydrodynamic Modelling of a Direct Drive Wave Energy Converter [J]. International Journal of Engineering Science, 43: 1377 – 1387.

Ferreira N M, Hadden M D. 2009. Experimental Methods for Power Take-off (PTO) Simulation of a Wave Energy Converter (WEC). Proceedings of the 8th European Wave and Tidal Energy Conference, Uppsala, Sweden.

Heath T, Whittaker T J T, Boake C B. 2000. The Design, Construction and Operation of the LIMPET Wave Energy Converter (Islay Scotland). 4th EWEC, Aalborg, Denmark.

Henderson R. 2006. Design, Simulation, and Testing of a Novel Hydraulic Power Take-off System for the Pelamis Wave Energy Converter[J]. Renewable Energy, 31: 271 – 283.

ITTC-Recommended Guidelines. 2011. Wave Energy Converter Model Test Experiments.

Mccormick M E. 1983. Ocean Wave Energy Conversion[M]. China Ocean Press.

M. Stalberg. 2005. Experimental Test Set-up for Wave Energy Converter Linear Generator. Uppsala University.

Ni X Y, Kuang X F, Miao Q M. 2011. Parameters Optimizing for Wave Energy Conversion System. 11th International Conference on Fluid Control, Measurements, and Visualization.

Ross D. 1995. Power from the waves [M]. Oxford: Oxford University Press.

Stahl A. 1892. The Utilization of the Power of Ocean Waves[J]. Trans Am Soc Mech Eng, 13: 438 – 506.

Sørensen H C, Hansen R, Friis-Madsen E, et al. 2000. The Wave Dragon-Now Ready for Test in Real Sea. 4th EWEC, Aalborg, Denmark.

Thorpe T W. 2000. Wave energy for the 21st century, Renewable Energy World, 7/8.

Washio Y, Osawa H, Nagata Y, et al. 2000. The Offshore Floating Type wave Power Device 'Mighty Wale': Open Sea Tests. 10th ISOPE, Seattle, USA.

Yemm R. 1999. The History and Status of the Pelamis Wave Energy Converter. Wave power—moving towards commercial viability. IMECHE Seminar, London, UK.

The Process Design of Liwan Gas Central Platform

Peilin liu, Wenfen Chen, Xifeng Zhao

Engineering Company, Offshore Oil Engineering Co., Ltd., Tianjin, China

Abstract Liwan Gas Central Platform is an offshore project and has the maximum platform area, weight and nature gas treatment capacity in China. The process design is very difficult since the complex production conditions and super type process treatment capacity. The paper introduces the process system design which gives a help for the big and super treatment capacity type nature gas center treatment platform.

Key Words Super Type Platform; Nature Gas Treatment; Process Design; Subsea; Compressor; Flare

INTRODUCTION

The Liwan area is located in South China Sea, with 1480m water depth. To exploit the nature gas, a shallow water central platform (Liwan Gas CEP) will be built at northwest of LW3-1 subsea facilities with 75km distance and 190m water depth.

The Liwan Gas CEP will receive and process the well fluid from two 22in, 75km subsea facilities of LW area or nearby gas fields, and the processed LW dry gas & dewatering condensate mixes together with the dry gas & dewatering condensate mixture coming from PY34-1CEP via a 14in, 30km subsea pipeline, to be pressurized and transported to the new built onshore terminal gas plant via a 30in, 261km export pipeline. See Fig. 1.

Fig. 1 Planned facilities

The development of Liwan Gas CEP contains two phases: Phase I (LW area gas production reaches $80 \times 10^8 m^3/a$, PY gas production reaches $20 \times 10^8 m^3/a$) and phase II (total gas production reaches $120 \times 10^8 m^3/a$).

Liwan Gas CEP has the maximum platform area, weight and nature gas treatment capacity in China. The process design is very difficult since the complex production conditions and super type process treatment capacity. The paper introduces the process system design which gives a help for the big and super type nature gas center treatment platform.

1 CHARACTERISTIC OF LIWAN GAS CEP

Liwan Gas CEP contains the following main process systems: Receiving & Gas/Liquid Separation System, Wet Gas Compression System, Gas Dehydration and Regeneration System, Condensate Treatment System, Dry Gas Compression System, Exportation System, MEG Regeneration Unit System. The platform has the characteristic:

(1) High design pressure and temperature. The maximum design pressure of process equipment is 246barg and the maximum design temperature is 170℃.

(2) The maximum duplex stainless steel reaches 36in, 600lb, the maximum grade is 2500lb, 22in. The maximum process pipe is 48in.

(3) $120 \times 10^8 m^3/a$ nature gas treatment and compress capacity. But the production is only $50 \times 10^8 m^3/a$ when it be put into production.

(4) Tree Gas/Liquid Separation, three TEG trains, and each trains has $40 \times 10^8 m^3/a$ capacity.

(5) There are ten nature gas turbines and eight nature gas compressors in Phase I, and five nature gas turbines and four nature gas compressors will be installed in Phase II.

(6) 26in high pressure flare tip, 16in low pressure flare tip, the flare boom is 90m.

2 SUBSEA PRODUCTION SYSTEM LONG-DISTANCE CONTROL

Liwan Gas CEP is a manned facility with fully equipped Living Quarters (LQ). In the LQ, a Central Control Room (CCR) will be provided for monitoring and controlling the Liwan Gas CEP facilities and the SSTB (Subsea Tieback). Therefore the primary control of the SSTB facilities will be from the Liwan Gas CEP CCR. See Fig. 2.

The SSTB facilities are monitored and controlled by the Master Control Station (MCS) located in the Liwan Gas CEP CCR. The MCS can enable the operator to monitor and control the surface and subsea systems associated with the Subsea Control System (SCS) including subsea flow metering, valve and choke control, sand monitoring, wellhead, downhole pressure-temperature and numerous other monitoring and control function.

The MCS communication to Liwan Gas CEP Process Control System (PCS) is peer to peer as the both the system will be a node in the PCS control system network. MCS system function is to gather the SSTB facilities instrumentation data for monitoring and to provide the commands to operate subsea valves. The MCS (PCS node) communicates with the subsea control unit (SPCU) via a dual high speed data link such as Ethernet TCP/IP Sockets with a suitable protocollayer.

In addition the Liwan Gas CEP topside houses all the surface-installed SSTB equipment such as

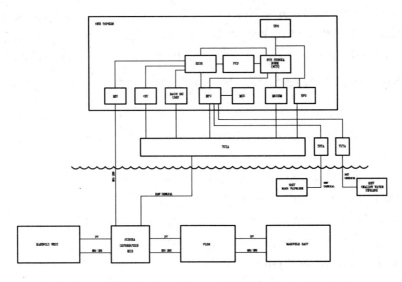

Fig. 2 Control system block diagram

the EPU, chemical Injection skid, and HPU for the interfaces and control of the SSTB facility.

3 MULTI-TRAINS DRY GAS COMPRESSION SYSTEM

Liwan Gas CEP has $120 \times 10^8 m^3/a$ nature gas treatment and compress capacity. But the earlier production is only $50 \times 10^8 m^3/a$. To meet the complex production conditions, the process systems are designed to sever trains. The typical is the dry gas compression system.

The dry gas compression system is required beginning from the first production year. During phase I, four dry gas compression trains are installed, three trains for running and one train for standby. Among which, two small dry compression trains have the same design capacity of $21.43 \times 10^4 \sim 29.76 \times 10^4 m^3/h$, the other two big dry gas compression trains have the same design capacity of $29.76 \times 10^4 \sim 39.28 \times 10^4 m^3/h$. When the total dry gas flow rate reaches $66 \times 10^8 m^3/a$, two small dry gas compression trains will be removed which could be used as the wet gas pre-compression trains. Another two new big dry gas compression trains with the same specification will be installed to satisfy the phase I design capacity ($100 \times 10^8 m^3/a$). And four big dry gas compression trains can also meet pressurization requirement of phase II design capacity ($120 \times 10^8 m^3/a$). During phase II, train E with the same specification will be added as standby one. See Table 1.

Table 1 Main Process Parameter of Dry Gas Compressor

Case	Suction Pressure (kPaG)	Discharge Pressure (kPaG)	Quantity Of Dry Gas Compressor Running (set)
First Year	7,130	11,300	2
$66 \times 10^8 m^3/a$	7,130	14,300	2
$86 \times 10^8 m^3/a$	7,130	17,600	3
Design capacity of phase I	7,130	19,500	3
Design capacity of phase II	7,130	22,800	4

Like it, to meet the aim of step explores and adapt to the big gap between the design capacity and early production, it has four Receiving & Gas/Liquid Separation trains, five Wet Gas Compression trains, three Gas Dehydration and Regeneration trains, two Condensate Treatment train.

4 DECLINE BLOW DOWN RATE AND OPTIMIZATION OF FLARE SYSTEM

Normally, when the fire is detected, all BDVs should blow down at the same time, the relief flow rate will reach $11.2 \times 10^5 m^3/h$ which is the maximum flare relief case after emergency cases are compared and analyzed. Based on it, the flare boom length is 104m which will be a big challenge to the structure design, and a separately flare platform is required. So it is very useful and valuable to reduce the blowdown rate and decrease flare boom length.

The following methods are adopted during the process design.

4.1 High Integrity Protection Systems (HIPS)

HIPS are used at the outlet of each deepwater subsea pipeline to isolate the high pressure production stream and decrease the blowdown rate of blockage case.

4.2 BDVS in Different Fire Zone Stage Blow Down

The platform has three decks including upper deck, middle deck and lower deck and nine fire zone. The most process separators are located in the middle deck which is the largest fire zone.

After discussed with HAZOP and ABS, the method is adopted: When detecting fire, BDVs in the fire zone will be blowdown firstly, and BDVs in the adjacent fire zone will blowdown delay 5 minutes respectively. If the equipment crosses two decks or different fire zones, or one system equipment locates at different decks or fire zones, and no SDV for isolation, BDV on the equipment is considered firstly.

The maximum relief rate will appear in upper deck fire first. See Fig. 3.

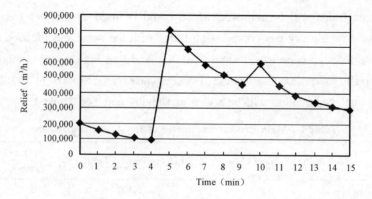

Fig. 3 Relief Profile during Upper Deck Fire First

4.3 Guard Against All BDVS Open Automatically When Instrument Air Losses

A separate instrument air drum is installed for each BDV to avoid all BDVs blowdown in loss instrument air case, and a check valve is also installed at the instrument air inlet line. See Fig. 4.

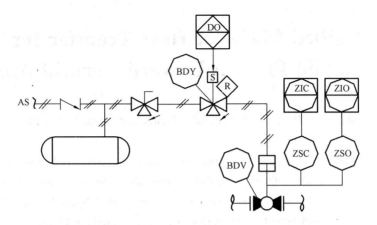

Fig. 4 Legend of BDV

After adopted the three optimize methods, the relief flow rate of flare system at emergency case is reduced from $11.2 \times 10^5 m^3/h$ to $8.1 \times 10^5 m^3/h$, the flare boom length is reduced from 104m to 90m, and the weight decreases 40tons.

CONCLUSIONS

Liwan Gas Central Platform is the key process production and control center of Shallow Water Facilities of the South China Sea Deepwater Gas Project.

It adopts the subsea production system long-distance control method to monitor and control the subsea production, multi-trains to meet the big extent production fluctuation, optimization design of the emergency relief system to decrease the flare system load.

The process design experience will give a help for the future super treatment capacity type nature gas center treatment platform.

REFERENCES

American Petroleum Institute. 2007. ANSI/API STANDARD 521, Pressure-relieving and depressuring systems. Fifth Edition.

Unified Model of Heat Transfer for Gas-Liquid Flow in Upward Vertical Annuli

Bangtang Yin[1], Xiangfang Li[2], Hongquan Zhang[3], Lei Zhu[4]

[1] College of Mechanical and Transportation Engineering, China University of Petroleum, Beijing, China
[2] College of Petroleum Engineering, China University of Petroleum, Beijing, China
[3] McDougall School of Petroleum Engineering, The University of Tulsa, Oklahoma, United States
[4] Development Research Dept. CNOOC RC, Beijing, China

Abstract A unified model of multiphase heat transfer is developed for different flow patterns of gas-liquid flow in upward vertical annuli. The required local flow parameters are predicted by use of the unified hydrodynamic model for gas-liquid flow recently developed by Yu and Zhang (2010). The model prediction of the pipe inside convective heat transfer coefficients are compared with unified zhang model (2006) for a gas-liquid system in upward vertical flows by using the hydraulic diameter and good agreement is observed.

Key Words Gas-liquid flow; Vertical annuli; Heat transfer

NOMENCLATURE

A', A'_{LFd}, A'_{LFc}, A'_C, B_{LFd}, B'_{LFd}, B''_{LFd}, Φ_a, Φ'_a, Φ'_d, Φ_{LFd}—the local constant defined in Eqs. 1-15;

c_{Pa}—specific heat;

d—diameter;

dl—the length of wellbore segment;

f—friction factor;

h—enthalpy;

h_d—the convective-heat-transfer coefficient of tubing fluid;

H—liquid holdup;

k—conductivity;

q_1—the heat influx to the annuli;

q_2—the heat influx to the formation;

r—outside radius of tubing;

T—temperature;

T_D—dimensionless temperature;

U_d—overall heat-transfer coefficient of wellbore;

v—velocity;

W—the mass flow rate;

ρ—density;

η—Joule-Thomson coefficient;
μ—viscosity;

Subscripts

a1—inlet of annuli;
a2—outlet of annuli;
ai—inside of casing;
ao—outside of casing;
C—gas core;
cem—cement
d—tubing;
di—inside of tubing;
do—outside of tubing;
dp1—outlet of tubing;
dp2—inlet of tubing;
ei—formation;
G—gas;
LFd—tubing liquid film;
LFc—casing liquid film;
L—liquid;
M—mixture;
R—representative;
wb—wellbore.

INTRODUCTION

As oil and gas production moves to deep and ultradeep waters, multiphase flow occurs during the production and transportation. The flow normally occurs in horizontal, inclined, or vertical pipes and wells. Gas-liquid two phase flow in an annulus can be found in a variety of practical situations. In high rate oil and gas productions, it may be beneficial to flow fluids vertically through the annulus configuration between well tubing and casing. For surface facilities, some large, under-utilized flow lines can be converted to dual-service by putting a second pipe through the large line (flowing produced water in the inner line and gas in the annulus). During gas production, liquids may accumulate at the bottom of the gas wells during their later life. In order to remove or "unload" the undesirable liquids, a siphon tube is often installed inside the tubing string, which would form a gas-liquid two-phase flow in the annulus. Flow-assurance issues such as wax deposition and hydrate formation in annulus become very crucial in this situation. These flow-assurance problems are strongly related to both the hydraulic and thermal behaviors (such as liquid holdups, local fluid velocities, pressure gradient, slug characteristics, and convective-heat-transfer coefficients corresponding to different phases and flow patterns) of the multiphase flow (Zhang, 2006). Therefore, multiphase hydrodynamics and heat transfer in an annulus need to be modeled properly to optimize the design and operation of the flow system.

Compared to experimental and modeling studies of multiphase hydrodynamics, very limited research results can be found in the open literature for multiphase heat transfer. Davis et al. (1979) presented a method for predicting local Nusselt numbers for stratified gas/liquid flow under turbulent-liquid/turbulent-gas conditions. A mathematical model based on the analogy between momentum transfer and heat transfer was developed and tested using heat-transfer and flow-characteristics data taken for air/water flow in a 63.5mm(ID) tube.

Shoham et al. (1982) measured heat-transfer characteristics for slug flow in a horizontal pipe. The time variations of temperature, heat-transfer coefficients, and heat flux were reported for the different zones of slug flow. Substantial difference in heat-transfer coefficient was found to exist between the bottom and top of the slug.

Most previous modeling studies were aimed at developing heat-transfer correlations for different flow patterns (Shah, 1981). Kim et al. (1997) evaluated 20 heat-transfer correlations against experimental data collected from the open literature and made recommendations for different flow patterns and inclination angles. However, these recommended correlations did not give satisfactory predictions when compared with experimental results by Matzain (1999). Manabe (2001) developed a comprehensive mechanistic model for heat transfer in gas/liquid pipe flow. The overall performance was better than previous correlations in comparison with experimental data; however, some inconsistencies in the hydrodynamic model and the heat-transfer formulations for stratified (annular).

Zhang et al(2006) developed a unified model of multiphase heat transfer for different flow patterns of gas-liquid pipe flow at all inclinations −90° to +90° from horizontal. And the required local flow parameters are predicted by use of the unified hydrodynamic model for gas/liquid pipe flow developed by Zhang et al(2003a,2003b). However it is not fit for the gas-liquid flow in annulus. Because the flow patterns in annuli are different from pipe-flow patterns, as seen in Fig. 1 (Yu and Zhang,2010; Caetano,1986). A new heat transfer model for gas-liquid flow in vertical annuli needs to be established.

A mechanistic model is developed to predict flow patterns, pressure gradient, and liquid holdup for gas-liquid flow in upward vertical annuli (Yu and Zhang, 2010). The major advantage of this model compared with previous mechanistic models is that it's developed on the basis of the dynamics of slug flow, and the film zone is used as the control volume. Multiphase heat transfer depends on the hydrodynamic behaveior of the flow. The objective of this study is to develop a unified heat-transfer model for gas-liquid flow that is consistent with the unified hydrodynamic model in vertical annuli.

1 MODELING

Bubbly Flow and Dispersed-Bubble Flow. When fluids flow through the tube and the surrounding temperature is colder than the fluids, heat is lost from the fluids to the annuli and the formation, resulting in a decline in temperature, as seen in Fig. 2. For the element of wellbore.

The conservation of mass is:

$$\frac{d}{dl}(\rho_L v_L) = 0 \qquad (1)$$

The conservation of momentum is:

$$\frac{d}{dl}(\rho_L v_L^2) = -\frac{dp}{dl} - \rho_L g\sin\theta - \frac{\tau \pi d_R}{A} \quad (2)$$

And the energy equation is:

$$\frac{d}{dl}\left(\rho_L v_L\left(e + \frac{1}{2}v_L^2\right)\right) = -\frac{d}{dl}(pv_L) - \rho_L v_L g\sin\theta - \frac{q_1}{A} \quad (3)$$

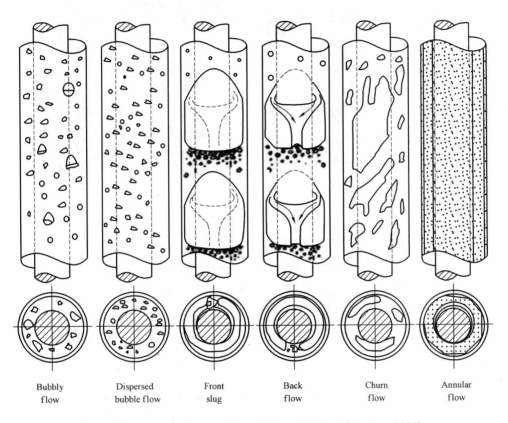

Fig. 1 Flow patterns in upward vertical-annulus flow (Caetano, 1986).

Using the mass balance, we can reduce Eq. (2) and Eq. (3) further:

$$\frac{dp}{dl} = -\rho_L v_L \frac{dv_L}{dl} - \rho_L g\sin\theta - \frac{\tau \pi d_R}{A} \quad (4)$$

$$\rho_L v_L \frac{d}{dl}\left(e + \frac{p}{\rho_L}\right) = -\rho_L v_L v_L \frac{dv_L}{dl} - \rho_L v_L g\sin\theta - \frac{q_F}{A} + \frac{q_D}{A_D} \quad (5)$$

Or

$$\frac{dh}{dl} = -v_L \frac{dv_L}{dl} - g\sin\theta - \frac{q_1}{A\rho_L v_L} = -v_L \frac{dv_L}{dl} - g\sin\theta - \frac{q_1}{w_L} \quad (6)$$

There is no fluid flow in annuli and the heat loss is very small. So $q_1 = q_2$. The energy equation is as follows:

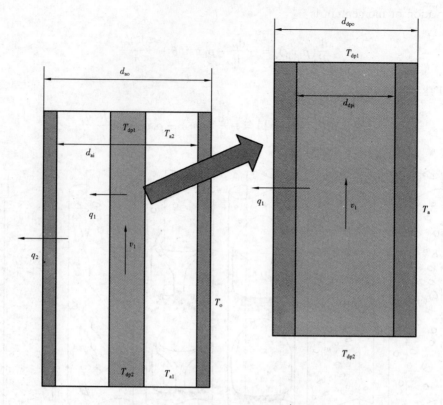

Fig. 2　Heat transfer from tube to the formation

$$\frac{dh}{dl} = -v_L \frac{dv_L}{dl} - g\sin\theta - \frac{q_2}{w_L} \tag{7}$$

The Hasan and Kabir model (1991) can be used for calculating the heat influx to the formation q_2:

$$q_2 = \frac{2\pi k_e (T_{wb} - T_{ei})}{T_D} = \frac{w_{Ld} c_{pa}}{A'}(T_{ei} - T_{dp}) \tag{8}$$

Where the dimensionless temperature, T_D, can be easily estimated from the following models:

$$T_D = \begin{cases} 1.1281\sqrt{t_D}(1 - 0.3\sqrt{t_D}) & 10^{-10} \leq t_D \leq 1.5 \\ (0.4063 + 0.5\ln t_D)\left(1 + \frac{0.6 t_D}{t_D}\right) & t_D > 1.5 \end{cases} \tag{9}$$

$$A' = \frac{w_{Ld} c_{pa}}{2\pi}\left[\frac{k_e + r_{do} U_d T_D}{r_{do} U_d k_e}\right] \tag{10}$$

U_d is the overall heat-transfer coefficient, depends on the resistances to heat flow through the tubing fluid, tubing wall, annuli and the wellbore.

$$\frac{1}{U_d} = \frac{r_{do}}{r_{di} h_d} + \frac{r_{do} \ln(r_{do}/r_{di})}{k_{dp}} + \frac{r_{co} \ln(r_{wb}/r_{co})}{k_{cem}} \tag{11}$$

Eq. (7) and Eq. (8) can be combined to yield:

$$\frac{dh}{dl} = -v_L \frac{dv_L}{dl} - g\sin\theta - \frac{c_{pa}}{A'}(T_{ei} - T_{dp}) \tag{12}$$

The enthalpy gradient can be written in terms of the temperature and pressure gradients:

$$\frac{dh}{dl} = c_{pa}\frac{dT}{dl} - \eta c_{pa}\frac{dp}{dl} \tag{13}$$

Combining Eq. (9) and Eq. (10):

$$\frac{dT_{dp}}{dl} + \frac{1}{A'}(T_{ei} - T_{dp}) + \frac{1}{c_{pa}}\left(v_L\frac{dv_L}{dl} + g\sin\theta - \eta c_{pa}\frac{dp}{dl}\right) = 0 \tag{14}$$

Defining a dimensionless parameter Φ_a, as:

$$\Phi_a = \left(\rho_L v_L \frac{dv_L}{dl} + \rho_L g\sin\theta - \rho_L \eta c_{pa}\frac{dp}{dl}\right)\bigg/\frac{dp}{dl}$$

We can write Eq. (14) as:

$$\frac{dT_{dp}}{dl} + \frac{1}{A'}(T_{ei} - T_{dp}) + \frac{1}{\rho_L c_{pa}}\frac{dp}{dl}\Phi_a = 0 \tag{15}$$

In bubbly flow and dispersed-bubble flow, the gas holdup is small and the gas superficial velocity is low, the gas phase is distributed as small discrete bubbles in a continuous liquid phase. So bubbly flow and dispersed-bubble flow can be treated as pseudosingle-phase flow. The fluid physical properties are adjusted on the basis of liquid holdup. Zhang correction (2006) for bubbly flow will be modified based on "representative diameter", introduced by Omurlu and Ozbayoglu (2007).

$$d_R = \sqrt{d_{ai}^2 - d_{dpo}^2} \tag{16}$$

Then the convective-heat-transfer coefficient of Eq. (15) for bubbly or dispersed-bubbly flow is obtained from:

$$h_d = \frac{N_{Nu}^M k_{LM}}{d_R} \tag{17}$$

Where N_{Nu}^M is mixture Nusselt number.

$$N_{Nu}^M = \frac{\left(\frac{f_M}{2}\right)N_{Re}^M N_{Pr}^M}{1.07 + 12.7\sqrt{\frac{f_M}{2}}(N_{Pr}^{M2/3} - 1)}\left(\frac{\mu_L}{\mu_{LW}}\right) \tag{18}$$

Where N_{Re}^M, N_{Pr}^M is the mixture Reynolds number and Prandtl number.

$$N_{Re}^M = \frac{\rho_M v_M d_R}{\mu_{LM}}$$

$$N_{Pr}^M = \frac{c_{pM}\mu_{LM}}{k_{LM}}$$

Annular Flow. As shown in Fig. 3, δ_d and δ_c are the thickness of tubing film and casing film, which is different from the annular flow in pipes. T_d and T_a are the temperatures of the fluid in tubing and the gas core with small liquid droplets in annuli.

Assume there are no temperature and heat transfer changes in vertical direction. Heat distribution in annuli includes three parts:

In casing film, tubing film and gas core. The heat transfer models in annuli are obtained by the similar method above.

The model for the tubing film:

$$\frac{dT_{LFd}}{dl} + \frac{1}{A'_{LFd}}(T_C - T_{LFd}) + \frac{1}{B_{LFd}}(T_{dp} - T_{LFd}) + \frac{1}{\rho_L c_{pLFd}} \frac{dp}{dl} \Phi_{LFd} = 0 \quad (19)$$

For the tubing,

$$\frac{dT_{dp}}{dl} + \frac{c_{pLFd} w_{LFd}}{c_{pd} w_d}(T_{dp} - T_{LFd}) + \frac{1}{\rho_L c_{pd}} \frac{dp}{dl} \Phi_d = 0 \quad (20)$$

For the gas core,

$$\frac{dT_C}{dl} + \frac{1}{A'_C}(T_{LFC} - T_C) + \frac{1}{B_C}(T_{LFd} - T_C) + \frac{1}{\rho_C c_{pC}} \frac{dp}{dl} \Phi_C = 0 \quad (21)$$

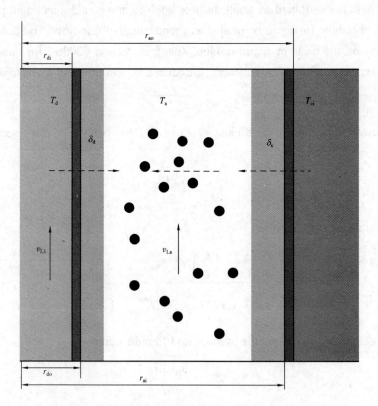

Fig. 3 Control volume and temperatures in annular flow

For the casing film,

$$\frac{dT_{LFc}}{dl} + \frac{1}{A'_{LFc}}(T_{ei} - T_{LFc}) + \frac{1}{B_{LFc}}(T_C - T_{LFc}) + \frac{1}{\rho_L c_{pLFc}}\frac{dp}{dl}\Phi_{LFc} = 0 \qquad (22)$$

Compounding Eq. (13) and Eq. (14):

$$A'_{LFd}B'_{LFd}\frac{d^2T_d}{dl^2} + B''_{LFd}\frac{dT_d}{dl} - T_{dp} + T_C + \left(\frac{\Phi'_d}{\rho_L c_{pd}} + \frac{\Phi'_{LFd}}{\rho_{LFd} c_{pLFd}}\right)\frac{dp}{dl} = 0 \qquad (23)$$

Where

$$A'_{LFd} = \frac{w_{LFd}c_{pLFd}}{2\pi(r_{do} + \delta_d)U_{LFd}}$$

$$A'_C = \frac{w_{Cd}c_{pC}}{2\pi r_{ci}U_C}$$

$$A'_{LFc} = \frac{w_{LFc}c_{pLFc}}{2\pi}\left[\frac{k_e + r_{co}U_{LFc}T_D}{r_{co}U_{LFc}k_e}\right]$$

$$\frac{1}{U_{LFd}} = \frac{1}{h_{LFd}}; \qquad \frac{1}{U_C} = \frac{1}{h_C}$$

$$\frac{1}{U_{LFc}} = \frac{r_{co}}{r_{ci}h_{LFc}} + \frac{r_{co}\ln(r_{co}/r_{ci})}{k_{cas}} + \frac{r_{co}\ln(r_{wb}/r_{co})}{k_{cem}}$$

$$B_{LFd} = \frac{w_{LFd}c_{pLFd}}{2\pi U_d}$$

$$B'_{LFd} = \frac{w_d c_{pd} B_{LFd}}{c_{pLFd} w_{LFd}}$$

$$B''_{LFd} = A'_{LFd} - B'_{LFd} - \frac{A'_{LFd}B'_{LFd}}{B_{LFd}}$$

$$\Phi'_d = A'_{LFd}B'_{LFd}\left(\frac{1}{A'_{LFd}} + \frac{1}{B_{LFd}}\right)\Phi_d$$

$$\Phi'_{LFd} = A'_{LFd}\Phi_{LFd}$$

$$\Phi'_a = \left(\rho v_L \frac{dv_L}{dl} + \rho g\sin\theta - \rho\eta c_{pa}\frac{dp}{dl}\right)\bigg/\frac{dp}{dl}$$

$$\Phi_{LFd} = \left(\rho_{LFd}v_{LFd}\frac{dv_{LFd}}{dl} + \rho_{LFd}g - \rho_{LFd}\eta_{LFd}c_{pLFd}\frac{dp}{dl}\right)\bigg/\frac{dp}{dl}$$

Eq. (19), Eqs. (21) to Eq. (23) are the heat transfer models for gas-liquid annular flow in annuli. Then the fluid temperature in wellbore is calculated by weighted average method based on holdup.

$$T_{ta} = H_{LFd}T_{LFd} + (1 - H_{LFd} - H_{LFc})T_C + H_{LFc}T_{LFc} \tag{24}$$

The convective-heat-transfer coefficients for the casing film, the tubing film and gas core are obtained by using $h_{LFd} = \dfrac{N_{Nu}^{LFd} k_{LF}}{d_{LFdR}}$, $h_{LFc} = \dfrac{N_{Nu}^{LFc} k_{LF}}{d_{LFcR}}$, $h_C = \dfrac{N_{Nu}^{C} k_C}{d_{CR}}$, where k_{LF}, k_C are the thermal conductivities of the liquid film and gas core. d_{LFdR}, d_{LFcR}, d_{CR} are the "representative diameter" of the tubing film, the casing film and gas core.

$$d_{LFdR} = \sqrt{(d_{do} + \delta_d)^2 - d_{do}^2}$$

$$d_{LFcR} = \sqrt{d_{ci}^2 - (d_{ci} - \delta_c)^2}$$

$$d_{CR} = \sqrt{(d_{ci} - \delta_c)^2 - (d_{do} + \delta_d)^2}$$

The Nusselt numbers for the liquid film and gas core are calculated using the correlations for single-phase convective heat transfer. The Petukhov (1970) correlation is used for turbulent liquid-film flow:

$$N_{Nu}^{LFi} = \dfrac{\left(\dfrac{f_{LFi}}{2}\right) N_{Re}^{LFi} N_{Pr}^{LFi}}{1.07 + 12.7 \sqrt{\dfrac{f_{LFi}}{2}} (N_{Pr}^{LFi\,2/3} - 1)} \left(\dfrac{\mu_{LFi}}{\mu_{LWi}}\right)^{0.25} \tag{25}$$

Where i is tubing film or casing film, i = d or c. f_{LFi} is the friction factor at the wall in contact with the liquid film.

$$N_{Re}^{LFi} = \dfrac{\rho_L v_{LFi} d_{LFiR}}{\mu_{LF}}$$

$$N_{Pr}^{LFi} = \dfrac{c_{PLFi} \mu_{LFiR}}{k_{LF}}$$

The Dittus and Boelter (1930) correlation is used for turbulent gas-core flow:

$$N_{Nu}^{C} = 0.023 (N_{Re}^{C})^{0.8} (N_{Pr}^{C})^{0.33} \tag{26}$$

For fully developed laminar flows of the liquid film and gas core, the Nusselt number approaches a constant value. According to Zhang et al. (2006), the Nusselt numbers for fully developed laminar flows are calculated by:

$$N_{Nu}^{LFi} = 3.657 + \dfrac{7.541 - 3.657}{0.5} \left(0.5 - \dfrac{\delta_i}{\sqrt{d_{ci}^2 - d_{do}^2}}\right) \tag{27}$$

$$N_{Nu}^{C} = 3.657 \tag{28}$$

The associated hydraulic parameters are calculated by Yu and Zhang model (2010).

Slug Flow. Film Region. The flow characters of film region are similar to the annular flow. The difference is the gas core with liquid droplet change into Taylor bubble. So the overall heat transfer coefficients of Eq. (11) ~ Eq. (15) should change from U_C to U_T. Then the heat transfer of the film region can be calculated.

$$\frac{1}{U_T} = \frac{1}{h_T} \tag{29}$$

$$h_T = \frac{N_{Nu}^T k_T}{d_{TR}} \tag{30}$$

where k_T is the thermal conductivities of Taylor bubble. d_{TR} is the "representative diameter" of Taylor bubble.

$$d_{TR} = \sqrt{(d_{ci} - \delta_c)^2 - (d_{do} + \delta_d)^2}$$

Slug Region. There are small discrete bubbles in a continuous liquid phase. The flow characters are similar to bubbly flow. So the heat transfer can be calculated by the bubbly flow model.

Slug Unit. The fluid temperature in wellbore is calculated by weighted average method based on holdup.

$$T_{TS} = \frac{[H_{LFd}T_{LFd} + (1 - H_{LFd} - H_{LFc})T_T + H_{LFc}T_{LFc}]l_F + T_{LS}l_S}{l_U} \tag{31}$$

2 SOLUTION PROCEDURE

Fig. 4 show the overall solution flow chart for the present model. Flow pattern is first determined on the basis of the input variables, and then all the flow conditions, such as flow pattern, liquid holdups, local fluid velocities of the liquid film and gas core, and slug characteristics, are predicted by use of the unified hydrodynamic model for gas-liquid flow in annuli recently developed by Yu and Zhang(2010), as seen in Fig. 4. Finally, heat transfer will be calculated according to the hydrodynamic behaviors of the gas-liquid flow. Fig. 5 is the flow chart for the present annular flow heat transfer model.

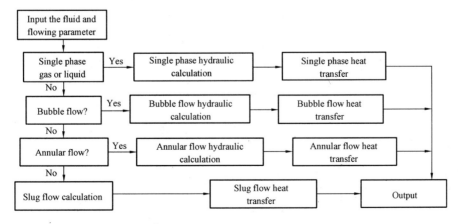

Fig. 4 Overall flow chart for present model

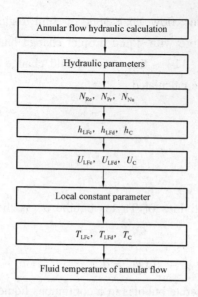

Fig. 5 Flow chart for present annular flow heat transfer model

3 COMPARISONS WITH UNIFIED ZHANG MODEL

The heat transfer of single-phase gas or liquid flow can calculated by the present model if the thermal conductivities of the tubing wall equals to infinite. Fig. 6 and Fig. 7 show the comparisons between models predictions and experimental measurements of the convective-heat-transfer coefficients for single-phase gas and liquid flows, respectively. The Dittus and Boelter (1930) and Petukhov (1970) correlations are used to predict the (turbulent) single-phase gas and liquid convective-heat-transfer coefficients, respectively. The good agreement between models predictions and experimental measurements for single-phase gas and liquid flows indicates that the instruments are reliable and the selected correlations are appropriate.

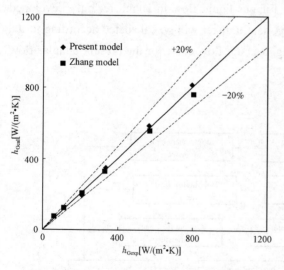

Fig. 6 Comparison of single-phase gas flow model predictions and measured convective-heat-transfer coefficient

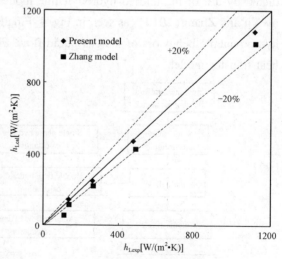

Fig. 7 Comparison of single-phase liquid flow model predictions and measured convective-heat-transfer coefficient

There are seldom experiment researches results can be found in the open literature for multi-phase heat transfer in annuli. The unified zhang model (2006) is verified by comparison with Manbe's experimental results for different flow patterns in a crude-oil/natural-gas system. And good agreement has been observed in the comparison. Based on Caetano's (1986) and Manabe's (2001) experimental data, the wellbore temperature is calculated by unified zhang model based on the "representative diameter" and the present unified heat-transfer model for gas-liquid flow in annuli. Although the errors may be big by the unified zhang model, there is no other better way to compare.

Fig. 8 and Fig. 9 show the comparison of convective-heat-transfer coefficient for bubble flow and annular flow in vertical annuli predicted by the present model and unified zhang model (2006). It is seen that most of the data points are located inside the 30% error band. The present model over predicts the two-phase convective-heat transfer coefficients by approximate 20% ~ 30% compared to the unified zhang model (2006). It may because the predictions by unified zhang model are normally developed by using the hydraulic diameter. It ignores the influence of the flow pattern and annulus geometry.

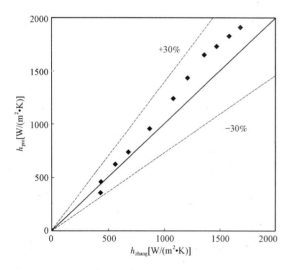
Fig. 8 Comparison of bubble flow models predictions convective-heat-transfer coefficient in vertical annuli

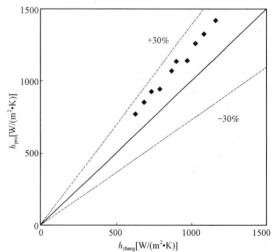
Fig. 9 Comparison of annular flow models predictions convective-heat-transfer coefficient in vertical annuli

Comparisons between models predictions of the temperature gradient in the flow direction are not demonstrated in this study. They are almost the same as the comparisons for the inside two-phase convective-heat-transfer coefficients.

CONCLUSIONS AND DISCUSSION

A unified heat-transfer model for gas-liquid flow in vertical annuli is developed in conjunction with the unified hydrodynamic model of Yu and Zhang (2010), which can predict flow-pattern transitions, liquid holdups, pressure gradient, and slug characteristics in gas-liquid flow in vertical annuli. The heat-transfer modeling is based on energy-balance equations and analyses of the temperature differences and variations in the tubing film, casing film, gas core, Taylor bubble and slug body.

The unified heat-transfer model for single gas or liquid flow is verified by comparison with Manabe's (2001) experimental results. For differ ent flow patterns, it is compared with unified zhang model modified based on "representative diameter". Good agreement has been observed in the comparison.

Experimental investigation of heat transfer in vertical are required to improve the model performance.

ACKNOWLEDGEMENTS

This paper was sponsored by the Key Technologies R & D Programs (2011ZX05056-001-01-2,2011ZX05056-001-01-3,2011ZX05021-005). We recognize the support of China University of Petroleum (Beijing) for the permission to publish this paper.

REFERENCES

Caetano E F. 1986. Upward Vertical Two-Phase Flow through An Annulus[D]. Ph. D. Dissertation, U. Tulsa, Tulsa, oklahoma, USA.

Davis E J, Cheremisinoff N P, and Guzy C J. 1979. Heat Transfer with Stratified Gas-Liquid Flow[J]. AIChE J,25 (6):958 – 966.

Dittus F W and Boelter L M K. 1930. Heat Transfer in Automobile Radiators of the Tubular Type, U. California (Berkeley) Pub. Eng:461.

Kim D, et al. 1997. An Evaluation of Several Heat Transfer Correlations for Two-Phase Flow with Different Flow Patterns in Vertical and Horizontal Tubes[C]. Proc. Natl. Heat Transfer Conference, Baltimore, Mary-land.

Manabe R. 2001. A Comprehensive Mechanistic Heat Transfer Model for Two-Phase Flow with High-Pressure Flow Pattern Validation[D]. PhD dissertation, U. Tulsa, Oklahoma, USA.

Matzain A B. 1999. Multiphase Flow Paraffin Deposition Modeling [D]. PhD dissertation, U. Tulsa, Oklahoma, USA.

Omurlu C M and Ozbayoglu M E. 2007. Analysis of Two-Phase Fluid Flow through Fully Eccentric Horizontal Annuli[C]. Proceedings BHRG Multiphase Technology Conference, Edinburgh, UK.

Petukhov B S. 1970. Heat Transfer and Friction in Turbulent Pipe Flow with Variable Physical Properties [M]. Advances in Heat Transfer, J P Hartnet and T V Irvine (eds.) Academic Press, New York City:505-564.

Shah M M. 1981. Generalized Prediction of Heat Transfer during Two Component Gas-Liquid Flow in Tubes and Other Channels[J]. AIChE Symp. Ser. ,77(208):140 – 151.

Shoham O, Dukler A E, Taitel Y. 1982. Heat Transfer During Intermit-tent/Slug Flow in Horizontal Tubes [J]. Ind. Eng. Chem. Fundamentals,21(3):312 – 319.

Yu T T and Zhang H Q. 2010. A Mechanistic Model for Gas/Liquid Flow in Upward Vertical Annuli. J SPE Production & Operations, SPE 124181, 285 – 295.

Zhang H Q, Wang Q, Sarica C, et al. 2006. Unified Model of Heat Transfer in Gas/Liquid Pipe Flow. J SPE Production & Operations, SPE 90459, 114 – 122.

Zhang H Q. et al. 2003a. Unified Model for Gas-Liquid Pipe Flow via Slug Dynamics—Part 1: Model Development [J]. J. Energy Res. Technol,125 (4):266 – 273.

Zhang H Q, et al. 2003b. Unified Model for Gas-Liquid Pipe Flow via Slug Dynamics—Part 2: Model Validation [J]. J. Energy Res. Technology,125 (4):274 – 283.

Wave Loading Uncertainties and Structural Fatigue Reliability Researches for Semi-Submersible

Peng Yang, Xuekang Gu

China Ship Scientific Research Center, Wuxi, China

Abstract Uncertainties of wave loading have large proportions in analysis of structural fatigue strength for ocean platform. The assessing methods and formulas are built for wave spectrum, sea wave height, wave loading transfer function, cross-zero period, sea state occurring possibility and wave direction due to investigate the uncertainties of wave loading during calculating fatigue loading. Meanwhile, wave loading model trial is taken for assessing the uncertainty of wave loading transfer function. The correcting formulas of significant wave height and mean cross-zero period in Northwest Pacific are regressed statistically after specially calculating the sea states in Northwest Pacific and South China Sea. Furthermore the influence factors of difference season sea states and principle of wave direction distribution to wave loading and fatigue life of semi-submersible are investigated. Finally, the fatigue reliability index of typical structures for semi-submersible is researched on the basis of uncertainties of wave loading and the other uncertainties during structural fatigue analysis.

Key Words Semi-submersible; Wave loading; Uncertainty; Fatigue; Reliability.

INTRODUCTION

Traditional and certain strength assessing method has a certain imperfection on safety design and check of marine structure due to manufacture technology, material property, some hypothesis on strength and loading calculation and so on. So, more reasonable approaches for assessing structure and loading must base on statistics theory for guaranteeing enough safety reliability of marine structure. Furthermore, advanced design rules of marine structure demand to make reliability design for marine platform. Guedes Soares(1984) has researched various uncertainties of short-term and long-term wave loading in detail mainly for North Atlantic and Global sea states. But, the author has not taken any analysis for Northwest Pacific which is very important for Chinese ship and ocean engineering designs. In addition, Guedes Soares has mainly analyzed short-term and long-term uncertainties of total ship wave loading, and, not taken any uncertainties of fatigue wave loading. Moan et al. (2005) have analyzed the influence of sea state annual variation in North Atlantic to long-term wave loading; meanwhile, researched the uncertainties of vertical bending moment for FPSO, local structure stress and fatigue loading. Wu(2011) and Yang et al. (2012) have investigated the uncertainties of short-term and long-term wave loading predictions for semi-submersible including the uncertainties of wave spectrum, wave height and wave loading transfer function.

For studying the uncertainties of wave loading in fatigue life calculation and structural fatigue reliability, the assessing expressions should be established after considering the uncertainties under random wave loading effect. The model test result of wave loading is utilized in assessing the uncertainty of wave loading transfer function. The paper investigates the regressing formulas of wave height and period under various sea states with the statistical datas of buoy and ship observation in Northwest Pacific for wave height and period given by Fang et al. (1996); meanwhile, studied the effect of wave period uncertainty to structural stress amplitude and fatigue life. The influence of sea state occurring possibility to structural fatigue life is studied using long-term scatter of S4 area in South China Sea in which the wave direction distribution rules are used to investigate the uncertainty of wave direction and the effect to structural fatigue life. Finally, the fatigue reliability index of typical structures for semi-submersible is researched on the basis of uncertainties of wave loading and the other uncertainties during structural fatigue analysis.

1 UNCERTAINTIES IN FATIGUE LIFE ANALYSIS

Long-term statistical sea state characteristic is determined by wave scatter of the sea area during service life for marine platform which works in specific area. On the basis of fatigue damage linear cumulative rule, total fatigue stress parameter could be obtained by linear superposition through every short-term sea state's. After considering rain-drop and low stress range corrections, when short-term stress characteristic obeys Rayleigh distribution, the equation of structural fatigue stress parameter is (Hu et al., 2010):

$$\Omega_j = (2\sqrt{2})^m \Gamma(m/2 + 1) \sum_{i=1}^{k} \lambda(m,\varepsilon_i) \Lambda_i f_{0i} \gamma_i \sigma_i^m \tag{1}$$

where Ω_j is stress parameter in some direction. Due to ABS (2003) rules, total fatigue stress parameter reads:

$$\Omega = \sum_{j=1}^{q} p_j \Omega_j \tag{2}$$

where q is totality of wave direction, p_j is wave direction occurring possibility, σ_i is alternating stress amplitude in No. i sea state, Γ is gamma function, f_{0i} is cross-zero period of stress response for each short-term sea state, γ_i is occurring possibility of each combination of H_s and T_z, k is sum of short-term sea state in wave scatter, ε_i is band coefficient, $\lambda(m,\varepsilon_i)$ is rain-drop correction coefficient (Wirsching and Light, 1980), Λ_i is fatigue damage correction coefficient considering low stress range (Hu et al., 2010). Substituting Eq. (1) into Eq. (2), the total fatigue stress parameter is

$$\Omega = (2\sqrt{2})^m \Gamma(m/2 + 1) \sum_{i=1}^{k} \sum_{j=1}^{q} \lambda(m,\varepsilon_i) \Lambda_i f_{0i} \gamma_i p_j \sigma_i^m \tag{3}$$

After considering the uncertainties of fatigue cumulative damage and stress parameter, structural fatigue life with $S-N$ curve method reads

$$T_f = \frac{\Delta A}{B^m \Omega} \tag{4}$$

where A is random for $S-N$ curve parameter, Δ is random considering fatigue cumulative damage, uncertainties in fatigue loading calculation is accounted by random B.

2 UNCERTAINTY ANALYSIS OF WAVE LOADING

The uncertainties of wave loading in structural fatigue life reliability mainly include: (1) sea state occurring possibility; (2) wave spectrum; (3) significant wave height and cross-zero period; (4) wave direction occurring possibility; (5) wave loading transfer function, and so on. Accounting for the influence of uncertainty, Eq. (3) becomes

$$B_W^m \Omega_s = \phi \eta (2\sqrt{2})^m \Gamma(m/2+1) \sum_{i=1}^{k} \sum_{j=1}^{q} \lambda(m,\varepsilon_i) \Lambda_i f_{0i} \gamma_i p_j (\phi_T \sigma_i^2)^{\frac{m}{2}} \quad (5)$$

where Ω_s is standard stress parameter not including uncertainties, B_w associating with wave loading uncertainties obeys lognormal distribution, then B_W^m also obeys lognormal distribution which associates with stress parameter uncertainty, ϕ is the bias of uncertainty of sea state occurring possibility, η is the bias of uncertainty of wave direction, $\phi_T(\alpha, H_s, T_z)$ considering spectrum style, wave loading transfer function, wave height and cross-zero period is the bias of variance of short-term stress amplitude response, in the other word is 0-order moment of stress response which reads

$$R(\alpha, H_s, T_z) = \sum_{i=1}^{n} S(\omega_i, H_s, T_z) \cdot H^2(\omega_i, \alpha) \cdot \Delta \omega_i \quad (6)$$

where $S(\omega_i, H_s, T_z)$ is wave spectrum, $H(\omega_i, \alpha)$ is stress amplitude transfer function in wave direction α degree.

2.1 Uncertainty of Sea Wave Spectrum

Uncertainty of sea wave spectrum could simply take as that, for a short-term sea state (H_s, T_z) if taking different wave spectrum to describe sea state, the corresponding calculation is various. For actual engineering calculation uses theoretical spectrum formula which is different to real sea state, a series of uncertainty has been taken in. At present, some theoretical spectrums such as $P-M$ spectrum, JONSWAP spectrum, double peeked spectrum are recommended in engineering world.

The 0-order moment of wave loading mean square response is:

$$R(\theta_i, \alpha, H_s, T_z) = \sum_{i=1}^{n} \bar{S}(\omega_i, \theta_i, H_s, T_z) \{1 + \varepsilon_i(\omega_i, H_s, T_z)\} \cdot H^2(\omega_i, \alpha) \cdot \Delta \omega_i \quad (7)$$

where $H(\omega_i, \alpha)$ is wave loading transfer function, θ_i is spectrum style, ε_i is random of zero mean value normal distribution.

The uncertainties due to variations of spectrum could be represented by 0-order moment of some loading response spectrums and each spectrum has one conditional probability, then mean value and variance of total bias respectively are:

$$\bar{R} = \int \bar{R}(\theta) f_\theta(\theta) d\theta \quad (8)$$

$$\mathrm{Var}[R] = \int \{\mathrm{Var}[R(\theta)] + [\overline{R}(\theta) - \overline{R}]^2\} f_\theta(\theta) \mathrm{d}\theta \qquad (9)$$

where θ is different spectrum, f_θ is occurring possibility of spectrum. The Eq. (9) shows that the uncertainty of R is from itself uncertainty of spectrum and the differences among them. The loading mean square response value with standard spectrum reads:

$$R_S = \int_0^\infty S(\omega) \cdot H^2(\omega) \mathrm{d}\omega \qquad (10)$$

where $S(\omega)$ could be $P-M$ spectrum. The bias is relative to standard spectrum is:

$$\phi_S = \frac{R}{R_S} \qquad (11)$$

where ϕ_S has the same distribution type with R, also normal distribution. The mean value and COV of bias respectively are:

$$\overline{\phi}_S(\alpha, H_s, T_z) = \frac{\overline{R}(\alpha, H_s, T_z)}{R_S(\alpha, H_s, T_z)} \qquad (12)$$

$$V_{\phi_S}(\alpha, H_s, T_z) = V_R(\alpha, H_s, T_z) \qquad (13)$$

The paper chooses $P-M$ and JONSWAP spectrums to synthetically simulate real sea state for analyze the effect of wave spectrum representation to uncertainty of spectrum. According to the result of Guedes Soares(1984), the occurring possibility of $P-M$ spectrum in North Atlantic is $p(H_s) = 0.86 - 0.06H_s$ when H_s not more then 14.33m, $p(H_s) = 0$ when $H_s > 14.33$. The variance of ε_i equals to 0.525. $P-M$ spectrum is chosen as standard spectrum. The uncertainty results of three main loading of semi-submersible (DNV-RP-C103, 2005) in surviving condition ($H_s =$ 13.7m, $T_z = 11.89$s) are shown in Table 1.

Table 1 Mean value and COV for spectrum bias

Loading type	Longitude vertical moment		Transverse vertical moment		Transverse splitting force	
Wave direction(°)	$\overline{\phi}_H$	COV	$\overline{\phi}_H$	COV	$\overline{\phi}_H$	COV
0	0.927	0.193	0.930	0.168	0.916	0.172
50	0.933	0.188	0.954	0.186	0.990	0.187
90	0.904	0.165	0.946	0.174	0.965	0.175

Table 1 shows bias mean value of spectrum is between 0.90 and 0.99, COV is between 0.17 and 0.19. It is concluding that bias is approach 1.0 and COV is comparatively large, that implies uncertainty of spectrum mainly induces discreteness.

2.2 Uncertainty of Wave Height

Based on observing datas and theoretical analysis, Soares(1984) has given the regression equation between real significant wave height and observing value in North Atlantic, as following:

$$H_s = 2.33 + 0.75H_v + \varepsilon, s = 1.59 \qquad (14)$$

where H_v is chosen from sea state datas, ε is normal random that mean value and standard deviation are respectively zero and s. Then, bias mean value and COV of wave height respectively are:

$$\overline{\phi}_H = \frac{\overline{H_s}}{\overline{H_v}} = 0.75 + \frac{2.33}{\overline{H_v}} \tag{15}$$

$$V_H = \frac{\sqrt{\mathrm{Var}(\phi_H)}}{E(\phi_H)} = \frac{1.59}{0.75\overline{H_v} + 2.33} \tag{16}$$

It can be known that mean square response of wave loading is proportional to wave height from Eq. (6). Bias mean value of mean square response of wave loading caused by the uncertainty of wave height is:

$$\overline{\phi}(H_s) = \overline{\phi}_H^2 \tag{17}$$

From Eq. (15) and Eq. (16), it is concluded that mean value and COV of wave height bias respectively are shown in Table 2.

Table 2　Result in Soares's formula

H_v(m)	1	3	5	7	9	11	13	13.7
$\overline{\phi}_H$	3.080	1.527	1.216	1.083	1.009	0.962	0.929	0.920
V_H	0.516	0.347	0.262	0.210	0.175	0.150	0.132	0.126

Based on the sea state set of significant wave height in Northwest Pacific, the relationship between H_s and H_v could be regressed using means of H_s and H_v, as following:

$$H_s = 0.45 + 0.76H_v + \varepsilon \tag{18}$$

where ε is normal random which has a zero mean value and standard deviation s.

From statistics, relationship of variance between H_s and H_v is:

$$\sigma_{H_s}^2 = 0.76^2 \sigma_{H_v}^2 + \sigma_\varepsilon^2 \tag{19}$$

It is obtained that $s = 0.560$ by weighting σ_ε^2, then

$$V_H = \frac{0.560}{0.76\overline{H_v} + 0.45}$$

So bias mean and COV of wave height in Northwest Pacific could be calculated and the corresponding result is shown in Table 3.

Table 3　Result in northwest Pacific

H_v(m)	1	3	5	7	9	11	13	13.7
$\overline{\phi}_H$	1.210	0.910	0.850	0.824	0.810	0.801	0.795	0.793
V_H	0.463	0.205	0.132	0.097	0.077	0.064	0.054	0.051

After comparing Table 2 with Table 3, it is revealed that the COV of wave height formula in this paper is less than Soares's. The potential reason maybe formula in this paper is regressed for Northwest Pacific, but Soares has regress for North Atlantic. The latter's sea state is more abominable; meanwhile, statistical discreteness and standard deviation are larger.

2.3 Uncertainty of Mean Cross-Zero Period and the Corresponding Effect to Stress Amplitude

There is a certain difference between observing and measuring values. The relationship between measuring and ship observing is investigated for Northwest Pacific using the cross-zero period data given by Fang et al. (1996), and the meaning of the result lies on supporting marine structural design for this area. The regressing formula between measuring and observing datas is:

$$T_m = AT_v + B + \varepsilon \tag{20}$$

where A and B are constants and relation to sea state set. , ε is normal random that mean value and standard deviation are respectively zero and σ_ε. Then bias mean value and COV of period respectively are:

$$\overline{\phi}_p = \frac{\overline{T}_m}{\overline{T}_v} = A + \frac{B}{\overline{T}_v} \tag{21}$$

$$V_p = \frac{\sigma_{\phi_p}}{E(\phi_p)} = \frac{\sigma_\varepsilon}{A\overline{T}_v + B} \tag{22}$$

The result of linear regress is:

$$T_m = 3.59 + 0.62 T_v + \varepsilon \tag{23}$$

From statistics, the variance between T_m and T_v is:

$$\sigma_{T_m}^2 = 0.62^2 \sigma_{T_v}^2 + \sigma_\varepsilon^2 \tag{24}$$

where σ_{T_m} and σ_{T_v} are standard deviation of T_m and T_v, respectively. It is obtained that $s = 0.533$ by weighting σ_ε^2, then $V_p = \frac{0.533}{0.62\,T_V + 3.59}$. After considering the effect of uncertainty of period, the 0-order moment of stress amplitude response is:

$$R(\alpha, H_s, T_z) = \sum_{i=1}^{n} S(\omega_i, H_s, T_z, \phi_p) \cdot H^2(\omega_i, \alpha) \cdot \Delta \omega_i \tag{25}$$

When $\phi_p = 1.0$, it is obtained the unbiased R_{p0} from Eq. (25). When $\phi_p = \overline{\phi}_p$, mean value of R is $\overline{R}_p$ only considering uncertainty of cross-zero period. Then the bias mean value ϕ_z of R induced by uncertainty of period is:

$$\overline{\phi}_z(\alpha, H_s, T_z) = \frac{\overline{R}_p}{R_{p0}} \tag{26}$$

When $\phi_p = \overline{\phi}_p + \overline{\phi}_p \times V_p$, R becomes R_{p+}; $\phi_p = \overline{\phi}_p - \overline{\phi}_p \times V_p$, R becomes R_{p-}. The variance of ϕ_z is:

$$\sigma_{\phi_z}(\alpha, H_s, T_z) = \frac{|R_{p+} - R_{p-}|}{2R_{p0}} \tag{27}$$

It could be proved that the Eq. (26) and Eq. (27) have 1-order precision. Then the COV V_z of ϕ_z is:

$$V_z(\alpha, H_s, T_z) = \frac{|R_{p+} - R_{p-}|}{2\bar{R}_p} \tag{28}$$

The Eq. (27) and Eq. (28) suppose that mean value and COV of ϕ_z have no relation to significant wave height. The calculating result of some semi-submersible is shown in Table 4.

Table 4 Mean and COV of ϕ_z

$T_z(s)$	Wave direction(°)					
	0		90		180	
	Mean	COV	Mean	COV	Mean	COV
4	1.731	0.144	1.842	0.081	1.320	0.073
8	1.128	0.083	1.031	0.014	1.134	0.090
12	0.791	0.150	0.982	0.002	0.767	0.180

2.4 Uncertainty of Wave Loading Transfer Function

For rationally investigating the uncertainty of wave loading transfer function, the wave loading model trial should be carried out. CSSRC 05 wave keeping Tanker has taken one model trial for some semi-submersible with scale 1∶40. Transverse vertical moment and horizontal splitting force loading transfer function of middle longitudinal section are measured; meanwhile, trial value is taken as real value and uncertainties induced by various factors in trial is 5%. In direction 90°, numerical and trial results of the transverse vertical moment and horizontal splitting force loading transfer function around peaked frequency are shown in Fig. 1 and Fig. 2 which suppose the numerical result comparatively coincides with trial's. Assuming taking trial and numerical result as real and theoretical value, the assessing result of bias mean value and COV of loading transfer function is shown in Table 5. For there are some systematic and incidental errors during trial and structural stress induced by transverse vertical moment is remarkably larger than horizontal splitting force's. So, measurable signal of horizontal splitting force is less and the measuring errors may be larger than transverse vertical moment which just be proved by the results in Table 5.

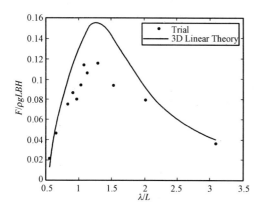

Fig. 1 Horizontal splitting force

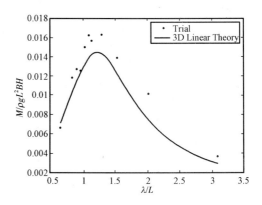

Fig. 2 Transverse vertical moment

Table 5 Mean and COV for uncertainty of load transfer function bias

Loading style	Mean	COV	COV(5%)
Horizontal splitting force	0.76	0.17	0.19
Transverse vertical moment	1.12	0.10	0.11

After considering the 5% uncertainty in trial, COV of transverse vertical moment transfer function is: $\sqrt{0.10^2 + 0.05^2} = 0.11$, COV of horizontal splitting force transfer function is: $\sqrt{0.18^2 + 0.05^2} = 0.19$. Taken bias mean value and COV in Table 5 as references, it is rational to suppose that wave loading transfer function bias B_F's mean value is 0.76 ~ 1.14 and COV is 0.11 ~ 0.19. Then the paper sets bias mean value and COV 0.90 and 0.15, respectively; meanwhile extends this conclusion to every direction. Based on Eq. (6), it is supposed that the mean square response is proportional to loading transfer function's square.

2.5 Uncertainty of Sea State Occurring Possibility and the Corresponding Effect to Stress Parameter

Generally, the sea state set is from statistical regressing analysis to many years' sea state datas. Since sea state set have seasonal and annual variation, the sea state sets obtained by various time have comparatively large differences. Using the sea state set in North Atlantic with various statistical periods such as 1, 2, 4, 29 years, Moan et al. (2005) have investigated the influence to long-term extreme forecast and fatigue life assessment. If taken 1 year as a statistical period, it is obtained 29 fatigue damage values. The ratio of maximum and minimum of fatigue damage is 4.3 for FPSO and 1.9 for semi-submersible which shows that sea state occurring possibilities in different statistical period have comparatively large influence to fatigue life assessing result. In Addition, research of Nolte(1973) reveals that significant wave height obviously varies by year and Guedes Soares & Ferreira(1995) have analyzed annual variation of sea state.

For investigating the uncertainty of sea state occurring possibility in sea state scatter, the paper assesses fatigue life of semi-submersible by the sea state set of S4 area in South China Sea given by Fang et al. (1996). Assuming each wave direction occurring possibility is equal and considering the influence of uncertainty of occurring possibility to stress parameter, real stress parameter could be expressed as:

$$\phi \Omega_s = \frac{1}{4q}(2\sqrt{2})^m \Gamma(m/2+1) \sum_{n=1}^{4} \sum_{i=1}^{k} \sum_{j=1}^{q} \lambda(m,\varepsilon_i) \Lambda_i f_{0i} \gamma_i \sigma_i^m \qquad (29)$$

where Ω_s is unbiased nominal value of stress parameter, k is short-term sea state number, q is direction number, ϕ is bias and is random which is used to express uncertainty of occurring possibility, n express four seasons. The bias of occurring possibility to fatigue stress parameter is 0.955 using Eq. (29).

2.6 Uncertainty of Wave Direction and the Corresponding Effect to Stress Parameter

Marine platform's bow direction has a certain random and wave direction generally doesn't obey uniform distribution that causes difficult to establish heading angle, and, only introducing random could solve this dilemma. It is usually assuming heading obeys uniform in wave loading assessment for actual engineering design. The paper uses whole year sea state statistical datas to investigate the uncertainty of wave direction and consider the effect of wave direction uncertainty to fatigue life assessment. It shows that the sea state scatter is very complex from the datas of S4 area (Fang et al., 1996).

(1) Uncertainty of wave direction occurring possibility.

Setting stress parameter Ω under uniform heading distribution as nominal stress parameter Ω_s, and, stress parameter under actual heading is Ω_D. The this kind of bias is defined as η_1, showed in Eq. (30), and, the result is 0.824.

$$\eta_1 = \frac{\Omega_D}{\Omega_s} \tag{30}$$

(2) Uncertainty of sea state scatters under various wave directions.

In the sea state sets of S4, there are some differences between sets of each single wave direction and sets including all wave directions. Generally, sea state sets including all directions are used to assess fatigue life, but more exact structural fatigue damage should accumulate fatigue damage of each sea state sets under different wave directions, therefor it induces bias of fatigue life, also including stress parameter. Now the bias η_2 of stress parameter which caused by the uncertainty of sea state sets under various wave directions and the expression is:

$$\eta_2 \Omega_s = \frac{(2\sqrt{2})^m \Gamma(m/2+1) \frac{1}{q} \sum_{n=1}^{q} \sum_{j=1}^{q} \sum_{i=1}^{k} \lambda(m,\varepsilon_i) \Lambda_i f_{0i} \gamma_{in} P_n \sigma_i^m}{\sum_{n=1}^{q} P_n} \tag{31}$$

where Ω_s is nominal value of stress parameter calculated by the sea state sets including all direction under uniform heading distribution, k and q respectively are the number of short-tern sea state and wave direction, P_n is wave direction occurring possibility. Assuming heading obeys uniform distribution and the occurring possibility of sea state scatter of each wave direction is uniform, the result of bias is 0.963.

In conclusion, the total bias of stress parameter due to the uncertainty of wave direction is:

$$\eta = \eta_1 \eta_2 \tag{32}$$

The result of η is 0.794 less than 1.0 which reveals the method directly using the sea state scatter of full direction and supposing uniform heading to assess fatigue life is too conservative.

2.7 Uncertainty of Short-Term Stress Response

Former research in the paper has introduced random ϕ_S to describe the uncertainty of sea wave spectrum, ϕ_L to describe the uncertainty of wave loading transfer function, ϕ_H to describe the uncertainty of wave height, ϕ_Z to describe the uncertainty of mean cross-zero period, and ϕ_T to describe the total uncertainty of short-term response. Assuming these uncertainties is independent of each other, it is concluded that sea wave spectrum is proportional to the square of wave height. And, it only counts the uncertainty of wave loading transfer function which is included in hotspot stress transfer function. Then, according to Eq. (6) the total bias of 0-order moment of stress response under short-term sea state is:

$$\phi_T = \phi_S \cdot \phi_L^2 \cdot \phi_H^2 \cdot \phi_Z \tag{33}$$

The mean value and COV respectively are:

$$\overline{\phi_T} = \overline{\phi_S} \cdot \overline{\phi_L}^2 \cdot \overline{\phi_H}^2 \cdot \overline{\phi_Z} \tag{34}$$

$$V_T^2 = V_S^2 + 4(V_L^2 \cdot \overline{\phi_L}^2 + V_H^2 \cdot \overline{\phi_H}^2) + V_Z^2 \tag{35}$$

Based on the former result of the paper, uncertainty analysis is carried out for Northwest Pacific. The calculating uncertainty result of total 0-order moment under the sea state ($H_s = 4.5$m, $T_z = 8.0$s) for some semi-submersible, which has the highest contribution to fatigue damage, is showed as Table 6.

Table 6 Mean value and COV of bias ϕ_T for short-term response

Wave direction(°)	$\overline{\phi_T}$	COV
0	0.579	0.445
90	0.538	0.409
180	0.570	0.448
270	0.528	0.411

2.8 Uncertainty of Long-term Stress Response

While using $S-N$ curve method, the stress parameter considering uncertainty factors in fatigue life calculation is:

$$B_w^m \Omega_s = \phi \eta (2\sqrt{2})^m \Gamma(m/2+1) \sum_{i=1}^{k} \sum_{j=1}^{q} \lambda(m, \varepsilon_{ij}) \Lambda_{ij} f_{0ij} \gamma_i p_j (\phi_T \sigma_{ij}^2)^{\frac{m}{2}} = \phi \eta \phi_Q \Omega_s \tag{36}$$

where Ω_s is unbiased nominal value of stress parameter. The paper introduces the following expression for Eq. (36).

$$\Omega = \phi_Q \Omega_s = (2\sqrt{2})^m \Gamma(m/2+1) \sum_{i=1}^{k} \sum_{j=1}^{q} \lambda(m, \varepsilon_{ij}) \Lambda_{ij} f_{0ij} \gamma_i p_j (\phi_T \sigma_{ij}^2)^{\frac{m}{2}} \tag{37}$$

When ϕ_T equals mean value $\overline{\phi_T}$, the corresponding stress parameter in Eq. (37) is $\overline{\Omega}$; on the other hand, ϕ_T equals 1.0, unbiased stress parameter in Eq. (37) is Ω_s. Then the mean value of bias of stress parameter is:

$$\overline{\phi_Q} = \frac{\overline{\Omega}}{\Omega_s} \tag{38}$$

The result is of $\overline{\phi_Q}$ is 0.438. From Eq. (36) it is obtained that:

$$B_w^m = \phi \eta \phi_Q \tag{39}$$

The mean value of B_w^m can be gotten by expression:

$$\overline{B_w^m} = \phi \eta \overline{\phi_Q} \tag{40}$$

The mean value is 0.332 after calculation.

Through Taylor series expansion, the mean square variance of Ω is:

$$\sigma_\Omega = \left| \frac{d\Omega}{d\phi_T} \right| \sigma_{\phi_T} \tag{41}$$

Substituting Eq. (37) into Eq. (41), it's obtained:

$$\sigma_\Omega = \frac{m}{2}(2\sqrt{2})^m \Gamma(m/2+1) \sum_{j=1}^{q} \sum_{i=1}^{k} \lambda(m,\varepsilon_{ij}) \Lambda_{ij} f_{0ij} \gamma_i p_j (\phi_T)^{\frac{m}{2}-1} \sigma_{\phi_T} \sigma_{ij}^m \tag{42}$$

From Eq. (37), the mean square variance of ϕ_Q could be calculated by $\sigma_{\phi_Q} = \frac{\sigma_\Omega}{\Omega_s}$ and the result is 0.309. COV of B_w^m is:

$$C_{B_w^m} = \frac{\phi \eta \sigma_{\phi_Q}}{\phi \eta \phi_Q} = \frac{\sigma_{\phi_Q}}{\phi_Q} \tag{43}$$

From the above equation, it shows that COV of B_w^m only is relative to short-term uncertainty and is independent of occurring possibility of sea state and wave direction. The COV $C_{B_w^m}$ is 0.722. Assuming B_w^m obeys lognormal distribution, the middle value of B_w^m is 0.269 which is calculated by:

$$\widetilde{B_w^m} = \frac{\mu_{B_w^m}}{(1+C_{B_w^m}^2)^{0.5}} \tag{44}$$

Because both of B_w and B_w^m obey lognormal distribution, the middle value and COV of B_w respectively are 0.646 and 0.39 calculated by:

$$\widetilde{B_w} = (\widetilde{B_w^m})^{\frac{1}{m}} \tag{45}$$

$$C_{B_w}^2 = (1+C_{B_w^m}^2)^{\frac{1}{m}} - 1 \tag{46}$$

3 FATIGUE RELIABILITY ANALYSIS FOR SEMI-SUBMERSIBLE

The uncertainties in fatigue stress calculation could be described by random B which is expressed as:

$$B = \prod_i B_i \quad (i = M, W, N, H) \tag{47}$$

where M represents structure fabricating technology, W represents sea wave description, N represents nominal stress calculation, H represent hotspot stress coefficient. B_i and B all obey lognormal distribution, meanwhile, the values of B_i are showed in Table. 7 and B_w adopts the former result in the paper.

Table 7 Values of B_i

B_i	$\tilde{B}_i$	C_{B_i}
B_M	1.0	0.10
B_N	1.0	0.20
B_H	1.0	0.10

The limit state equation of fatigue life with the method of $S - N$ curve is:

$$Z = \ln T_f - \ln T_D = \ln\left(\frac{\Delta A}{B^m \Omega}\right) - \ln T_D \tag{48}$$

where all of Δ, A and B are randoms with lognormal distribution, T_D is design life and for this platform is 30 years, Ω is stress parameter and adopts certainty value referencing Eq. (3) and for the paper is 616.

Wirsching (1984) has recognized Δ obeys lognormal distribution after lots of trials and researches and sets $\tilde{\Delta} = 1.0$ and $C_\Delta = 0.30$. The D curve parameters in DNV(2005) rules are $\tilde{A} = 3.81 \times 10^{12}$, $m = 3$, $C_A = 0.51$.

The reliability index is calculated by following equation and result is 2.2.

$$\beta = \frac{\mu_Z}{\sigma_Z} = \frac{\ln\tilde{\Delta} + \ln\tilde{A} - m\ln\tilde{B} - \ln\Omega - \ln T_D}{[\ln(1 + C_\Delta^2) + \ln(1 + C_A^2) + m^2\ln(1 + C_B^2)]^{1/2}} \tag{49}$$

CONCLUSION

The paper has supposed a suit of method to assess the uncertainties of fatigue wave loading mainly including the uncertainties of wave spectrum, wave loading transfer function, wave height, mean cross-zero period, occurring possibility of sea state and wave direction, as well as the influence to wave loading and fatigue life. Moreover, the fatigue reliability index is investigated according to the uncertainties of loading and strength. It is concluded that, first, there are some comparatively large differences among various wave loading transfer function. Second, there are some comparatively large distinctions of mean cross-zero periods between South China Sea and North Atlantic. Third, there are some comparatively large differences among sea state scatters of various seasons which could induce a certain uncertainties of fatigue life. For South China Sea the sea states in spring and summer have smaller contributions to fatigue damage than in autumn and winter. Fourth, the distribution principle of wave direction has distinguishing influence to fatigue life, meanwhile, the normal assumption of uniform wave direction and using the sea state sets of full direction make the calculating result is too conservative. So the actual distribution principle and uncertainties of wave direction are worth of study and concern. Fifth, the mean value and COV of wave loading uncertainties respectively are 0.646 and 0.39 during assessing fatigue reliability.

REFERENCES

ABS. 2003. Guide for the Fatigue Assessment of Offshore Structures.

DNV. 2005. Fatigue Design of Offshore Steel Structures.

DNV-RP-C103. 2005. Column-stabilized Units. DET NORSKE VERITAS.

Fang Z S, Jin C Y, Miao Q M. 1996. Wave Statistical Sets in Northwest Pacific[M]. National Defense Industry Press, Beijing, China.

Guedes Soares C. 1984. Probabilistic Models for Load Effects in Ship Structures[D]. Norwegian Institute of Technology. NTH, Trondheim, Norway.

Guedes Soares, C Ferreira A M. 1995. Analysis of the Seasonality in Non-stationary Time Series of Significant Wave Height[J]. Computational Stochastic Mechanics. Rotterdam: A A Balkema;559-78.

Hu Y R, Li D Q, Chen B Z. 2010. Ship and Ocean Engineering Structural Fatigue Reliability Analysis[M]. Harbin Engineering University Press, Harbin, China.

Moan T, Gao Z, Ayala-Uraga E. 2005. Uncertainty of Wave-induced Response of Marine Structures Due to Long-term Variation of Extratropical Wave Conditions[J]. Marine Structures, 18: 359-382.

Nolte K G. 1973. Statistical Methods for Determining Extreme Sea states[C]. POAC Conference, Reykjavik, Iceland.

Wirsching P H, Light M C. 1980. Fatigue under Wide-band Random Stresses[J]. Journal of Structural Division, ASCE, 16 (7).

Wirsching P H. 1984. Fatigue Reliability for Offshore Structures[J]. Journal of Structural Engineering, ASCE, 110 (10) :2340-2356.

Wu D W. 2011. Research on Reliability Assessment Method for Structural Strength of Semi-submersible Platform [C]. CSSRC, Wuxi, China.

Yang P, Gu X K, Wu D W. 2012. Research of Load Uncertainty and Structural Reliability for Semi-submersible [J]. Journal of Ship Mechanics, 16 (1): 108-117.

INSTALLATION AND RISKS

Development of Virtual Simulation System for Subsea Hardware Installation Procedure

Zhangquan Sun[1], Xiaoyao Ren[2], Mingjie Li[2] and Caixia Li[2]

[1] Offshore Oil Engineering Co. ,Ltd. ,Tanggu,Tianjin,China
[2] Offshore Oil/Gas Research Center,China University of Petroleum,Beijing,China

Abstract Subsea production system is widely used in the development of offshore oil fields owe to its unique advantages. The South China Sea is rich in oil and gas reservoir where the water depth ranges from 500m to 2000m, but the installation capacity in China is up to 300m water depth. Based on the analysis of the installation procedures for deep water facilities, this paper focuses on the development of virtual simulation system for the installation procedure of subsea production facilities by using commercial softwares such as MultiGen Creator and Vega Prime. This system is developed to realize the designing, previewing and training for deepwater installation, meanwhile, it will provide a virtual offshore platform for the development of oil fields located in deepwater.

Key Words Subsea production system; Installation; MultiGen Creator; Vega Prime; Simulation system

INTRODUCTION

Virtual Reality (VR) simulation technology provides virtual environment of simulate real events for users by integrating computer with graphics processing and generation, information synthesis, sensing, data visualization, and other advanced technology in 1993, Burdea G first summarized and presented characteristic triangle of VR simulation technology at Electro 93 international conference, as shown in Fig. 1, and this is the main difference between VR and animation.

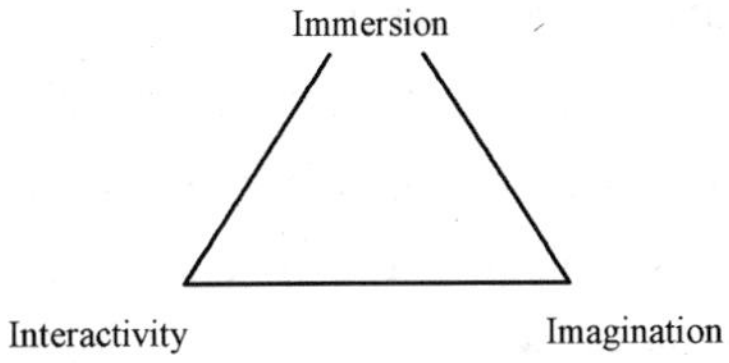

Fig. 1 Characteristic triangle

In recent years, VR simulation technology has been widely used in various fields. In the international oil industry, as an effective technology way of onshore oil and gas development, now it's being used in seismic data visualization explanation of complex exploration area, design optimization of well location and well trajectory, development plan design of mature exploration area, construction of reservoir model, simulation of oil reservoir value, virtual drilling and so on. The main examples of application in marine engineering are:

(1) In 2004, the Ormen Lange gas field development simulation system of Norway Hydro Company, which was developed by Switzerland ABB Company, has guaranteed installation and pipeline laying of underwater facility successfully.

(2) In 2004, the first internal large sailing manipulation simulator, which was developed by Britain Transas Company, was installed in Maritime School of Wuhan University of technology,

and then it was conducted training of ship maneuvering, ship collision avoidance, pilot operation, sailing comprehensive internship, crew assessment and so on.

(3) In 2010, the offshore oil collection simulation system, which was developed by Belgian Barco Company and University of Sao Paulo, Brazil, solved the problem of training of underwater oil spill collection.

Because of the special nature environment of ocean, for example, water depth, wave, high-pressure, dark, low-temperature and so on, the development of offshore oil becomes very difficulty. As relatively backward of marine science and technology, lack of advanced technology and marine equipment, combined with the South China Sea water depth is between 500m and 2000m, our country hasn't had the technology of oil and gas exploration and production at this level of water depth currently. In this paper, a VR simulation system applied in simulating installation and retrieve process of underwater production facilities is constructed by using the VR technology in offshore oil development, the system is used in South China Sea, played its unique "3I" feature advantage, bringing a new breakthrough in the development of technology for deep-sea oil development. In VR simulation system:

(1) The operator can well-known the environment of area of the target oil and gas field visually, such as, wind, wave, flow, seabed topography and so on.

(2) Simulating the installation process of underwater production real-time, conducting simulation exercises, providing the operator works cognitive.

(3) Allowing operators to manipulate simulation station real-time, such as, ship, cranes and so on, operators will grasp a more realistic, diverse operating experience in a short term, and they can be performed examinations.

(4) The harsh marine environment leads to high requirements of underwater equipment, so it can shorten the development cycle and reduce costs by designing and testing in virtual marine environment under the premise of don't have to manufacture prototype.

(5) Designing, optimization, testing the installation program with a variety of interactive tools, ensuring the safe operation of the marine.

So this article can achieve the goals of reduce risk and save money, promote the rapid development of marine technology, by using VR simulation system to auxiliary complete the installation of the subsea production facilities.

1 THE INSTALLATION PROCESS ANALYSIS OF SUBSEA PRODUCTION FACILITIES

Subsea production facilities has the features of wells flexible layout, small influence by disasters, parts of equipment can be recycle, the installation doesn't require the construction of experience offshore platforms, so its economic benefits is very considerable, it's reliability and widely used in various countries. The subsea production facilities of subsea production system mainly includes: subsea tree, cross-over, underwater manifolds, umbilical, underwater base plate, pipeline terminal and so on. Fig. 2 shows a layout plan of a typical subsea production system.

There is much oil and gas resources stored in South China Sea, its water depth is 500 ~

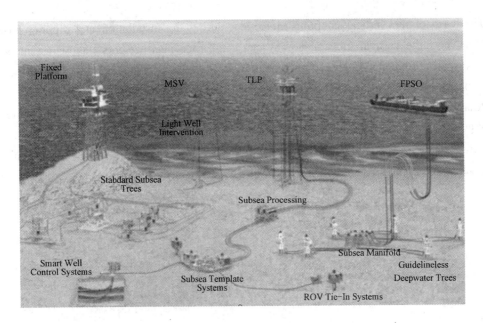

Fig. 2 Typical subsea production system

2,000m, but development of underwater production system in our country is relatively limited, its deepest water depth is 333m. After research and analysis the successful project of foreign deepwater installation, its water depth is 3,000m, according to the differences between different installation implements, there are four installation methods: installation with drill pipe, installation with crane or winch, installation with pulley, and swing installation.

When we select installation method, we should mainly consider the following questions:

(1) Climatic conditions;

(2) Target install water depth;

(3) Upgrading and decentralized system: the dynamic response of the upgraded load, decentralization, upgrading ability and so on;

(4) Load control and positioning: the accurate positioning, the stability of the seabed.

Considering the above factors, according to the actual working conditions of Liwan 3 - 1 gas field and the deepwater installation ability currently grasp by our country, the VR simulation of subsea production facilities installation and retrieve process will based on specific installation vessels and equipment of our country, under good sea conditions, mainly use the installation method which with the crane or winch. Now, the selections of a variety of installed equipments have been completed.

2 INSTALLING CALCULATION OF SUBSEA PRODUCTION FACILITIES

The theoretical calculation of the dipping process ensure the installation of subsea equipments safety and effectively. Different installation vessel and equipment has numerical calculated for its lifting and installing process, including the torque load calculation, power calculation, dynamic response of the enhanced load, the installation and position calculation and stress analysis. Simulation system has made the theoretical analysis visualization, enhanced system immersive, and verified the reasonable of the installation process.

The harm of the vertical coupling movement during the installation and dipping process in deep-sea. The large vertical movement may cause the rope near weights flabby instantaneously, when the rope is tightened again, an enormous mutation load is generated and shorten the fatigue life of the rope, and can lead to the structure destruction and device loss in severe cases.

Analyzing and calculating the boat-cable-body (here, slings, umbilical, drill pipe simplify for the cable) coupling when design the installation program. According to the ship motion theory, established the three orthogonal coordinate systems, as shown in Fig. 3, solved the movement mathematical model, and obtained the relationship between terminal movement and tension by applying

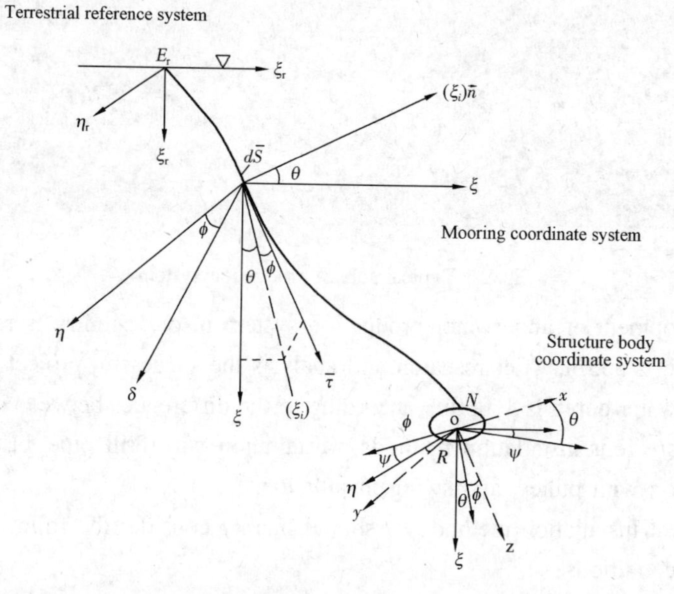

Fig. 3 Multi-body coupling coordinate system

Fig. 4 The real time calculation in visual simulation

analysis software, and established a database. Coupled motion response simulated in the three-dimensional simulation system, forecast the natural frequency of the dipping system, and ensure that the frequency of the control system away from its natural frequency.

The vortex module of the simulation software could not reflect the physical characteristics of the physical movement; the third-party software is still needed to develop collaboratively for establishing the complete and engineering significance simulation system. The system will write simulation program with complex calculation by using the VC + + and monitoring force characteristics, as shown in Fig. 4.

3 SIMULATION FOR INSTALLATION AND RETRIEVE PROCESS

3.1 Simulation Modeling

The virtual simulation system for subsea production facilities includes the installation and recycle procedure of subsea X-mas tree, manifold, jumper, PLET and so on. The simulation model library should include at least:

(1) The main subsea production facilities: subsea X-mas tree and others.

(2) Dedicated equipments for installation and recycle of subsea production facilities, such as X-mas tree installation requires core dedicated equipment and hydraulic installation equipment, PLET installation requires PLET bracket and A-frame.

(3) Facilities required in installation and retrieve process, such as installation vessels, cranes, winches, ropes.

(4) Auxiliary equipments, such as lifting beam (connection for ropes and subsea facilities to prevent rotate), three orifice plate (complete power conversion from crane to winch of subsea facilities), floating body module (alleviate ropes tension).

(5) ROV, Remotely Operated Vehicle, high ability in deepwater operations, real-time status monitoring of subsea facilities, lock and unlock implementing agencies, orientation, and measurement.

The three-dimensional modeling is the basis of the simulation system. The models is required high fidelity, but also to meet the real-time requirements. This paper use MultiGen Creator, develop by American MultiGen-Paradigm, to complete modeling. Creator is polygon modeling tools, provides user interface with hierarchy database and users could quickly generate three-dimensional models.

Creator generates facet model, focusing on real-time (such as the lod skill) and authenticity (set textures, materials), so for three-dimensional models will be more realistic, it is needed to also use three-dimensional CAD software SolidWorks, focusing on accurate modeling, to complete parametric modeling. Simulation models are generated by steps as shown in Fig. 5.

Simulation models generated by faces rotated and stretch in SolidWorks are required generating surface bounding volume by importing to 3DMAX. Only after this step they could be accepted by Creator. Ruled surface and bounding surface could be accepted by Creator directly. In addition, units are required consistent in SolidWorks and Creator to prevent the model is enlarged or reduced, see Fig. 6 and Fig. 7.

3.2 Process Simulation Design

This paper use Vega Prime system to develop the virtual simulation system for subsea production facilities. Vega Prime provides LynX Prime graphical environment with advanced cross-platform scene rendering API (Application Program Interface), and is compatible with C++ STL (Standard Template Library). Therefore, Vega Prime provides advanced simulation capabilities and also has an easy-to-use advantages, so that users can focus time and effort on solving the problem of application areas.

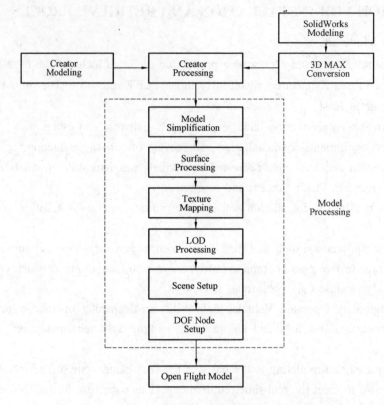

Fig. 5 Simulation modeling steps

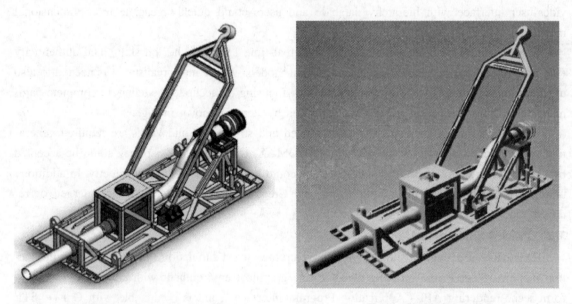

Fig. 6 Three-dimensional model Fig. 7 Simulation model

In addition, Vega Prime provides collision detection algorithm meeting requirement. VP FX modules can easily build and customize the particle special effects. VP Vortex module can add movement performance of real-time interaction, based on the physical characteristics. Marine module is capable of simulating realistic ocean state and the ship's posture in the water.

Vega Prime application development framework is shown in Fig. 8.

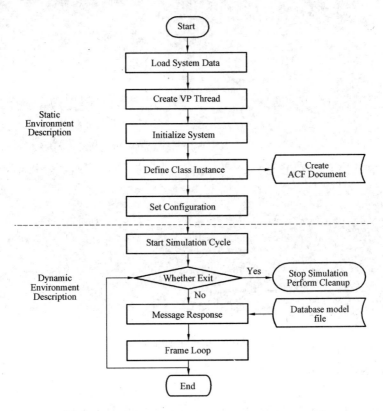

Fig. 8　Vega prime application development framework

According to the actual installation process to write simulation movement script, we get the following simulation operation program (simulation example of subsea manifold installation process):

For manifold weight of less than 250 tons, the crane will rise manifold from the deck of the barge, adjust to above the mounting orientation. We can use the crane to complete the entire process of installation if the work depth range is within the crane work depth. When installation in deepwater, using the crane down to a certain depth, then power conversion to winch after stable equilibrium to complete the whole process. During the down process, monitor subsea production facilities installation path by monitoring device. When down to the seabed, install facilities after deployed by deepwater signal positioning system. Manifold weighing more than 300 tons is usually installed by folding arm-type crane.

Even though we can use crane or winch to install when manifold is heavier, in the traditional installation methods generally use the drilling catheter method. At the moment, there are non-traditional installation methods used in ultra deep-water area internationally: sheave method and PIM (pendulum installation method). Sheave method requires two supply vessels, as well as a semi-submersible drilling platform. PIM need the help of the two work vessels and use of pendulum action to complete the installation, see Fig. 9.

Fig. 9 Installation visual simulation

CONCLUSIONS

Virtual reality simulation technology used in offshore oil development is a new attempt, and has a broad prospect. The paper has achieved the three-dimensional virtual simulation for installation process, modeling the three-dimensional model by using the MultiGen Creator, and using Vega Prime to roaming real-time.

Development of Motion Simulation in the installation process of subsea production facilities is the difficulty, using VC + +, assist to develop the calculation process for complex movements is safety and feasibility.

The article has developed the software of the virtual simulation system, without the design of the hardware platform. Increasing the simulation console emulator of the operating equipment, such as ships, cranes, winches and other equipments, could increase the sense of immersion and make the training more significance in engineering.

REFERENCES

Butler P Hine III, Carol Stoker, Michael Sims, Daryl Rasmussen, Phil Hontalas. The Application of Telepresence and Virtual Reality to Subsea Exploration[EB/OL]. http://daleandersen. seti. org/Dale_Andersen/Science_articles_files/Application_of_Telepresence_and_VR_to_Subsea_Exp - 1. pdf.

Dang-BaoSheng. 2007. Development of Virtual Reality [J]. China Modern Educational Equipment, 4.

Driscoll F R, Lueck R G, Nahon M. 2000. The Motion of a Deep-Sea Remotely Operated Vehicle System. Part 1: Motion Observations[J] Ocean Engineering, 27:29 - 56.

Korkealaakso P M, Rouvinen A J, Moisio S M, Peusaari J K. 2007. Development of a real-time simulation environment [J]. Multibody Syst Dyn 17:177 - 194.

Wang Chunsheng. 2010. Application of Subsea Production System in Liuhua 4 – 1 Oilfield [J]. Shipbuilding of china,25 Special2.

Wang Wei,Sun Liping,Bai Yong. 2009. Investigation on subsea production systems [J]. China Offshore Platform, 24 (6).

Zhang Jin,Xie Yi. 2011. Research and Development of Subsea Manifold Installation Methods [J]. The Ocean Engineering,29 (1).

Zhang Zhongwei. 2003. Application of Virtual Reality in Marine Technology[C]. Innovation academic conference about IT,network,information technology,electronic instruments in Tianjin.

Zhao Dangqian,Gao Pengfei,Gong Ying. 2007. Virtual Environment Simulation Based on the MultiGen Creator and Vega [J]. China Science and Technology Information,21.

Zhao Gaishan. 2002. Virtual Reality in E&P: Applications andprospects [J]. Progress in Exploration Geophysics,25 (4):9 – 20.

Gas Hydrate Problems during Deep Water Gas Well Test

Yonghai Gao[1], Baojiang Sun[1], Changsheng Xiang[1],
Linlin Yan[1], Chunlei Chen[1], Qinghua Zheng[2]

[1] College of Petroleum Engineering, China University of Petroleum, Qingdao, China
[2] CNOOC Research Institute, Beijing, China

Abstract Gas hydrate is an essential problem which must be considered in deepwater well testing. Pressure and Temperature are two key factors of hydrate formation. Based on deepwater testing procedure, the flowing and heat transfer equations were established on flowing test and shut in conditions, and the flowing was also simulated at throttle position. Temperature and pressure distribution in testing pipes were obtained. On this basis, the hydrate formation sections in pipes were predicted. The results indicated that at the stage of circulating test, gas production is an important factor to hydrate formation, the less output, the larger hydrate generation region, yet near the throttle in pipes, hydrate is easily formed at large production due to the throttle effect. Another indication of the result is that serious hydrate forming problem may take place as the high pressure and low temperature at the stage of shut in. Based on the results, three inhibitor injection positions were put forwarded: mud line, 650 meters below mud line and choke positions.

Key Words Well test; Deepwater; Gas hydrate; Prediction

INTRODUCTION

The low environmental temperatures encountered in deepwater well test increase the likelihood of gas hydrates formation in testing pipes, which is negatively impacting the deepwater well test operations by reducing the effective cross-sectional area or even plugging the test pipes (Pedroso, 2009; Chen, 2010; Peavy, 1994; Dalmazzone, 2003). Pressure and temperature are closely related to the formation of gas hydrate. In this paper, the pressure and temperature distributions in test pipes and chokes were obtained through theoretical analysis and numerical simulation. The potential hydrate formation regions were analyzed, and the preventive measures to control the formation of hydrate were put forwarded as well.

1 PRESSURE AND TEMPERATURE COMPUTATION METHODS IN PIPES IN DEEPWATER WELL TEST

To obtaining the fluids' temperature and pressure expressions in the test pipes, the pressure and temperature computing models in circulating conditions were set up considered the flow in pipes as gas-liquid biphase flow (majority is gas) on the basis of the mass, energy, and momentum conservation laws (Hasan, 1991, 1994; Sun, 2008; Romero, 1998).

$$\frac{dp}{dZ} = -\frac{\partial}{\partial t}(\sum_{i=1}^{2}\rho_i E_i v_i) - \frac{\partial}{\partial Z}(\sum_{i=1}^{2}\rho_i E_i v_i^2) - g\cos\alpha(\sum_{i=1}^{2}\rho_i E_i) - \frac{dF_r}{dZ} \qquad (1)$$

$$\frac{dT_t}{dZ} = \frac{1}{A}(T_t - T_e) - C_t \qquad (2)$$

Where p is pressure in pipe; Z is positive in the downward direction; t is time; ρ_i are densities of gases and liquids; E_i are volume fractions of gases and liquids, and v_i are velocity of gases and liquids; α is deviation angle; F_r is friction loss; T_t is mixed fluids' temperature in pipe, and T_e is the environmental temperature. A and C_t can be expressed as,

$$A = \frac{w_t c_{pt}}{2\pi r_{to} U_t} \qquad (3)$$

$$C_t = \frac{v_t \frac{dv_t}{dz} - g\cos\alpha - \frac{C_J c_{pt} \cdot dp}{dz}}{c_{pt}} \qquad (4)$$

Where w_t is the mass flow rate of mixed fluid; c_{pt} is the specific heat of fluids in pipes; r_{to} is the outside radius of pipes; v_t is the velocity of mixed fluid; C_J is the Joule-Thompson coefficient; U_t is the whole heat transfer coefficient, which can be calculated on two cases (above and below mud line).

At the stage of shut in, the pressure in pipes is the summation of hydrostatic pressure and wellhead pressure. The temperature can be calculated by the following equation:

$$\rho c_p \frac{\partial T_t}{\partial t} \pi r_{ti}^2 = \frac{k(T_e - T_t) \cdot 2\pi r_{to}}{r_{to} - r_{ti}} \qquad (5)$$

Where ρ is the density of fluid in pipes; r_{ti} is the inner radius of pipes.

Pressure and temperature often decrease sharply when liquids flow in pipes of variable diameter or chokes due to throttling phenomena, which can result in frost or hydrate generation (Perkins, 1993). It is difficult to calculate the pressure and temperature distribution using theoretical methods as the complexity of flowing. So, in this paper, the Computational Fluid Dynamics (CFD) method is applied to simulating the distribution of pressure and temperature when gas flows through the throttle.

2 TMPERATURE AND PRESSURE DISTRIBUTION UNDER DIFFERENT CONDITIONS

2.1 Different Testing Production

According to the established pressure and temperature calculation method of pipe in deep water test, a deepwater gas well LW-X in South China Sea was calculated and the distribution of temperature and pressure during testing operation were obtained, as well as the hydrate formation areas were predicted.

The basic data of the well are as follows: the water depth is 1350m; testing layer depth is 3150~3190m; pipe's outer diameter is 114.3mm.

The distribution of temperature and pressure inside the testing tube under the conditions of different production while well opened time-2 hours are shown in Fig. 1 and Fig. 2. From the curves it can be found that at the same conditions, the temperature of tubing fluid increases, while the pressure decreases with flow rate.

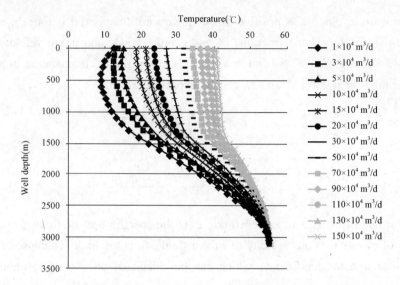

Fig. 1　Temperature distribution inside testing tube under different production

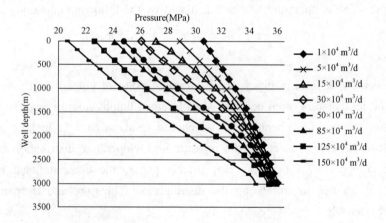

Fig. 2　Pressure distribution inside testing tube under different production

2.2　Shut-In Time

When some problems are encountered such as downhole testing tool failure, the gas well can only be shut in at surface. Fig. 3 shows the inside temperature changes of the testing tube with time at which the production rate is $15 \times 10^4 m^3/d$. After the surface shut-in, the inside pressure of the testing tube is approximately equal to the formation pressure, and with the shut-in time increasing, the inside temperature of the testing tube will be decreased to the surrounding temperature.

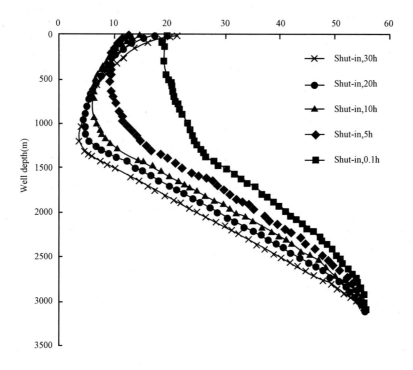

Fig. 3　Temperature distribution inside testing tube changes with time while surface shut-in

2.3　Influence of Variable Diameter Pipe

A model of natural gas flowing through a 3in – 2in – 3in variable diameter pipe was established, and the flow and temperature field were analyzed using computational fluid dynamics software. Computational conditions: inlet pressure, 11MPa; outlet pressure, 10MPa; inlet temperature, 300K. The results are shown as Fig. 4 to Fig. 6.

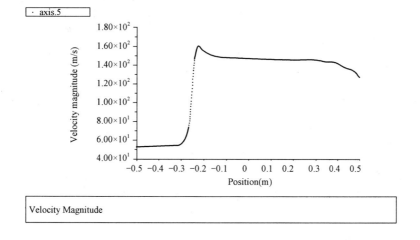

Fig. 4　Velocity of axis of variable diameter pipe

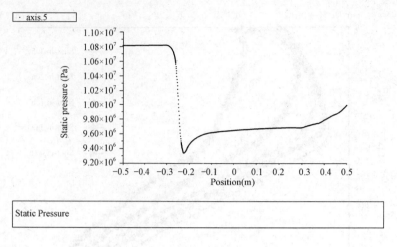

Fig. 5 Pressure of axis of variable diameter pipe

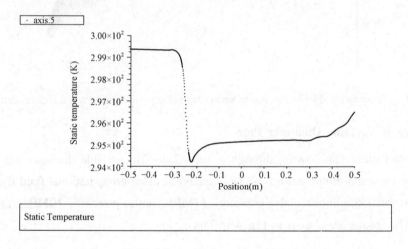

Fig. 6 Temperature of axis of variable diameter pipe

From the calculated results, it can be seen that when methane gas flow through the variable diameter pipe, the tube pressure along axis decreases from 11MPa to the minimum 9.3MPa, and then increases gradually to the outlet pressure 10MPa; meanwhile, the temperature decreases from 300K(27℃) to 294K(21℃) quickly, and then increases gradually. Whether hydrate forms depends on the pressure and temperature of the liquids. The lowest pressure is approximately 9.3MPa, and the lowest temperature is approximately 294K(21℃) in this case. Based on natural hydrate formation conditions, It can be conformed that if the inlet pressure is higher than 20MPa or the inlet temperature is lower than 290K(17℃) in designed conditions, there will be no hydrate formation in the full string, otherwise, there might have hydrate formation when gas-liquid fluids flow through variable diameter pipe because of the throttle effect.

2.4 Influence of Choke Throttle

Fig. 7 shows a choke physical model which gases flow through in it. Computational conditions: inlet pressure is 10MPa; outlet pressure is 5MPa; inlet temperature is 300K(27℃). The prediction of hydrate formation is done by analyzing the pressure and temperature before and after throttling. The results are shown as Fig. 8 to Fig. 10.

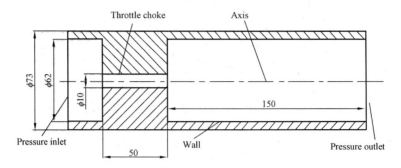

Fig. 7　Surface throttle choke physical model

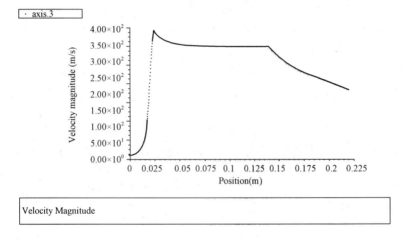

Fig. 8　Velocity of choke axis

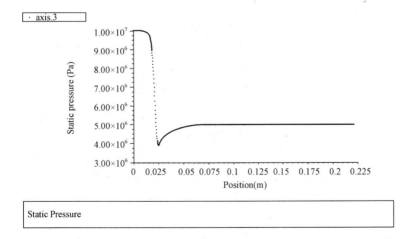

Fig. 9　Pressure of choke axis

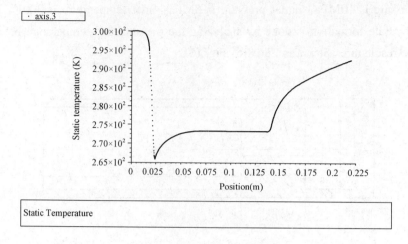

Fig. 10 Temperature of choke axis

The results indicate that with the velocity of fluids increase quickly from 15m/s to 300m/s when they flow through the choke throttle, the pressure decreases from 10MPa to 3.9 MPa. Meanwhile, because of the influence of throttle effect, the temperature decreases sharply from 300K(27℃) to 265K(-8℃), which is far lower than the hydrate formation temperature under the condition of choke pressure and even lower than freezing point, which will result in hydrate formation or freeze.

3 PREDICTION AND PREVENTION MEASURES OF HYDRATE FORMATION

The conditions of hydrate formation can be determined from the Vander Waals model (Vander Waals J. H. ,1959):

$$\frac{\Delta\mu_0}{RT_0} - \int_{T_0}^{T}\frac{\Delta H_0 + \Delta C_p(T - T_0)}{RT^2}dT + \int_{p_0}^{P}\frac{\Delta V}{RT}dp = \ln(f_w/f_w^0) - \sum_{i=1}^{2} v_i\ln(1 - \sum_{j=1}^{N_C}\theta_{ij}) \quad (6)$$

Where $\Delta\mu_0$ is the chemical potential difference between the empty hydrate lattice and pure water under standard conditions, T_0 and p_0 are the temperature and pressure under standard conditions, $T_0 = 273.15K$, $p_0 = 0$, ΔH_0、ΔV、ΔC_p denotes, respectively, specific enthalpy difference, specific volume difference and specific heat capacity difference between the empty hydrate lattice and pure water, and also $\ln(f_w/f_w^0) = \ln x_w$.

3.1 Hydrate Formation Region In Circulating Test

The Analysis from above indicated that the rate of flow can affect the temperature and pressure distribution within the string during the circulating, and these two parameters would determine the hydrate formation. Fig. 11 shows the hydrate formation regions under different flow conditions.

It can be seen from the above results that, the smaller the production of gases, the greater the hydrate formation region. This is mainly due to the high string pressure and low string temperature conditions at low flow rate. When the flowing rate is 1×10^4 m³/d, the pressure and the tempera-

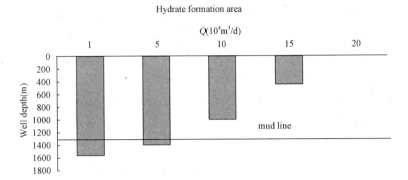

Fig. 11　The hydrate formation region under different rate of flow conditions

ture up to their highest and lowest point during the whole string respectively, at which the hydrate formation region will spread from 200m below the mud line (depth 1,550m) to the wellhead. Therefore, it will definitely cause hydrate formation and finally bring difficulties to the implementation of the test, if no inhibition measures are taken at low production, with increasing of the production, the string pressure is gradually reduced, at the same time the minimum temperature gradually increased, so the hydrate formation region decreases. Under the test conditions we calculated, when the production equals to 5×10^4 m^3/d, from the mud line to the wellhead is the most risk area for hydrate formation. As production increases to 10×10^4 m^3/d, the region from 1000m below the surface to the wellhead will probably hydrate. When production reaches 15×10^4 m^3/d, hydrate will appear near the wellhead only, while no hydrate will come into being at the production higher than 20×10^4 m^3/d. This is due to the transmission of high reservoir temperature. At high production flowing conditions, the borehole fluid temperature is much higher than the hydrate formation temperature, and hydrate is impossible to form. The smaller the production, the higher the wellhead pressure, so that the pressure is relatively high throughout the test string. Furthermore, string flow exchanges heat with the low temperature environment as much as possible, thus the flow temperature is low and the hydrate is easy to generate.

3.2　Hydrate Formation Region During the Surface Shut-In

During the surface shut-in, the main factor affecting the hydrate formation region is shut-in time. Fig. 12 describes the hydrate formation region within the test string at different shut-in times when the production is 15×10^4 m^3/d.

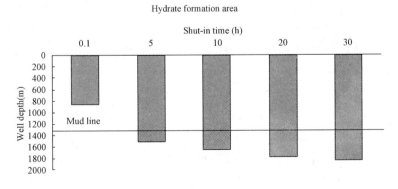

Fig. 12　Hydrate formation region within string at different shut-in time

The Figure reveals that the surface shut-in will lead to a serious hydrate plugging in deep-water test. With the increase of shut-in time, the string temperature is gradually close to the ambient temperature, while the pressure has maintained high value. The hydrate formation region gradually increased from 1,000m below surface at initial moment of shut-in to 1950m below surface or 600m below the mud line. Because hydrate is difficult to decompose as long as they came into being, once a long string hydrate, the consequences would be extremely serious and bring difficulties to the deep water test. Therefore, surface shut-in should be avoided as much as possible. If surface shut-in is necessary due to unforeseen circumstances, we must be sure to keep injecting inhibitors to ensure the circulation of the fluid within the string, in order to avoid serious hydrate plugging problems.

3.3 Positions of Choke and Nozzle

From the calculation results above, both the changed diameter and the nozzle choke will affect the string environment and lead it transforming to a state which the hydrates are easily to generate. As deep water test relates to high pressure, high production, as well as low environment temperature, we must take appropriate measures in the choke to avoid hydrate formation and decrease adverse effects.

3.4 Prevention Measures of Hydrate Formation

Inhibitors injection is an effective measure for hydrate formation prevention in deepwater gas well test. Generally, hydrate inhibitors can be divided into thermodynamic inhibitors and kinetic inhibitors. The thermodynamic inhibitors (alcohols, salts) can effectively reduce the hydrate formation temperature under a certain pressure. However, kinetic inhibitors do not contribute to the decreasing of formation temperature much but can reduce the hydrate formation rate obviously.

3.5 Inhibitors Choosing

The advantage of alcoholic inhibitors is obvious. There is no environmental polution because of its sufficient burning with the produced gas. The alcoholic inhibitors should be the first priority applying to avoid the of natural gas hydrate (NGH) forming in site operation.

3.6 Inhibitors' Injecting Position

Besides injecting the chemical inhibitors in the lowest point which is possible to generate the NGH, the inhibitors should be injected in the position which is close to the seabed mudline to avoiding the unfavourable influence of the low temperature, and the wellhead also should be considered due to the gas expansion.

The lowest injecting point determine: As the generating condition of NGH, the hydrocarbons hydrate would not be generated when temperature is higher than 21℃. With the water deeper than 1,500m the temperature is about 2~4℃ at the mudline. Usually, the temperature gradient below the mudline is more than 3℃/100m. The critical temperature appears in the 650m below the mudline, and the temperature would be higher than 21℃ in the deeper stratum. According to the critical temperature of NGH, there should be an injecting point at 650m below the mudline.

Nearby the mudline: The temperature is approximately 3℃ nearby the mudline in deepwater well. There should be a chemical inhibitors injecting point close to the mudline to avoid the hydrate generating.

At the choke position of wellhead: Because of the throttle effect in the chokes, the pressure drops, the gas volume expands and the temperature reduces in the downstream flow. The chemical injecting point should be set and the heat exchanger should be applied for heating at the well head to avoid the generating of NGH.

CONCLUSIONS

(1) With the increase of surface shut-in time, the string temperature is gradually close to the ambient temperature, while the pressure maintains high value. Serious hydrate forming problems may take place in deepwater well test.

(2) Production is the most important factor of hydrate formation at the stage of circulating, the less output, the larger hydrate generation region.

(3) Alcohols are effective in inhibiting hydrate formation. Three inhibitor injection points should be set: mud line, 650 meters below mud line and choke position, and they will do a good work in NGH problems prevention.

ACKNOWLEDGEMENT

This work is supported by the Major Program of National Natural Science Foundation of China (No. 51034007), the National Science and Technology Major Project of China(No. 2011ZX05026 – 001 – 02), the National Natural Science Foundation of China (No. 50874116, and No. 51004113), the Natural Science Foundation of Shandong Province of China(No. ZR2009FQ005) and the Fundamental Research Funds for the Central Universities, China.

REFERENCES

Chen Shing-Ming . 2010. Un-planned Shut-in and Deepwater Gas Hydrate Prevention. OTC 20436.

Dalmazzone C, Herzhaft B, Rousseau L, et al. 2003 . Prediction of Gas Hdrates Formation With DSC Techniue. SPE 84315.

Hasan A R , Kabir C. 1994. Aspects of Wellbore Heat Transfer During Two-Phase Flow. SPE Production & Facilities, 8:211 – 216.

Hasan A R, Kabir C S . 1991. Heat Transfer during Two-Phase Flow in Wellbores: Part 11 – Wellbore Fluid Temperature . SPE 22948, presented at the Annual Technical Conference and Exhibtion of the SPE held in Dalllas, TX.

Peavy M A, Cayias J L. 1994. Hydrate Formation/Inhibition During Deepwater Subsea Completion Operations [S]. SPE 28477.

Pedroso C A , Denadai N, Pinheiro C. 2009. The Use of Saturated Fracturing Fluid to Cope With Hydrate in Gas Wells Located in Ultradeep Waters. SPE 122185.

Perkins T K . 1993. Critical and Sub-critical Flow of Multiphase Mixture Through Chokes. SPE 20633. SPE Drilling & Completion.

Rornero J, Touboul E. 1998. Temperature Prediction for Deepwater Wells: A Field Validated Methodology. SPE 49056, presented at the 1998 SPE Annual Technical Conference and Exhibition held in Naw Orleans, Louisiana, 27 – 30.

Sun Baojiang, Gao Yonghai, Wang Zhiyuan, et al. 2008. Temperature Calculation and Prediction of Gas Hydrates Formed Region in Wellbore in Deepwater Drilling. The Proceedings of the Eighteenth (2008) International Offshore and Polar Engineering Conference, Volume I: 44 – 48.

Vander Waals J H, Platteeuw J C. 1959. Clathrate solutions, Adv. Chem. Phys. ,2 : 1 – 57.

Installation Technology on Riser Clamps for Platform Upgrading

Baoying Xia, Kai Yu, Lan Yu, Shouqiang Lin, Haiyun Shen

Maintenance Operations Department, COOEC Subsea Technology Co. ,Ltd, ShenZhen, China

Abstract Nowadays, Installation of riser and cable conduit accounts for a large part of offshore oil engineering work. In this paper, we mainly introduce the installation technology on riser clamps for platform upgrading, including the different forms of clamps, process of fabrication and installation. The equipment which be used usually in the process of clamps installation is be summarized simply.

Key Words Riser; Cable-Conduit; Clamp; Installation; Platform; Method; Column; Winch

INTRODUCTION

With the rapid development of offshore oil industry, many old platforms, which are outdated production and structure form, cannot meet the demands of existing application. The corresponding reform is necessary. And the installation of riser and cable conduit in the transformation is very important. This paper mainly introduces the clamps installation and fabrication, including the installation position of different clamps in different forms. The more attention needs to be focus on the process of making some special items of clamps, and on the process of installation operational methods and localization. The paper also summarizes the usual equipment which would be used in the procedure of clamp installation.

1 FUNCTION AND FORMS

Pipe clamps are essential and necessary for the riser pipe and cable conduit. The main function of pipe clamps is to fix the riser or cable to platform in order to avoid the accident that the riser or cable conduit became slack and fall off the platform. According to the different position, the clamps generally are classified as two forms. One kind is installed on the structure horizontal beam of the platform and the characteristic of them is that the angle between the both ends of the clamp is 90 degrees. We can see it in Fig. 1.

Another is installed on the diagonal support beam of platform, and the characteristic of them is that the angle between the both ends of the clamp is not fixed. It is adjusted according to the actual requirement. It can be seen in Fig. 2.

The large difference of clamps is the internal surface, the member clamp which will be fixed on riser needs protection and seal by Chloroprene Rubber, with the clamp on jacket having 10mm round bar for internal fixation.

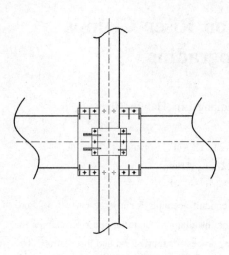 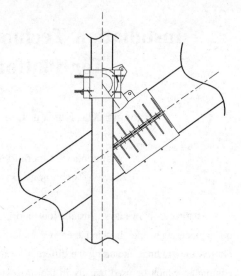

Fig. 1 Clamp on the horizontal beam   Fig. 2 Clamp on the diagonal support beam

2 FABRICATION

Clamps are generally produced on land, and transported to offshore for installation. So in the land production process, the most important thing is to make sure the size of the clamps. In order to ensure that the offshore installation of clamps can be smooth and successful, the method of trial installation on land for clamps is adopted usually (Fig. 3). It needs to roll a short pipe in advance as auxiliary structure, which has same diameter like the tube structure at the offshore position of clamps installation.

(a)   (b)

Fig. 3 Trial installation on land

Each clamp has fixed number, with in-between connection pipe. The axial lines of two clamps at each end are different in space, which needs more attention during fabrication.

3 INSTALLATION PROCEDURE OF CLAMPS

3.1 First-phase Preparations

3.1.1 Strengthening the base of the winch and fixed in place

As an usual equipment, winch is an essential part in the clamp installation process, and the quantity can be decided according to actual situation of clamp installation. When the clamp is installing, the winch is on the bottom deck of platform, and generally speaking, it is walkway for works to keep the platform in repair (see Fig.4). But there is a problem, the walkway is full of grating and it cannot meet the requirement of strength when the winch is working. So the base which is manufactured by H shape steel to make sure the requirement of strength is necessary.

Fig.4 The position of winch

3.1.2 The installation of the cantilever beam

We know that the winch is based on the walkway, and we need to put the clamps in right place along the legs of platform. So this needs to install the cantilever beam in the position of legs. Cantilever beam usually is made from H shape steel, if necessary, we can increase the support structure, and weld the lifting point on it for the fixed pulley. Corresponding to every installation position of clamp, there is a lifting point upper on it. There has a note the structure engineer need to check the strength of cantilever beam and the strength of the lifting point.

3.1.3 Pulleys connection

The winch and clamps are connected by block and tackle. Before the installation of pulley we need to install a fixed pulley on the legs of platform and a pulley on the cantilever beam for hanging clamps. We need to pay attention to the selection of pulley's installation points, which needs to consider the path of wire rope. It is necessary that the wire rope avoid to crash and conflict the other structures.

3.2 The Installation of Clamp Offshore

According to the actual situation, the offshore installation program of clamps can be dicided into:

(1) When the installation position is located in the area of crane hoisting on the platform, the clamps can be installed directly with the crane.

(2) When the installation position beyond the scope of crane hoisting on platform, the clamps can be installed indirectly with the crane.

3.2.1 The installation position in the scope of crane

The crane on platform will lift the clamp down to the cantilever beam position. The winch on steel base will catch it and lift it down to the specified location in water, and then divers will close the clamp.

Two sets of pneumatic winch could respectively locate on walkway. As mentioned, the walkway that stops the pneumatic should be reinforced. Cantilever beam and pulley systems can be welded with the column. The cantilever beam welds three lifting point which respectively corresponding to three clamps in the vertical position on the same column. See Fig. 5.

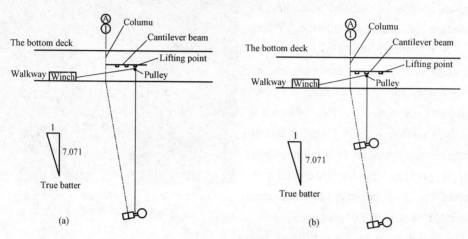

Fig. 5 Clamp lifting procedure

The crews on the platform must follow the instructions from the divers in the process of reaching the installation position in case of the clamps in the water meet catheter frame. There is a distance when the clamps arrive at the specify installation area, which requires lever blocks chain to do some adjustment.

When the clamps are installed on the column, divers will be shut off the clamps. Divers will mount bolt with the cooperation of the winch. But, do not fully tighten, that is for the clamps could do a coaxial adjustment. Sequence of clamps installation is from the bottom to the up.

In the installation of the second clamps on this column, firstly the pulley on the cantilever beam should be moved to the next lifting point, and so on. Repeat this step.

3.2.2 The installation position out of the scope of crane

If platform crane cannot reach the installation position of clamps, the platform crane will lift down into water nearby the clamps installation position. The assistance of divers and the floatation pocket will pull the clamps to the position that the hoist could reach. The divers will join the clamps to the winch. The winch will put down the clamps to the designated installation position. See Fig. 6.

Fig. 6 Flotation pocket

4 INSTALLATION OF CLAMPS UNDERWATER

The divers should install the clamps in the specified position after the clamps put down into the water.

4.1 Preparation for Underwater Work

Before the start of clamps installation, the first thing is to complete the following work:

(1) Diving equipment should be in a connective state and on debugging;

(2) Necessary scaffolding should be put up;

(3) Related crews should be familiar with the hydrological data.

4.2 Underwater Cleanup Work

Steel wire rope or steel metric ruler is measured in advance to make sure the installation position of clamps for finding out the center elevation mark of the underwater installation.

Divers clean the installation position of clamps with shovels and grinding wheel, which requires the installation position of clamps cleaned up and down respectively 50 mm. The same treatment should be employed to clean the fishing nets and the reproductive sea creatures in the area.

4.3 The Installation of Clamps

4.3.1 Installation of clamps under the water

When the clamps move from the surface to underwater installation position, the divers should give assistance and install the clamps underwater.

Divers in the water should observe the descend position of the stand pipe clamps in a safe area when it is sinking; when the stand pipe clamps arrives at the installation position, the divers should inform the crews to stop putting down. Make sure the dive guide cables intact during the process of putting down the stand pipe clamps.

When the stand pipe clamps closes to the column, it should cut off the rope behind. With the gravity itself and guide chain traction, close the clamps. And then, it should immediately lead a bolt wear through bolt hole, loading on nuts.

It should install of all the bolts, and constantly adjust and rectify the bolt, until completely fixed the stand pipe clamps.

Welding 4 to 8 supports at the bottom of the stand pipe clamps is for the prevention of the clamps sliding when matching the coaxial.

4.3.2 Installation of clamps above the water

After the underwater stand pipe clamps installation, it is the part of the installation of the stand pipe clamps above water. The installation of the stand pipe clamps over the water is the same to other product structures, which only requires welding the clamps on the column.

5 CLAMPS OF THE COAXIAL DEGREES ADJUSTMENT

In order to guarantee the clamps in a straight line, it should make a coaxial tolerance of opposites adjustment after the clamps installed in place. It's generally fall into two cases for coaxially adjustment:

5.1 The Column has Installed Cable Conduit or Riser

If the same column has installed cable protect tube or stand pipe, they could be regarded as reference. Adjustment the coaxial tolerance means the arrangement each clamp so that they could keep the same distance with the installed cable conduit and riser on the same column, which guarantees the installed clamps in a straight line and assures the installation of the riser and the cable conduit smoothly.

5.2 The Column has not Installed Cable Conduit or Riser

If the same column has not installed a cable protect tube or stand pipe yet, it will be marked on the rubber in the wall coating before the installation in the water. Marks should be made in the middle of the clamps as well as the either end of the clamps, which could apply the disposable paint brush:

(1) Platform staff should put a piece of wire rope in the surface of the clamps, and the diameter of wire rope should have enough strength;

(2) The divers should get into water and carry back the wire rope;

(3) The divers should get into water and put wire rope through the clamps one by one;

(4) The wire rope should be fixed at the bottom of the clamps by the divers. Then, platform staff tightening wire rope;

(5) The divers should check and confirm that whether the wire rope is close to the wall of the clamps or not.

Whether the wire rope is seated in the middle of the clamps or not should be checked by divers. The clamps marks should be seen on the straight line which tensed by wire rope line. But if not, the position of the clamps should be adjusted by putting down the steel wire until all the position of the clamps meet the requirements.

CONCLUSIONS

This paper introduced the installation technology of neutral pipe clamps in upgrading process of the old platform, describing the form and installation of clamps steps in detail, which hopes could provide a reference for the development of offshore oil industry.

Investigation on Active Model Test Methods Based on the Deepwater Riser Installation Process

Dongyan Zhao[1], Mao Ye[1], Menglan Duan[1], Zhigang Li[2],
Wei Yang[2], Tao Kuang[1] and Xiaonan Jiao[1]

[1] Offshore Oil Engineering Co., Ltd., Tianjin, China
[2] Offshore Oil/Gas Research Center, China University of Petroleum, Beijing, China

Abstract As the current hybrid model testing's environmental load and the boundary conditions at the truncated position of active truncation method are hard to control, the concept that simulates the process of installing deepwater riser on land is proposed for the first time.

The similarity theory and the boundary conditions at the truncated position are deduced. Base on the study of model test for deep water projects, an active truncation approach of hybrid model testing for the installation of deepwater risers is introduced. The results show that the test device can simulate the SCR installation process well. It also lays the foundation for future works on deepwater projects.

Key Words Active truncation method; SCR installation; Similarity; Model test

INTRODUCTION

Model test is regarded as the most effective way to guide the deep-sea project. For the high-risk, high investment, high technology of deep-sea oil development, hybrid model testing is considered the best way to solve the problem. Basically, hybrid model test methods are divided into passive truncation method and active truncation method. At present, passive truncation approach is used more frequently, but it requires numerical reconstruction to get the dynamic characteristics of the system.

Active hybrid model testing is a very challenging frontier technology. It requires advanced and accurate real-time control system which can simulate a wide range of six degrees of freedom motion. Active simulation uses the real-time computer control unit and servo mechanism to simulate the movement and stress of the truncated position of the riser at the bottom of the pool, and it can obtain the dynamic characteristics of the actual system directly.

In this article, we propose the active truncation method to simulate the installation of SCR in deepwater based on the research of SCR installation method when using HYSY201 in South China Sea. During the test, 6DOF motion platform is used to simulate the motion of HYSY201 and Servo control systems are utilized to apply the boundary conditions at the truncated position, as it solved the problem that the active model test requires precise control.

1 THE THEORETICAL FOUNDATION

The following sections provide a detailed presentation of the similarity deduction, and the boundary conditions of SCR at the truncated position, which is calculated by solving the catenary equations as well as numerical analysis and fitting of the boundary condition.

1.1 Similarity Theory

When it comes to constructing the simulation test system, model tests should be based on the similarity theory. Dynamic similarity is the foundation of model test, it is to say that if the prototype system is fully similar with the model system, they must meet geometric similitude, material similitude and kinematic similitude. In this paper, the test is mostly conforming to dynamic similarity.

Geometric similitude is that the ratio of the line which consists of the corresponding points between prototype system and the model system is a constant.

After scaled by 1∶10:

$$L_P/L_M = 10 A_P/A_M = 100 v_P/v_M = 1000$$

Material similitude is that the ratio of the mass at corresponding points between prototype system and the model system is a constant.

$$m_P/m_M = \rho_P v_P/\rho_M v_M = v_P/v_M = 1000$$

Kinematic similitude: During the test, the conditions of HYSY201 is scaled which must strictly abide by Newton's second law:

$$F_P = F_r F_M; m_P = m_r m_M; v_P = v_r v_M; t_P = t_r t_M \tag{1}$$

$$F_P = m_P \frac{dv_P}{dt_P} = m_r m_M \frac{dv_r v_M}{dt_r t_M} = \frac{m_r v_r}{t_r} m_M \frac{dv_M}{dt_M} \tag{2}$$

According to Eq.(1) and Eq.(2), we can get:

$$F_r F_M = \frac{m_r v_r}{t_r} m_M \frac{dv_M}{dt_M} = \frac{m_r v_r}{t_r} F_M \tag{3}$$

On the basis of Eq.(3) the following equation can be:

$$\frac{F_r t_r}{m_r v_r} = 1 \tag{4}$$

Similarity constant of quality are defined as:

$$m_r = \frac{m_P}{m_M} = \frac{\rho_P v_P}{\rho_M v_M} = \rho_r v_r = \rho_r L_r^3 \tag{5}$$

The relationship between velocity and time similarity constant are described as follows:

$$v_r = \frac{L_r}{t_r} \Rightarrow t_r = \frac{L_r}{v_r} \tag{6}$$

In accordance with Eq.(4) ~ Eq.(6), following equation can be acquired:

$$\frac{F_r}{\rho_r L_r^2 v_r^2} = 1 \Rightarrow \frac{F_P}{\rho_P L_P^2 v_P^2} = \frac{F_M}{\rho_M L_M^2 v_M^2} \tag{7}$$

Velocity similarity constant is:

$$v_r = \frac{v_p}{v_M} = \sqrt{\frac{F_p \rho_M L_M^2}{F_M \rho_p L_p^2}} = \sqrt{\frac{F_r}{\rho_r L_r^2}} \qquad (8)$$

Through the deduction and calculation of the similarity constant, we can get time and stress similarity according to Eq. (6).

$$t_r = \frac{L_r}{v_r} \qquad (9)$$

$$\sigma_r = \frac{F_r}{A_r} \qquad (10)$$

As the length similarity constant equals 10, it shows that the length similarity constant between the model system and the prototype system is 1∶10. Similarity constants of other physical quantities in the test are shown in Table 1.

Table 1 Similarity constants

Physical	Code	Similarity principle	Length similarity constant(λ)	Similarity constant
Length	C_L	λ		0.1
Area	C_A	λ^2		0.01
Volume	C_V	λ^3		0.001
Density	C_D	1		1
Mass	C_M	λ^3	0.1	0.001
Force	C_F	λ^3		0.001
Stress	C_σ	λ		0.1
Time	C_T	$\lambda^{1/2}$		0.32
Velocity	C_S	$\lambda^{1/2}$		0.32

1.2 In the Lowering and Recovering Process, Deducing the Boundary Conditions at the Truncated Position of CSR (Without Considering Seastate)

For the deepwater installation simulation, the length of SCR is far larger than its diameter. So we have to truncate the SCR at a certain height. We can acquire the boundary conditions by solving the catenary equation. The configuration of catenary riser is shown below. See Fig. 1.

Through solving the catenary governing equation, the relationship between lowering altitude and tip angle at the truncated position is described as follows:

$$x_0 = \sqrt{y_0^2 + 2\frac{y_0 \omega_0 \cos\theta}{\omega_0(1-\cos\theta)} y_0} - \sqrt{y_1^2 + 2\frac{y_1 \omega_0 \cos\theta_1}{\omega_0(1-\cos\theta_1)} y_1} + \frac{y_1 \omega_0 \cos\theta_1}{\omega_0(1-\cos\theta_1)}$$

$$\cdot \ln\left[\frac{\omega_0(1-\cos\theta_1)}{y_1 \omega_0 \cos\theta_1} y_1 + 1 + \sqrt{\left(\frac{\omega_0(1-\cos\theta_1)}{y_1 \omega_0 \cos\theta_1} y_1 + 1\right)^2 - 1}\right] + \frac{\Delta y}{\tan\theta_1} \qquad (11)$$

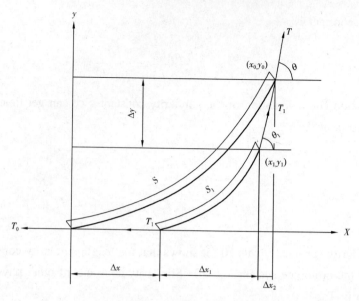

Fig. 1 Configuration before and after SCR decentralized

The tip angle is defined if the lowering depth is given, and then determine the morphology of the SCR.

The following formula can be used to obtain the axial tension:

$$T = \frac{y\omega_0}{1 - \cos\theta} \qquad (12)$$

We can get the x-coordinate in the new coordinate system as follows:

$$x' = \frac{T_1}{\omega_0}\ln\left[\left(\frac{\omega_0}{T_1}y_1 + 1\right)^2 - 1\right] \qquad (13)$$

Coordinate conversion formula:

$$x = x' + x_0 - \left(x_1 + \frac{\Delta y}{\tan\theta_1}\right) \qquad (14)$$

Since we solve the boundary conditions by using two catenary equations, coordinate transformation is needed. In a unified coordinate system, x coordinates at the truncated position are defined as follows:

$$x = \frac{T_1}{\omega_0}\ln\left[\frac{\omega_0}{T_1}y_1 + 1 + \sqrt{\left(\frac{\omega_0}{T_1}y_1 + 1\right)^2 - 1}\right] + x_0 - \left(x_1 + \frac{\Delta y}{\tan\theta_1}\right) \qquad (15)$$

Taking the installation of 16 in SCR for instance, the installation depth is 3,000m and the initial tip angle is 80°. 16 in SCR with its basic parameters containing pipe diameter 0.4064m, wall thickness 0.02032m, quality in the water per meter is 60.08kg.

We truncate at the position where $y = 2,970$m with its lowering altitude at 30m, when the decentralization of 5m, 10m, 15m, 20m, 25m, 30m, the top angle, axial force and x-coordinate at the truncated position can be solved from Eq. (11) to Eq. (14). In this way, the boundary conditions of the model riser are obtained as shown in Table 2.

Table 2 Boundary conditions of model riser

Lowering depth	Tip angle(°)	tension T(N)	x-coordinate(m)
Δy(m)	θ(°)	T(N)	x
0	79.9158	2,117.04402	153.05336
0.5	80.0068	2,113.03582	153.05819
1	80.0956	2,109.14228	153.06278
1.5	80.1842	2,105.26834	153.06724
2	80.2669	2,101.66488	153.07126
2.5	80.3552	2,097.83112	153.07545
3	80.4432	2,094.01598	153.07953

2 IN THE LOWERING AND RECOVERING PROCESS, THE MOVEMENT AND THE FORCE AT THE TRUNCATED POSITION

The study of the movement and the force is the most important part of active truncated test. Riser truncation boundary condition is acquired mainly by the superposition of two parts, the first part is that without considering sea state, the boundary conditions change with the lowing of SCR. The second part is with considering the case of sea conditions, the movement of the ship effect on the boundary conditions.

2.1 Simulation of Environmental Load

6 DOF platform (5m × 4m × 5.5m) is used to simulate sea conditions of HYSY201 under the influence of various working conditions. Under the influence of the waves, the ship will have six degrees of freedom of movement: three rotation(pitch, rolling, yawing), three translational motion(surge, sway, heave). Traditionally, large pool and wave exciter was used to simulate the environmental load during the model test. For the situation that large pools are not only expensive but also unable to simulate flow well, we use large 6DOF hydraulic platform to simulate the environmental load of the South China Sea. See Fig. 2.

Fig. 2 6 DOF platform

2.2 In the Case of Not Considering the Seastate, the Changes of Boundary Conditions

The relationship between lowering altitude and tip angle at the truncated position is described as Eq. (11), when using the formula to control the system, the operation is large, and the accuracy cannot meet the requirements of real-time control. In order to realize the precise control, at first, using Eq. (11) ~ Eq. (14) to calculate several finite discrete points, and then make data fitting, after that the boundary conditions at every moment can be obtained.

2.3 The Relationship Between the Tip Angle and the Lowering Altitude

The configuration of SCR mainly depends on SCR tip angle during the SCR recycling and lowering process. As the discrete points of tip angles in table 2 are fitted, we can get the following graph Fig. 3.

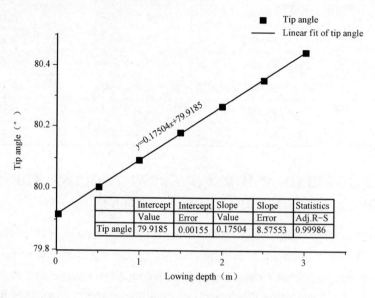

Fig. 3　Linear fit of tip angle

After fitting, the linear equation is $y = 0.17504x + 79.9185$ and the error is only 0.0015, so we can use this linear equation to replace Eq. (11).

2.4 The Relationship Between the Tension and the Lowering Depth

As the tension of table 2 is fitted, the relations between tension and lowering altitude are shown in Fig. 4.

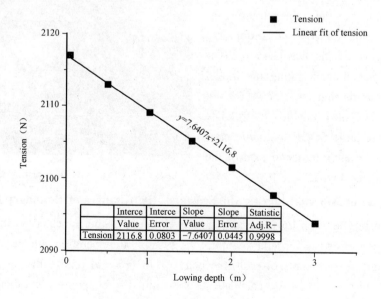

Fig. 4　Linear fit of tension

With the lowering depth increasing, the tension at the truncated position becomes much smaller. The fitted line equation is $y = 7.6407x + 2116.8$. We can use this linear equation to replace Eq. (12). The control of the tension just applying a fixed weight does not guarantee the tension changes with the lowering altitude's changes, so we must control the acceleration to control the tension changes.

In the initial state, applied weight is 216kg.

$$F = m(g + a)$$

When the axial tension changes, we can get the trend of the acceleration:

$$a = (F - mg)/m$$

The fitted acceleration applied on the weight and changes with the lowering altitude $y = -0.3537x$, as shown in Fig. 5.

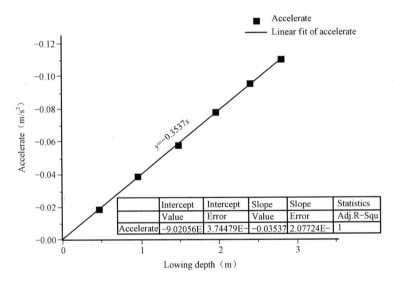

Fig. 5 Linear fit of accelerate

2.5 The Relationship Between X-coordinate and the Lowering Altitude

In the process of SCR recovering and lowering, we fit the discrete points of the displacement in x direction in the table 2, and then get the equation which related to lowering altitude and x-coordinate as shown below:

With the increase of lowering altitude, the displacement in x-direction increases. By fitting, the equation is $y = 0.00868x + 153.053$. According to the fitting equation, x-coordinate can be calculated at any lowering depth.

2.6 With Considering the Sea Conditions, the Ship's Movement Impact on the of Truncation Boundary Conditions

Under various sea conditions, the impact of the ship movement on the boundary conditions is large. And because of the existence of the phase angle, it is difficult to be calculated by theory. So we can use field test results to obtain the data. Before the experiment, test conditions are input

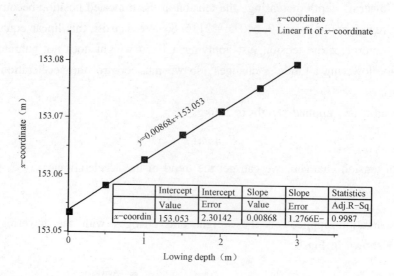

Fig. 6　Linear fit of x-coordinate

into 6DOF platform. When the platform is at the equilibrium position, the initial boundary conditions is imposed on the truncation of the SCR, then operate the platform and use the angle, displacement and acceleration sensors to measure the changes of angle, displacement, acceleration in real time.

2.7　The Boundary Conditions at Truncation

Without considering the sea conditions, we control lifting and lowering speed of SCR by using a servo system, according to the relationship between velocity and lowering height, we can turn the relationship between lowering height and the boundary conditions into the relationship between boundary conditions and time. Thus output the fitting function about time and boundary conditions, and take a point at every 0.001s to discretise the output function.

When considering the seastate, as the SCR is in the initial status of the truncated position, the axial tension, the tip angle and direction of the coordinates of x are: 2116.8N, 79.9185°, 153.053m. At the beginning, tension sensors, angle sensors, displacement sensors and acceleration sensors are arranged in the truncated position, and then calibrate sensors. Input the conditions of the HYSY201 to 6DOF platform, SCR start periodic movements under the influence of 6DOF platform's vibration, and the collection frequency of the sensor is 1000Hz. Acquisition board was used to collect real-time measurement of physical quantities, and record its variation.

In the time domain, the curve of the two situations is superimposed according to numerical analysis, and then new fitted equations about the boundary conditions are acquired at the truncated position.

3　DURING THE TENSION CONVERTING PROCESS, CONTROLLING OF TENSION CONVERTING PROCESS

During the tension converting process, in order to guarantee the motion path of the SCR head, the joint control of the speed of the A&R winch and the towing winch is needed. Tension converting process is shown in Fig. 7.

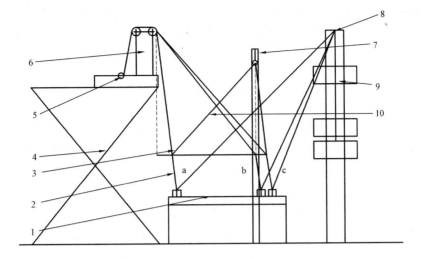

Fig. 7 Riser tension conversion diagram

1—Slideway；2—SCR model；3—SCR heading；4—6DOF platform；5—A&R winch；
6—SCR lifting system；7—Target platform；8—Servo motor；9—Weight；10—Towing winch

As the SCR is in positions a, b, c, the SCR heading will be in line. In order to calculate the relationship between the A & R winch and the towing winch, we divide the process into two phases: AB segment and BC segment.

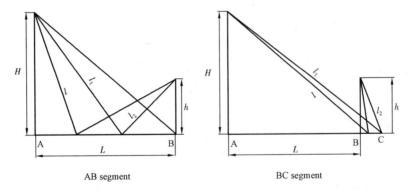

Fig. 8 AB, BC segment of riser tension conversion diagram

During the AB phase, we assume A & R winch is lowered at a constant speed v.

$$l_1 = l + vt$$

The length of the A & R cable is:

$$l_2 = \sqrt{h^2 + [L - \sqrt{(l+vt)^2 - H^2}]^2} \tag{16}$$

The speed of the towing winch can be acquired by derivation of Eq. (16):

$$v' = \frac{[L - \sqrt{(l+vt)^2 - H^2}](l+vt)v}{\sqrt{h^2 + [L - \sqrt{(l+vt)^2 - H^2}]^2}\sqrt{(l+vt)^2 - H^2}} \tag{17}$$

According to the Eq. (17), we can get the relationship between the lowering speed of towing winch and the time in AB segment. Similarly, in the BC phase, we can also get the relationship between the lowering speed and the towing winch, which is shown below:

$$v' = \frac{[\sqrt{(l+vt)^2 - H^2} - L](l+vt)v}{\sqrt{h^2 + [\sqrt{(l+vt)^2 - H^2} - L]^2}\sqrt{(l+vt)^2 - H^2}} \quad (18)$$

where, l is initial length of A & R cable; H is distance between lowering depth and the SCR head; h is height of the target platform; L is distance between the HYSY201 and the platform; l_2 is length of the towing cable.

In accordance with the results derived above, SCM systems achieve the automatic control of the tension converting system.

4 TESTING PROGRAMS DESIGN

4.1 Testing Programs Review

Simulation of the installation process includes the riser lifting and lowering, tension converting and hanging up to the platform, etc. During the test, follow devices are contained: 6DOF platform, SCR lifting system, SCR model, and the simulation system for the movement and the force of the truncated position. See Fig. 9.

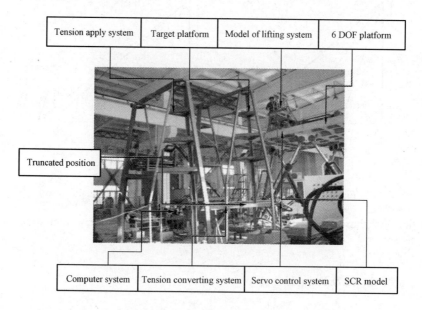

Fig. 9 Overall testing programs

4.2 Controlling of SCR Boundary Conditions at the Truncated Position

During the process of riser recovering and lowering, the movement and the force at the SCR truncation are controlled by a computer program and servo motors. SCR truncation top angle and horizontal coordinates are controlled by the A & R winch and servo motors, and the axial tension is controlled by the weight and servo motors. The motions of A & R winch and servo motor are controlled by the C + + program for precisely control. Control flow chart is shown in Fig. 10. See Fig. 11 for software controlling interface.

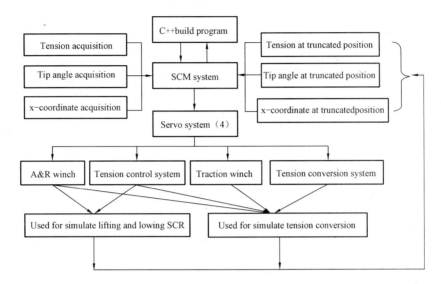

Fig. 10   Controlling flow chart

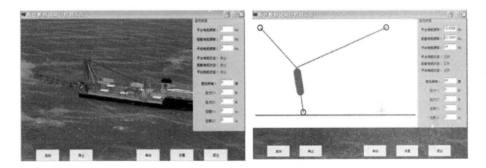

Fig. 11   Software controlling interface

During the process of tension transformation, in order to prevent the tension of A & R winch and the SCR bending strain near TDP overlarge, ship moving is needed in the process to control SCR's top angle within a certain range. Tension converting modeling system was used to accurately control the form and movement at the truncation during the installation process. The hardware of the device includes the servo motor, gear rack, roller wheel and angle sensors, etc. Software part is SCM process which complied by C + + program. SCR tension converting process contains SCR recovering, ship moving, and continuing recovering. The movement of 4 servo motors must be precisely controlled to complete installation process. One is used to simulate the A & R winch on the ship, another is used to simulate a platform towing winch, the other is used on the tension converting system and tension apply system which used to exert SCR axial tension. See Fig. 12.

Fig. 12 Boundary condition apply system

5 TEST RESULTS

In order to reduce the instability caused by underwater operation during the SCR installation process, fixing SCR on the fixture is needed for SCR head cutting, flexible joints welding and other operations. The SCR fixture is the main force component of the lifting system. This test simulates the dynamic response of the force of the fixture under the following three conditions. See Table 3.

Table 3 Sea conditions

Condition No.	H_s(m)	φ(°)	T(s)	Heave		Roll		Pitch	
				Amp(m)	Phase(°)	Amp(m)	Phase(°)	Amp(m)	Phase(°)
Seastate 1	3	90	2.84	0.169	90	2.07	0	0	0
Seastate 2	3	180	2.84	0.04	0	0	0	0.77	0
Seastate 3	3	270	2.84	0.238	0	2.73	0	0	0

The numerical results calculated by Abaqus are as follows (Fig. 13).

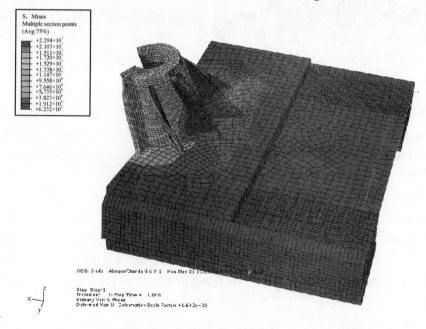

Fig. 13 Abaqus caculated results

The Fig. 14 shows the comparison of test values with numerical values computed by Abaqus:

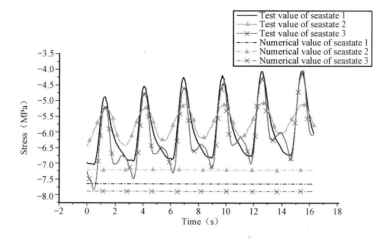

Fig. 14 Comparison of test and numerical values

It can be seen from above, this active truncation test can simulate the actual SCR installation well. The results acquired from the test, and that calculated values by Abaqus were close. See Fig. 15.

This active truncation test program can be used to simulate the riser installation process, including riser recovering, lowering, fixing, tension converting and hanging up to the platform. And also the test equipment can simulate the riser truncated boundary condition exactly. Thus, the test can be used to verify the security of the system and the feasibility of SCR installation process. Meanwhile, the improved test device also can be used to simulate the installation of PLET and PLEM.

Fig. 15 Active truncation test

ACKNOWLEDGMENTS

The authors are grateful for the financial support from the National 863 Program of China (Grant No. 2006AA09A105), the National Science and Technology Major Projects of China (Grant No. 2011ZX05027 - 002) and the COOEC Ltd.

REFERENCES

Andrew Pytel, Jaan Kiusalaas. 1994. Engineering Mechanics: Statics and Dynamics [M]. New York: Harper Collins College Publishers.

Baarholm R, Palazzo F G. 2004. Hybrid verification of a DICAS moored FPSO[C]. Proceedings of 14th ISOPE Conference, Toulon, France:307 - 314.

Chen Bangmin. 2010. The Research on Scaled Truncated Model Test of Installation of Deepwater Risers [D]. Beijing: China University of Petroleum.

Chen X, Zhang J, Johnson P, et al. 2000. Studies on the Dynamics of Truncated Mooring Line[C]. In: Proceedings, Vol. II, the 10th ISOPE Conference, Seattle, WA, USA:94 – 101.

CNOOC will develop the South China Sea to a "deepwater Daqing" in 20 years. XinMin Network, 2010 – 03 – 05. http://news.xinmin.cn/rollnews/2010/03/05/3871603.html.

Duan M L, Wang Y and Estefen S. 2010. Some Recent Advances on Installation of Deepwater Risers[J]. China Ocean Engineering, 24(4).

Duan M L. 2010. A Lifting System for Installation of Deepwater Risers and Subsea Hardware[C]. Invited presentation for the 2010 SUT technical conference, Society for Underwater Technology, Rio de Janeiro, RJ, Brazil: 23 – 24.

Huse E, MARINTEK, et al. 1998. Large Scale Model Testing Of Deep Sea Risers[C]. Offshore Technology Conference, Houston, Texas, USA.

Jiang X Z, Li Z G, He N, Wang Y, Duan M L. 2009. A New Lifting System for Installation of Risers in Deeper Water[C]. Proceedings of ISOPE – 14IDOT 2009, Beijing, China.

John Bouwman. 2007. Installation Challenges with Lifting and Pull-in of the 20 SCR[C]. Huston: Proceedings Offshore Technology Conference, No. 19061, Huston, Texas, USA:1 – 10.

Kai Kurea. 1981. Model Tests with Ocean Structures[J]. Applied Ocean Research, 3(4).

Mao D F, Duan M L, Wang Yi, et al. 2010. Model Test Investigation on An Innovative Lifting System for Deepwater Riser Installation[J]. Petroleum Science, 17 (4).

Sheng Z B, Xiao L F. 2003. Deep Sea Platform Hybrid Model Testing Technique[J]. Shanghai Shipbuilding, (1): 12 – 14.

Stansberg C T, Oritsland O, Kleiven G. 2000. Verideep: Reliable Methods for Laboratory Verification of Mooring and Station Keeping in Deep Water [C]. OTC 2000 Conference, Houston, TX, USA. Proceeding Paper, 12087.

Stansberg C T, Yttervik R, Oritsland O, et al. 2000. Hydrodynamic Model Test Verification of a Floating Platform System in 3000 m Water Depth[C]. Proceedings of ETCE/OMAE2000 Joint Conference Energy for the New Millenium, February 14 – 17, New Orleans, LA, OMAE2000/OSU OFT 4145, 2000. 1 – 9.

Testing Technique for Deep-Sea Platforms Based on Equivalent Water Depth Truncation[J]. CHINA Ocean Engineering, 21(3).

Watts S. 1999. Hybrid Hydrodynamic Modeling[C]. Offshore Technology Conference, Seattle, USA:13 – 17.

Watts S. 1999. Hybrid hydrodynamic modeling[J]. Offshore Technology: 13 – 17.

Watts S. 2000. Simulation of Metocean Dynamics: Extension of the Hybrid Modeling Technique to Include Additional Environmental Factors. The SUT Workshop: Deepwater and Open Oceans, the Design Basis for Floaters, Houston, TX, USA.

Xu T. 1982. Theory of Similarity and Model Test[M]. Beijing: China Agricultural and Mechanical Press.

Yang S X, Yang T, Xun Y T. 2003. The Development of the Digital Six-DOF Stewart Platform[J]. Hydraulics & Pneumatics, (8):46 – 4.

Zhang H M, Sun Z L, Yang J M, et al. 2009. Investigations on Optimization Design of Equivalent Water Depth Truncation[J]. Science in China, 39 (4): 523 – 536.

Zhang Huoming, Yang Jianmin, Xiao Longfei. 2007. Hybrid Model

Zhen Jianwen. 2008 – 09 – 17. CNOOC 201 begins to be built in Rongsheng Heavy Industries[J]. Nantong Daily.

Latest Progress in Deepwater Installation Technologies

Alan Wang, Yun Yang, Shaohua Zhu, Huailiang Li, Jingkuo Xu, Min He

Installation Division, Offshore Oil Engineering Co., Ltd., Tianjin, China

Abstract This paper gives a comprehensive overview of the current state-of-the-art deepwater installation technologies and categorizes various installation methods based on the latest advancements in the subsea constructions. Different deepwater installation technologies are described here to address the important aspects of subsea installation technologies, including different types of subsea construction vessels, various single-wire and multi-fall deployment systems, steel wire ropes, synthetic fiber ropes, anti-rotation systems, free-fall installations, dynamic amplification factors, motion compensation systems, buoyancy elements, duration of deepwater installations, etc.

Key Words Deepwater installation; FRDS; HMPE

NOMENCLATURE

AHC	Active Heave Compensation
DAF	Dynamic Amplification Factor
DCV	Deepwater Construction Vessel
DSV	Diving Support Vessel
FRDS	Fiber Rope Deployment System
HLV	Heavy Lift Vessel
HMPE	High Modulus Polyethylene
LBL	Long Base Line
PLEM	Pipeline End Manifold
PLET	Pipeline End Termination
ROV	Remotely Operated Vehicle
SDA	Subsea Distribution Assembly
SSCV	Semisubmersible Crane Vessel
USBL	Ultra Short Base Line

INTRODUCTION

Deepwater installation technologies may be classified into three major interrelated challenges, that is, lifting/lowering technology, dynamic responses to environmental conditions, positioning and control techniques. Subsea 7 has successfully installed the world's first wellhead Christmas tree in over 3,000m water depth using a newly developed fiber rope deployment system. The deepest subsea manifolds and boosting pumping stations were installed in August of 2009 by Technip's deepwater pipelay and construction vessel Deep Blue at depths in excess of 2,500 meters for the

Petrobras's Cascade and Chinook Subsea Development in the Gulf of Mexico US waters. Chung (2010) recently revealed that a 5,000m-long, 15in outer-diameter, full-scale pipe was deployed from the large Moon Pool (270ft ×70ft) of the Hughes Glomar Explorer while a deep-ocean mining system operated from the ship in the North Pacific Ocean in 1976 and 1979. This full-scale field test conducted by Hughes Glomar Explorer was a quantum jump in 1970s and still a great challenge to today's deepwater operation technology.

Up to now existing conventional deployment systems are unable to meet the requirements of deploying subsea structures more than 300 tonnes in water depths beyond 3,000 meters. Most of small Diving Support Vessels (DSV) and large Deepwater Construction Vessels (DCV) with conventional wire rope deployment systems, as well as drilling riser system of expensive day-rate drilling vessels, have their limited capability when installing 300 tonnes subsea hardware in a water depth up to 1,000 meters. Only the few dedicated heavy lift vessels and deepwater construction vessels, such as Balder, Saipem 7,000, Deep Blue, etc., can install 300 tonnes subsea structures beyond a water depth of 3,000 meters. The self weight of steel wire ropes makes wire rope lifting/lowering systems inefficient and impractical for water depths beyond 3,000 meters. With the advent of in-depth understanding the mechanical properties of various man-made fiber ropes and the industry's confidence in the Fiber Rope Deployment Systems (FRDS), the subsea constructions will therefore turn increasingly to deepwater fiber rope deployment system. Many unconventional deepwater installation methods, such as Sheave Installation Method, Pendulous Installation Method, Pencil Buoy Method, etc., have been also developed and successfully applied.

This paper gives a comprehensive overview of the current state-of-the – art deepwater installation technologies and categorizes various installation methods based on the latest advancements in the subsea constructions. Different deepwater installation technologies are described here to address the important aspects of subsea installation technologies, including different types of subsea construction vessels, various single-wire and multi-fall deployment systems, steel wire ropes, synthetic fiber ropes, anti-rotation systems, free-fall installations, dynamic amplification factors, motion compensation systems, buoyancy elements, duration of deepwater installations, etc.

1 DEVELOPMENT OF SUBSEA HARDWARE

The modern concept of subsea field development started in the middle of 1970s when offshore industry began placing wellhead and production equipment below the water surface and having associated subsea components encapsulated in a sealed chamber. Since then subsea hardware has advanced from manually operated systems in shallow water into remote operated systems in deep water up to a water depth of 3,000 meters. Benefiting from high oil prices and advanced subsea hardware solutions, a variety of subsea hardware have been developed for deepwater fields at increasingly remote locations and in extremely harsh environmental conditions. So far subsea structures and equipment cover a very wide range of subsea production facilities, including manifolds, templates, foundations, PLEMs, PLETs, in-line sleds, SUTAs, SDAs, jumpers and flying leads, subsea wellheads and Xmas trees, riser bases, subsea control system, subsea multiphase flow measurement, subsea compression modules, subsea boosting stations, subsea water injection modules, subsea separation modules, etc.

Subsea structures vary greatly in size, shape, and weight. Some of these subsea facilities can be huge structures reaching a weight of thousands of tons, a length and width of 50 meters, and a height of 30 meters or so. Fig. 1 shows a proposed concept of gas compression package for the Ormen Lange field which is the first large-scale offshore field development at water depth varying between 850 to 1,100 meters without surface platforms. The subsea compression system is a modularized system with separately retrievable separator, compressor, and pump modules, etc. Modular lifting/lowering method has to be employed to install such a large size of subsea facility. Table 1 presents typical range of weights and dimensions for various subsea structures and equipment modules, which are only an approximate indication and reproduced from Frazer et al. (2005).

Fig. 1 Courtesy of Aker Solutions: impression of subsea gas compression package proposed for Ormen Lange Gas Field

Table 1 Typical subsea structures & equipment modules

Hardware Unit	Weight (t)	Dimension $(L \times B \times H)/(\phi \times H)$ (m)
Processing Modules	200~400	up to $(15 \times 15 \times 8)$
Manifolds	50~400	$(5 \times 5 \times 4) \sim (25 \times 20 \times 8)$
PLEMs	50~400	$(5 \times 5 \times 4) \sim (25 \times 20 \times 8)$
Template	100~400	$(10 \times 10 \times 6) \sim (30 \times 20 \times 7)$
Riser Base	50~200	up to $(20 \times 20 \times 10)$
PLETs	30~100	$(5 \times 4 \times 3) \sim (10 \times 8 \times 6)$
SDAs	50~100	$(5 \times 5 \times 4) \sim (10 \times 10 \times 8)$
SUTAs	5~50	up to $(5 \times 5 \times 6)$
Pumping Modules	5~50	$(1 \times 1 \times 1.5) \sim (5 \times 5 \times 6)$
Vertical Jumpers	5~50	up to (50×6) $(L \times H)$
Horizontal Jumpers	5~50	up to (50×15) $(L \times B)$
Suction Pile	40~200	$(\phi 4.5 \times 15) \sim (\phi 10 \times 30)$
Drag Anchor	50~75	up to (15×5)
Subsea Tree	10~70	up to $(5 \times 5 \times 6)$
Mid-Water Arch	50~100	$(10 \times 6 \times 4) \sim (20 \times 9 \times 5)$

2　DEEPWATER INSTALLATION VESSELS

　　Different subsea structures and equipment modules may require different deepwater installation vessels. The typical deepwater installation vessels include expensive drill vessels, heavy lift vessels, deepwater construction vessels or called subsea construction vessels, pipe-lay and umbilical-lay barges, ROV support vessels, diving support vessels, field support vessels, AHTS tugs, etc. The worldwide megalift fleet includes four semisubmersible crane vessels (SSCVs) and one crane barge with lifting capacities exceeding 5,000 metric tons. These megalift vessels include the largest SSCV Thialf and S7000 with dual rotating cranes capable of lifting 14,200t at a 31.2m radius and 14,000t at a 42m radius, respectively, SSCV Hermod with a dual-crane lift capacity of 8,100t, SSCV Balder with a dual-crane lift capacity of 6,300t, mono-hull crane barge Lan Jing with a single crane cable of lifting 7,500t. There are 35 vessels with lifting capacities ranging from 1,500t to 5,000t, where much of the recent and anticipated newbuild growth is concentrated.

　　Among the floating production platforms, SEMIs, TLPs, and FPSOs don't require the super heavy lift vessels for the lifting installation of the topsides. Recently all these topsides modules are either installed by modular-lifting at quayside of fabrication yards or floatover installation inshore in a shelter area. Then integrated units are transported by wet tow or dry tow to the installation sites where the units are hooked up with the pre-installed mooring systems. Only fixed platforms and SPARs still need the super heavy lift vessels because the integration between topside modules and substructures cannot be completed at quayside. However more and more fixed platforms, especially the large integrated topsides, are in favor of floatover installations. Refer to Wang et al. (2010) for various floatover installation technologies. Even one of the 19 existing SPARs, the Kikeh SPAR located offshore Malaysia, enjoyed a twin-barge floatover installation. The majority of megalift crane vessels are actively used in the installations of deepwater pilings, moorings, and TLP tendons. This is why the recent and anticipated newbuilds are concentrated in the lifting range of 1,500t to 5,000t and mainly used for deepwater constructions and pipelays. A few dedicated heavy lift vessels and deepwater construction vessels, such as Balder, Deep Blue, Aegir, Fairplayer, etc., are worth mentioning here.

　　Heerema's DCV/HLV Balder: The Balder went through an upgrade conversion in 2002 into a full Class III DP deepwater construction vessel. This DCV is equipped with a series of multifunctional deepwater tools for deepwater field development up to 3,000 m water depth. The Balder is capable of a tandem lift of 6,300t, equipped with a 2,700t crane and a 3,600t crane, and operates a J-Lay system. The cranes provide for depth reach lowering capability as well as heavy lift capacity to install topside modules. The starboard auxiliary hoist has a lowering capacity of 400 short tons when plumbing a depth of 3,000 m below water surface at minimum radius. Subsea lowering capabilities are project specific engineered for the use of the starboard crane in combination with the A & R winch and the mooring line deployment winch. Currently the Balder's lowering capacity has been increased to 1,490t at 1,000 m, 600t at 2,000 m, and 340t at 3,000 m. The further upgrade allows for the addition of two more traction winches, thus increasing the lowering capacity up to 100%. These upgrades and combinations of the two cranes, the mooring line deployment winch, and/or the A&R winch, make its lowering capacities up to 1,000 metric tons in 3,000 m of water economically feasible.

This multi-functional dynamic positioned DCV is tailored for the installation of foundations, moorings, SPARs, TLPs and integrated topsides, as well as pipelines and infield flowlines. Refer to Fig. 2 for the photo of DCV Balder. The Balder maintains many deepwater construction records, which installed the heaviest subsea template with dimensions of 44.3m long × 32.7m wide × 14.9m high and a weight of 1,150t, at 850m water depth in the Ormen Lange field offshore Norway; installed the world's deepest foundation piles, 52m long and 156t, in water depths of 2,030 m at Block 31 offshore Angola; installed the world's largest, 1,039t, Submerged Turret Production

Fig. 2 Courtesy of Heerema, Deepwater Construction Vessel Balder

(STP) Buoy in the record breaking water depths of 2,515m in the Cascade & Chinook fields for the first FPSO in Gulf of Mexico; laid the world's deepest Pipe-in-Pipe flowline at a water depth of 2,691m in the Chinook field; installed the world's deepest and heaviest export SCR, producing the heaviest SCR loads of 800t when hooking up with the SEMI Independence Hub in Gulf of Mexico; installed the world's longest mooring lines at 3.86 km each for Independence Hub; deployed PLET in 2,740 m water depth when initiating Independence flowlines; installed the deepest mooring pile at 2,440 m water depth for Independence Hub; and so on.

Technip's DCV Deep Blue: The DCV Deep Blue is the world's largest purpose-built ultra deepwater pipelay and subsea construction vessel. The Deep Blue utilizes the reel-lay and J-lay methods for the installation of all types of flexible and rigid pipe, as well as umbilicals. The Deep Blue is one of the most advanced pipelay and construction vessels for deepwater installations and the flagship of the Technip fleet. This flagship vessel supports field developments in water depths up to 3,000 m.

The main crane of the Deep Blue has been designed for offshore lifts ranging from 60t at a 55m radius up to 400t at an 18m radius enabling efficient subsea equipment installation through its active heave compensation system. In addition to the main deck crane, a number of smaller utility cranes ranging from 10t to 30t capacity are provided in way of the pipelaying equipment. Two large winches are provided in support of pipelay operations and subsea constructions. They are the 327t SWL Traction and Storage Winch provided for abandonment and recovery of pipelines with tension range from 150t to 350t and the 150t Initiation/A&R Winch provided for initiation of pipelines used primarily for suction piles, diverless latch and pipe transfer, etc. , and also for abandonment and recovery of pipelines with tensions up to 150t. Refer to Fig. 3 for the photo and general layout of DCV Deep Blue. The Deep Blue created many deepwater construction records, including installation of 9 anchoring piles, 2 manifolds, 2 pump stations at water depths ranging from 2,400m to 2,700m at the Cascade & Chinook field in Gulf of Mexico; installation of the deepest

Pipe-in-Pipe (PiP) with reel-lay operation in 2,115m water depth, the deepest PiP with J-lay operation in 1,912m water depth, the deepest PiP Steel Catenary Riser in 1,938m water depth for the Na Kika field in Gulf of Mexico.

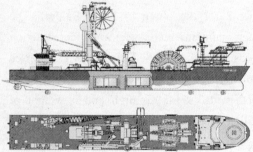

Fig. 3 Courtesy of Technip, Deepwater Construction Vessel Deep Blue

Heerema's DCV Aegir: Heerema's latest DCV, Aegir, is designed as a monohull crane and pipelay vessel with the capability to execute complex subsea infrastructure and pipe line projects in ultra deep water. The Aegir is fitted with a pipelay tower for J-Lay and Reel-Lay operation whose tension capacity is 2,000t. A crane with a revolving lift capacity of 4,000t is installed together with deepwater lowering equipment to reach a water depth of 3,500 meters. This DCV has sufficient lifting capacity in order to execute installation of fixed platforms in relatively shallow water.

Fig. 4 Impression of DCV, Aegir, courtesy of Heerema

The vessel is equipped with a Class 3 dynamic positioning system. The monohull configuration is specially designed for fast transit speed of 12 to 14 knots and optimum motion characteristics in operation whilst maintaining the maximum pipe payload of 4,500 metric tons. Refer to Fig. 4 for the impression of DCV Aegir. This DCV is equipped with a 2,000t A&R system which is capable of plumbing a depth of 3,500 m below water surface. Its active and passive heave compensation system can withstand a load up to 750 metric tons.

Subsea 7's DCV Borealis: The DCV Borealis is a 5,000t DP-III monohull derrick and pipelay vessel with 13-knot transit speed. The Borealis is fitted with state-of-the-art deepwater equipment to meet the exacting requirements of today's ultra-deep and deepwater projects in the world's harshest environments. A 5,000t offshore Huisman mast crane is heave compensated. The dual main hoist has a revolving lifting capacity of 4,000t at a 34m radius and 1,500t at a 78m radius. The AHC subsea hook, i.e. auxiliary hoist, has a lifting capacity of 1,200t at a 70m radius with 4 falls, 600t at a 103m radius with 2 falls, 300t with single fall when operating at 6,000m water depth. A 400t deepwater lowering system is equipped with an active heave compensation system to counter-act the vessel's heaving motion when landing a load on the seabed. These features make the vessel an efficient tool for the installation of heavy loads required for deepwater subsea production systems. In addition, the A&R capacity includes a 600t traction winch and a 200t CT drum winch capable of operating at water depth from 20m to 3,000m. This A&R system can be upgraded to 360t capacity when operating at 3,000m water depth. Refer to Fig. 5 for the impression of DCV Borealis.

Jumbo Offshore's HLV Javelin & Fairplayer: Jumbo Offshore's DP2 offshore heavy lift vessels Javelin and Fairplayer are able to provide a single solution for loading, transporting and installing subsea structures in water

Fig. 5 Impression of DCV, Borealis, courtesy of Subsea 7

depths up to 3,000 m. The DP2 class vessels are very versatile transport, subsea installation and construction vessels. Key features include the twin 900t Huisman mast cranes with an inshore lift capacity of 1,800t and an offshore lift capacity of 1,100t large and heavy subsea structures. Their transit speed can reach 17 knots. Its patented deepwater deployment system (DDS) can handle subsea structures and equipment modules from deck and place them on the seabed without the need for subsea transfer. This DDS covers a range of typical weights and depths, that is, 1,000t at 1,000m, 600t at 1,700m, and 200t at 3,000m. Both vessels can be fitted with a modular deepwater installation system capable of installing heavy structures in deep water. The dual crane offshore lift capacity combined with the large deck space makes the vessel ideal for the installation of large and heavy structures on the seabed and for the installation of mooring systems. The twin HUISMAN mast cranes have the subsea installation capacity of 1,000t in water depths up to 900m, 660t in water depths up to 1,500m, 280t in water depths up to 2,000m, 200t in water depths to 3,000m. The depth capability can be increased through the use of pennants in combination with a subsea transfer. Fig. 6 shows the impression of HLV Fairplayer where the top picture illustrates twin lifting installation of a subsea manifold and the bottom photo illustrates installation of a suction pile with two mast cranes.

(a) Twin lifting installation of subsea manifold

(b) Installation of suction pile with two mast cranes

Fig. 6 Pictures of HLV, Fairplayer, courtesy of Jumbo Offshore

3 DEEPWATER DEPLOYMENT TECHNOLOGY

Deepwater heavy structure installations face many challenges. The state-of-the-art deepwater deployment systems are still difficult to meet the requirements of deploying subsea structures more than 300 tonnes in water depths beyond 3,000 meters. Self weight of steel wires makes conventional wire rope systems inefficient and impractical on most of deepwater installation vessels. Synthetic fiber ropes are essentially neutrally buoyant in sea water, thus eliminating the impact of the self weight of the deployment ropes on the lifting/lowering capacity of an ultra-deep water deployment system. HMPE material has a very high strength-to-weight ratio, good elongation properties and dynamic toughness, which is emerging as the most appropriate material for ultra-deep water deployment applications. Some key issues of deepwater deployment technologies are addressed below.

Single Wire Deployment System: It is common to use a single wire deployment system with a simple drum winch to deploy small and medium size subsea hardware. The subsea structure is normally lifted off from the deck of the construction vessel using a ship crane, and then overboard and lower the structure through the wave slash zone. When being lowered at a convenient depth, say 50m to 100m below surface, the load is transferred to the winch wire for further lowering through the water column and landing on the target box at the seabed. However, as the water depth becomes very deep, it may require anti-rotation system to avoid the rotation of the load. There are three main types of steel wires, that is, ordinary wire, rotation resistant wire, and low rotation wire. The ordinary wire is not torque balanced and may rotate up to 360° every meter under a tension load of 20% MBL, which is not suitable for subsea installations. The rotation resistant wire comprises two layers of strands in opposite directions, thus providing a limited degree of rotation resistance. The low rotation wire is designed to counter balance the torque produced by their layers of strands. The rotation can be less than 2° per meter of wire under a tension load of 20% MBL. The free rotation of the wire under tension can cause wire damage or loss of the end termination. Especially when the tension is removed, the rotation tries to unwind. Only the low rotation wire can be used for deepwater installations. In addition, deepwater installations will increase loads on the drum flanges, which accumulates significantly at the flanges when the load is large, thus resulting in damage to the drum flanges.

A traction winch does neither suffer reduction of the in-line pull nor build up the load on the drum flanges of the storage winch as the stored wire length increases. Fig. 7 and Fig. 8 show a 400t deployment system which is equipped with a traction winch, a AHC system, and a storage winch, where the steel wire of ϕ109mm is reeving around the traction winch, the heave compensation system, the storage winch, and then through portside main crane or starboard A & R winch. This development system can be operated at a lowering speed of 500m/h and work at a maximum water depth of 2,000m. Single fall subsea winches can be operated in three different modes. The normal mode is to pay in or out the wire by directly following action of operator on the joystick. The AHC mode couples the winch with the vessel vertical motions and automatically compensates for heave motions. The Constant Tension (CT) mode is to set the winch to a pre-determined tension and automatically adjusts the wire length paid in or out and thereby maintaining the pre-set tension in the wire.

(a) Traction winch

(b) Active heave compensation system

(c) Storage winch

Fig. 7   400t deployment system

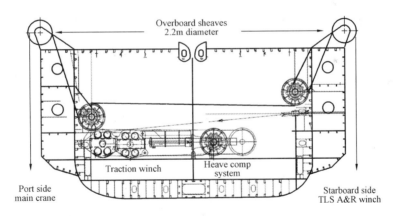

Fig. 8   400t deployment winch reeving arrangement

Modular Winch System: Steel wire ropes make singe wire deployment system lose lift capacity due to self weight. This technical constraint makes wire rope use deteriorate in ultra deep waters. Steel wire rope technology with multi-fall lowering systems are mature and durable. However it is difficult to manufacture sufficient long lengths of steel wire ropes for ultra-deep water installations. The current manufacturing capability is approximately 200t steel weight which can produce a single length of 2,900m, ϕ5in wire with SWL 350 tonnes. Even the low rotation designs can still have significant problems with the very long wires required for deepwater. This will further complicate the entangle problem of multi-fall systems. One of the solutions is to use modular winch system, so-called Module Handling System, which is used for lifting/lowering operations of special modules onto seabed. This system may comprise a total of six winch units ranging from 5t to 70t lifting capacity and an operational capacity down to 2,000 meters below surface. All the winches have electrical drives and have a built-in Active Heave Compensation (AHC) and Active Rope Tension (ART) controls. This system compensates for the vessel movements relative to the seabed. The total power consumption for this system can peak well above 3000kW. In addition a special control and operation system is developed in harmony with other systems onboard. Fig. 9 show an exemplary application of a dual winch tandem development system which has a 250t lowering capacity into water depths up to 3,000m.

Fig. 9 Dual winch tandem development system

Use of Synthetic Fiber Ropes: Synthetic fiber ropes provide a potential solution to the self-weight problems. The main fiber options for deepwater applications include aramid, polyester, and HMPE. Other possible candidate materials are high-strength zylon, vectran, and nylon, etc. Table A-1 shows the comparison of the mechanical properties of five deployment rope materials. Fig. 10 demonstrates the loss of the wet lowering capacity due to the self-weight of the steel wire. The steel wire diameter is ϕ4in whose weight in air is 44kg/m while weight in water is 36.5kg/m. The minimum break load (MBL) is 665t while the safety working load (SWL) is 214.5t based on the assumption of a safety factor = 3.1. If the dynamic amplitude factor (DAF) is 1.2, the lowering capacity is 178.8t at the surface. At the water depth of 3,000m, the wet lowering capacity will reduce to 69.3t, approximately 61% net loss without considering any hydrodynamic effect and dynamic resonant behavior. McKenna (2004) provided the rope weights and rope sizes of the five deployment ropes based on 1,000t MBL, see Table 2 for details.

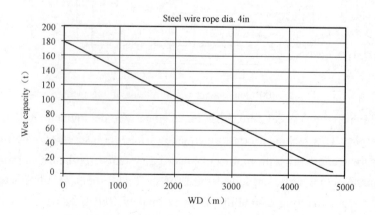

Fig. 10 Wet lowering capacity of steel wire (ϕ4in)

Table 2 Weights & sizes of deployment ropes based on 1,000t MBL

Parameter	Unit	HMPE	Aramid	Polyester	Nylon	Steel
Weight in air	kg/m	8.4	12.0	23	25	58
Weight in water	kg/m	Neutral	3.3	5.9	2.5	49
Overall diameter	mm	125	120	175	200	110
Rope size No	in	15½	15	22	25	14

The HMPE rope has become the most appropriate candidate for ultra-deep water deployment systems. The high-strength and high-modulus polyethylene HMPE fiber, essentially neutrally buoyant, has excellent mechanical properties with low density, thus resulting in high performance-on-weight basis and making it one of the strongest man-made fibers. The HMPE fibers have a high strength and a high modulus in the fiber direction, which yields resistance against deformation. The main characteristics include high strength, low density, low elongation at break, long fatigue life, and resistance to most chemicals and sea water, etc. The elongation at break is relatively low compared with other synthetic fibers. However, due to the high strength, the energy to break is very high. Like other synthetic fibers, the mechanical properties of HMPE fibers are influenced by temperature. The strength and modulus increase at sub-ambient temperatures but decrease at higher temperatures. For long duration exposure HMPE fibers can be used from cryogenic conditions up to a temperature of 70°C. The melting range is between 144~152°C. Unlike other synthetic fibers, the mechanical properties are not influenced by the presence of water. Subjecting HMPE fibers to long-term static loads leads to a permanent elongation, called creep. The HMPE fibers are flexible and have a long flexural fatigue life. The fibers also have a higher fatigue resistance than other fiber types when subject to repeated axial load or even partly compression bending.

Adequate bending fatigue life and low heat generation under repeated rapid cycling through a heave compensator and bending, or slipping, on sheaves, adequate coefficient of friction on sheaves and drums to operate the winch effectively, high axial stiffness and strength even after thousands of loading cycles, dynamic performance, high susceptibility to abrasion, rope crushing by compressive forces exerted on the rope successive layer reeling on the drum, field inspectability and reparability, and retirement criteria are the areas of main concerns.

Single-Fall FRDS System: A fiber rope deployment system and a conventional steel wire crane onboard can be used together for an ultra-deep water deployment. The conventional crane can be used to lift off and overboard the subsea structure from the deck, lower the load through slash zone with or without AHC system, and then transfer the load to the FRDS system at a certain water depth, say 50m below surface, for the final deployment to the seabed.

Thanks to the Deepwater Installation of Subsea Hardware (DISH) JIP, a 46t SWL prototype has been built and extensively tested. This single-fall FRDS system has been tested through a field pilot program at the Ormen Lange gas field in Norway and then worked at 2750 m water depth in Gulf of Mexico for installations of mudmats, second end of umbilicals, manifolds, jumpers, and spool pieces, etc. A blend of HMPE and Vectran (liquid crystal polymer) fibers has been used to provide good temperature and good creep properties. A special lubricant coating is applied to ensure very low coefficients of friction between fibers and thereby reducing internal wear and heat buildup due to high cyclic bending over sheaves. A 12×12 strand (Braid Optimized for Bending, BOB®) are high strength, low elongating single braided rope constructions with excellent long term creep resistance and superior cyclic fatigue performance, especially in bend-over-sheave applications. The braided construction assures a torque free rope which can be easily inspected internally and repaired offshore by trained riggers. Recently a ϕ100 large diameter rope braided from Dyneema® HMPE and ePTFE (expanded polytetrafluoroethylene) fibers has been tested at the DNV facility, Cyclic Bend Over Sheave test rig, in Oslo. Tornqvist et al. (2011) presented the test results based on a 12×3 strand design. The promising results demonstrate that the large diameter fiber ropes on small-diameter sheaves have a potential to be applied in deepwater deployment systems with active heave compensators for large subsea structures and equipment modules.

Torben et al. (2007) presented a 46t SWL prototype Cable Traction Control Unit (CTCU), refer to Fig. 11 for details. The CTCU unit comprises a series of sheaves with individual drives used to de-tension rope, thus controlling speed and torque on each individual sheave to avoid accumulated slip due to rope elongation and variations of diameters due to splices. A storage winch is designed to store the rope at low tension while maintaining a constant back tension for the CTCU. This de-tension avoids squeezing of rope into soft layers due to the large load difference between deployment of a heavy load and recovery of empty hook with neutrally buoyant lifting line. An inboard damping device is used to smoothen the tension between the CTCU and the storage winch. An outboard damping device is used for constant tension and pull limit control. The rope is guided over an overboarding device. HPU with accumulators supplies the system with high and low pressure hydraulic oil. A computer control system is used to for dynamic control of individual machines and interactions between machines and rope management system to monitor and manage the real condition of the rope and thereby determining the retirement criteria. The CTCU system has automatic lift-off function for transition from CT mode to AHC mode during lift off, automatic landing function for the transition from AHC mode to CT mode upon loading, and a crane function when being integrated with cranes, A-Frames or other overboarding devices. A ratio of sheave and rope diameters has a minimum requirement of 30:1 for cyclic bending over sheave of fiber rope through the AHC system.

(a) Impression of 46t fiber rope development system   (b) Setup for a field test

Fig. 11   The 46t FRDS system, courtesy of Odim Alitec AS

The low axial stiffness of fiber ropes causes axial resonance problems at more shallow water than for steel wires. The axial resonance will become significant due to very long length of the lowering rope, which depends on the rope stiffness (EA/L). The DAF is the ratio of the response acceleration of subsea structure and the vessel acceleration due to wave action. The dynamic load acting on the lowering system depends on the vessel acceleration, the dynamic response and the submerged weight of the lowering system, the submerged weight, hydrodynamic added mass and damping of the structure, etc. Frazer et al. (2005) presented the relationship between the dynamic amplitude factor and water depth shown in Fig. 12.

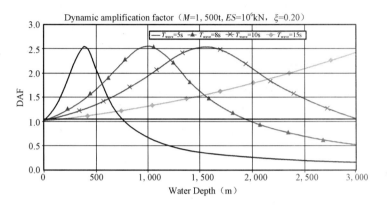

Fig. 12 Dynamic amplitude factor (DAF) vs water depth

Two-Fall FRDS System: A 125t FRDS system has been designed cased on the CTCU technology stated above and can be operated in two-fall configuration with a lifting capacity of 250t up to more than 3,000m water depths. Torben and Ingeberg (2011) presented the two-fall 250t FRDS prototype for a field pilot test. The components of the two-fall fiber rope deployment system include storage winch (1), spooling device (2), inboard damping device (3), cable traction control unit CTCU (4), storage winch electrical container (5), CTCU electrical container (6), accumulator units (7), fiber rope (8), cable counter (not visible in the figure). Refer to Fig. 13 for the labeled components.

The storage winch has a capacity of $\phi 88$mm $\times 7,000$m rope blended with HMPE & Vectran fibers, thus ensuring an adequate length for ultra-deep water operations in more than 3,000m water depths. The rope has a 12×12 strand construction. Each of the 12 primary braids contains 6 S-twist strands and 6 Z-twist strands. The rope is totally balanced by design to minimize twisting problem. In addition all the sheaves have been aligned to ensure a zero fleeting angle. This is very

Fig. 13 The 125/250t fiber rope development system

important for a two-fall lifting configuration to further minimize introduction of rope twist during handling. The flexible inboard damping device is designed to maintain robust and reliable synchronization between the CTCU and the storage winch. Any transient speed mismatch between the CTCU and the storage winch will activate the inboard damping device to either extend or retract accordingly. The accumulator system is used in AHC and CT modes to store hydraulic energy when available and release the energy when needed. The AHC system can perform at a stroke of ±2.4 m in 10 seconds with a peak speed of 1.5m/s at 125t load in a single-fall configuration while at a stroke of ±1.2m in 10 seconds with a peak speed of 0.75m/s at 250t load in a two-fall configuration. A specially formulated coating has been applied to maximize the durability during bending over sheaves. In the two-fall lifting configuration, the dead end of the deployment rope has to be fixed at a hang-off beam. The hang-off beam is installed on top of the skidding rails secured on the hatch of the moon pool. This field proven prototypes FRDS system will be available in the market very soon that would fulfill the ultra-deep water installation requirements. It should be pointed out that due to the two-fall configuration the deployment speed will decrease by half. Assuming that the winch is operated at a speed of 20m/min for deployment operation while at a speed of 10m/min for recovery operation, respectively, each operation in 3,000m water depth will take nearly 7 hours just for lowering and recovering, and more than 24 hours for the entire operation, including time to rig up the lifting gear at the surface, connect the line from storage reels, position and place the load onto the seabed in the target box, and recover the line.

4　VARIOUS INSTALLATION METHODS

Deepwater installation methods may be classified into conventional methods and non-conventional methods. The conventional methods include installations by cranes of heavy lift vessel, A-Frames of offshore support vessels, drilling riser of drill vessels, deepwater construction vessels, etc. The non-conventional methods include special construction vessels, Sheave Installation Method, Pendulous Installation Method, Pencil Buoy Method, Modular Winch System, etc. These deepwater installation systems and methods have been developed and successfully applied in deepwater applications.

Deepwater Construction Vessels: Take the Deep Blue as an exemplary application. Before offshore installations, the subsea hardware will be transported to site via a barge. Each of the hardware packages and suction piles will be pre-fitted with two high-performance synthetic slings. The Deep Blue's 400t outboard crane is used to lift the hardware packages using one of the two attached slings. The hardware will be lowered to 100 meters below surface, where a ROV captures a special grommet attached to the second lifting sling and ferries the eye of the sling to the hook of the Deep Blue's A & R winch through the moonpool. When the load is transferred to the A & R winch, the sling will be freed from the hook of the outboard crane and the hardware will be lowered onto the seafloor for installation. As Fig. 14 shows, Technip's riggers are pre-rigging the hardware packages with the ϕ88mm AmSteel® -Blue grommets. The HMPE ropes are rated at 190t safe working loads. The grommets are intended to use the wet handshake technique to transfer loads from the overboard crane to the A & R winch through the moonpool for deepwater deployment onto the seafloor. A high-strength and light-weight Dyneema® fiber rope is used to fabricate the slings. AmSteel® -Blue fiber material is as strong as steel wire for the same size ropes, which is neutrally buoyant or slightly positive in seawater. This makes handling by ROV a simple matter and requires a minimum of power. The ϕ88mm AmSteel® -Blue is spliced into a grommet of 50 meters in length.

MODU Drilling Risers: The drilling riser system of expensive day-rate drilling vessels are often used to install wellheads, Christmas trees and BOPs but rarely used to install the manifolds, etc. One of the reasons is due to

Fig. 14　Hardware packages pre-rigged with ϕ88mm HMPE grommets

its low availability and almost prohibitively expensive day rates. In addition, majority of drilling semisubmersibles have their limited lifting capability when installing 300t subsea hardware in a water depth up to 1,000 meters. It is inefficient to use the drilling riser to install the subsea hardware. As Fig. 15 shows, Petrobras has used a large SEMI riser system to install a 240t manifold in 940m water depth in December 1997.

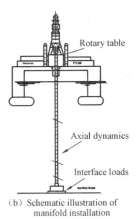

(a) Using large SEMI drilling riser to install 240t manifold
(b) Schematic illustration of manifold installation

Fig. 15　Lowering heavy subsea hardware with drill-strings from large SEMI, courtesy of Petrobras

Heavy Lift Vessels: Only the few dedicated heavy lift vessels, such as Heerema's Balder, Saipem 7,000, Jumbo Offshore's Javelin & Fairplayer, etc., can install 300 tonnes subsea structures beyond a water depth of 3,000 meters. These heavy lift vessels are specialized in deepwater installations. They are very capable and efficient. The main problems are due to low availability and high cost. Fig. 16 shows that a crane barge use conventional lifting slings to install 420t manifold in 620m water depth in 1995 offshore Brazil.

Offshore Support Vessels with A-Frame: Most of small offshore support vessels, such as ROV support vessels, diving support vessels, field support vessels, AHTS tugs, with A-Frames and conventional wire rope deployment systems have their limited capability when installing 200 tonnes subsea hardware in a water depth up to 1,000 meters. Most of these small offshore support vessels can be only used to install more compact subsea hardware in shallow waters. However, if a heavy duty FRDS system is used onboard, these small offshore support vessels with a capable A-Frame can be used to deploy 300 tonnes subsea structures beyond a water depth of 3,000 meters. Fig. 17 shows that an AHTS tug with A-Frame is used to install a manifold which can be launch into water.

Fig. 16 Crane boom & lifting slings for installing 420t manifold at 620m water depth, courtesy of Petrobras

Fig. 17 AHTS tug with A-Frame for manifold installation

Sheave Installation Method: The Sheave Installation Method (SIM) is a non-conventional method which was developed to deploy the 175t Roncador Manifold I into 1,885m water depth in 2002. The SIM method is based on the two-fall configuration of a conventional deployment system. The major difference is to relocate the fixed point for the dead end of the deployment rope from the same installation vessel to another vessel. Fig. 18 shows the schematic illustration for the SIM method. Three vessels are used to install the manifold. The semisubmersible Pride South America is a DP-3 drilling vessel whose crane has 300t lifting capacity and drilling riser can only reach the limit of 1,000m water depth. The semisubmersible is mainly used to lift off the manifold by its crane and deploy it through splash zone while providing heave motion compensation during lowering through splash zone and landing on the seabed. The AHTS Tug 1 provides the fixed point for the dead end of the wire rope developed from the semisubmersible when lowering the manifold through water column. The AHTS Tug 2 maintains a adequate distance from the AHTS Tug 1 and therefore provides assistance to orient the manifold, thus avoiding any potential twist induced by a two-fall configuration.

Pendulous Installation Method: The Pendulous Installation Method (PIM) is another non-conventional method developed by Petrobras due to the low availability and high cost of deepwater construction vessels and heavy lift vessels. The PIM method is to use conventional small vessels without special riggings and a FRDS system, which allows deployment of 300t heavy equipment in deep waters up to 3,000m. As Fig. 19 shows, the PIM method requires two small installation vessels to launch and deploy a subsea structure onto the seabed.

(a) Lift and overboard the manifold

(b) Lower the manifold through splash zone

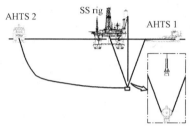
(c) Schematic illustration of sheave installation method

Fig. 18　Sheave Installation Method for 175t manifold at 1,885m WD

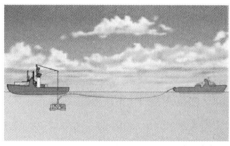
(a) Illustration of manifold overboarding

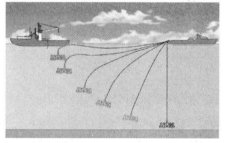
(b) Illustration of pendulous motion to lower manifold

(c) Dummy manifold for full-scale test (16.6m×8.5m×5.2m)

Fig. 19　Pendulous Installation Method for 280t manifold at 1,900m water depth, courtesy of Petrobras

Different from traditional vertical launch, the PIM method uses a conventional steel wire winch system as a launch line to launch the structure in a pendulous motion while using a polyester rope as deployment line to finally deploy the structure onto the seabed. The first installation vessel uses its crane to lift off and overboard the manifold into water and through slash zone, and then transfer the load from the crane to the launch winch wire at a certain water depth, say 50m below surface. The deployment line is pre-rigged with the lifting slings of the manifold and equipped with a number of buoyancy elements to reduce the winching capacity requirement for both the launch winching system and the deployment winching system. The deployment line is pre-deployed with a certain length which shall ensure that the manifold should maintain a vertical position approximately 50m above the seabed. The elongation of the deployment polyester rope under load tension shall be taken into account to avoid any premature touchdown. Upon transferring the load from the crane to the launch winch wire, the launch winch can start paying out the launch line while maintaining the deployment winch at the braking mode. Through a pendulous movement, the load will gradually transfer from the launch line to the deployment line. When the pendulous motion is completed and the manifold will swing from the bottom of the first vessel, 50m below water surface, to the bottom of the second vessel, 50m above the seabed. The polyester rope is gradually tensioned up to the full load. This will prevent axial resonance by using ropes much longer than the lengths that would fall into the resonance region when lowering through water column and allow ropes to undergo gradual tension after pre-paid. After the pendulous launch, the launch line can be disconnected by using ROV to pull the trigger sling. Then the deployment winch can continue deploy the manifold vertically, position and land it into the target box on the seabed with a conventional way. This PIM method is cost effective compared to utilization of scarce specialized deepwater installation vessels or drilling rigs. It should be pointed out that, due to complex geometry of the manifold, hydrodynamic instability may rise during launch operation.

Pencil Buoy Method: The Pencil Buoy Method (PBM) is a subsurface transportation and installation method developed Aker Solutions. The PBM method requires a crane barge to lift the structure from the transportation barge at a nearby inshore transfer location with sufficient water depth. Then transfer the structure weight from the crane barge to an top class AHTS vessel. After rigging the structure with the tug winch wire and the tubular buoyancy tank shaped as a pencil, the pencil buoy is then launched from the stern deck by paying out of the towing winch line while the tug moves slowly forward. The structure and the rigging weight are suspended from the Pencil Buoy during wet tow.

The Pencil Buoy is a steel structure with internal ring stiffeners. The buoy is subdivided into many watertight compartments to meet the requirement of one-compartment damage. Normally a wet tow speed of 3.0 to 3.5 knots is maintained. Upon arrival at the installation site the towing wire is winched in while transferring the structure weight from the Pencil Buoy back to the towing winch wire. Then the buoy can be disconnected. The AHTS tug shall move slowly forward during load transfer and therefore ensure no contact between the tug stern and the buoy. A passive heave compensator is used during lowering of structure to seabed. The lifting off and lowering through splash zone are performed inshore sheltered area rather than offshore operations. The subsurface wet tow is designed for typical unrestricted summer storms. The lowering operation is a standard offshore operation similar to other deepwater installations.

Fig. 20 Shows illustrative setup for the Pencil Buoy Method and the Pencil Buoy at pre-launch position. Only standard equipment onboard AHTS tugs, tailor-made special pencil buoy, and a passive heave compensator are required. The first generation of the Pencil Buoy has 150t cargo capacity, and the second generation and the recent third generation have 250t and 350t cargo capacities, respectively. The submerged cargo and the surface pencil buoy are subject to complex hydrodynamic problems due to wave and current action.

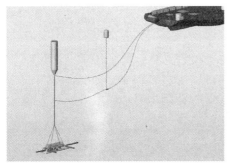 

(a) Setup for Pencil Buoy Method   (b) Photo of slender pencil shaped buoy at pre-launch

Fig. 20 Pencil Buoy Method, courtesy of Aker Solutions

Free-Fall Installation - Torpedo Anchor Piles: Torpedo anchors are used as foundations for mooring deep-water offshore facilities, including risers and floating structures. The free-fall torpedo pile dynamically penetrates the soil by the kinetic energy which acquires during free fall through the water column. The torpedo anchors are cone-tipped cylindrical steel pipes ballasted with concrete and scrap metal. The conical tip, stabilizing fins, ballast, and an omni-directional chain attachment on the pile top help the anchor penetrate the seabed within the target tolerance and thereby preventing fluttering and resultant unacceptable vertical tilt angles after penetration in the soil. The ballast into the bottom portion of the pile will keep the center of gravity low and stabilize the free fall. A mooring line is connected at the top of the anchor. Once the mooring line is loaded to cut in the embedded anchor chain, the chain forms an inverse catenary shape in the soil. This shape reduces the horizontal load on the top of the pile, thus increasing its lateral capacity. The design of such anchors involves estimation of the embedment depth as well as short-term and long-term pullout capacities.

Torpedo piles were first conceived as an inexpensive, easily installed anchor for riser flowline restraint. So far three type of Torpedo piles have been developed, including T-24 torpedo piles typically used for flowline restraint, T-43 for MODUs, and T-98 for FPSO permanent anchors. Torpedo piles work best in the soil conditions that work well for suction piles and plate anchors, such as soft to medium clay soils. Fig. 21 shows a torpedo an-

Fig. 21 Torpedo anchor piles in combination with polyester ropes

chor piles in combination with polyester mooring ropes, which are used in the Cascade and Chinook field for the first FPSO in Gulf of Mexico. A computational fluid dynamic analysis shall be performed to predict torpedo-anchor embedment depth and evaluate the resisting forces on the anchor. The CFD approach provides estimates of not only the embedment depth but the pressure and shear distributions on the soil-anchor interface and in the soil.

Fig. 22 shows the typical free-fall installation steps of the anchors:

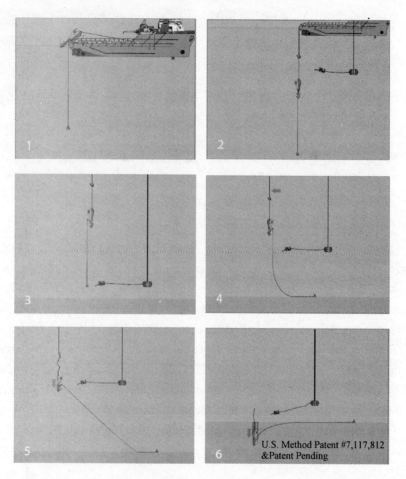

Fig. 22 Typical free-fall installation steps for anchors

Step 1: The anchor is deployed over the AHTS stern roller without the use of an A-frame.

Step 2: Once the anchor is overboard, ROV inspects the rigging arrangement.

Step 3: The anchor is lowered over the subsea connector mud-mat placement. ROV verifies the water depth, and re-inspects the rigging arrangement.

Step 4: The AHTS pays out the winch line and lowers the anchor as ROV observes the connector mud mat touching down. ROV monitors the mooring component being laid out toward the anchor drop location, which allows pre-alignment of the mooring arm.

Step 5: The AHTS finishes paying out and moves to the proposed drop location and drop height from the sea floor. ROV verifies the correct orientation of the mooring arm and the drop height, say 30m above seabed, and then observes the release hook from a safe distance. The drop coordinates are recorded with a LBL system. Then acoustic release-hook is triggered.

Step 6: ROV verifies the anchor's penetration depth. After free-fall installation, the top of the pile is typically 9m to 15m below the mudline. The anchor has been successfully deployed and is ready for attachment with the mooring leg of the platform.

CONCLUSIONS

This paper gives a comprehensive overview of the current state-of-the-art deepwater installation technologies and categorizes various installation methods based on the latest advancements in the subsea constructions. Different deepwater installation technologies are described here to address the important aspects of subsea installation technologies, including different types of subsea construction vessels, various single-wire and multi-fall deployment systems, steel wire ropes, synthetic fiber ropes, anti-rotation systems, free-fall installations, dynamic amplification factors, motion compensation systems, buoyancy elements, duration of deepwater installations, etc.

ACKNOWLEDGEMENTS

Very special thanks to Aker Solutions, DNV, DSM Dyneema, Heerema, Jumbo Offshore, Petrobras, Puget Sound, Subsea 7, and Technip for their invaluable experience and expertise, as well as their valuable photograph courtesy.

REFERENCES

Chung J S. 2010. Full-Scale, Coupled Ship and Pipe Motions Measured in North Pacific Ocean: The Hughes Glomar Explorer with a 5,000 – m-Long Heavy-Lift Pipe Deployed[J]. International Journal of Offshore and Polar Engineering, 20(1): 1 – 6.
Frazer I, Perinet D, and Vennemann O. 2005. Technology Required for the Installation of Production Facilities in 10,000 ft of Water[C]. Offshore Technology Conference, Paper OTC – 17317: 8.
McKenna, H A, Hearle, JWS, and O'Hear N. 2004. Handbook of Fibre Rope Technology[M]. Woodhead Publishing Ltd., CTR Press: xv + 416.
Torben S R, Ingeberg P, Bunes O, Bull S, Paterson J and Davidson D. 2007. Fiber Rope Development System for Ultradeepwater Installations[C]. Offshore Technology Conference, Paper OTC – 18932: 11.
Torben S R and Ingeberg P. 2011. Field Pilot of Subsea Equipment Installation in Deep Water Using Fibre Rope in Two-Fall Arrangement[C]. Offshore Technology Conference, Paper OTC – 21204: 10.
Tornqvist R, Strande M, Cannell D, Gledhill P, Smeets P, Gilmlore J. 2011. Development of Subsea Equipment: Qualification of Large Diameter Fibre Rope for Deepwater Construction Applications[C]. Offshore Technology Conference, Paper OTC – 21588: 6.
Wang A M, Jiang X Z, Yu C S, Zhu S H, Li H L, Wei Y G. 2010. Latest Progress in Floatover Technologies for Offshore Installations and Decommissioning[C]. Proceedings of the 20th International Offshore and Polar Engineering Conference, Beijing, China: 9 – 20.

ANNEX A

Table A−1 Mechanical Properties of Deployment Rope Materials

Parameter	Symbol	Rope Material		HMPE	Aramid	Polyester	Nylon	Steel
		Formula	Unit					
Density		ρ	t/m^3	0.97	1.44	1.38	1.14	7.85
Minimum Break Load	MBL	MBL/d^2	kN/mm^2	0.75	0.60	0.19	0.19	0.70
Stiffness	K	K/MBL		63	31	38	4.8	100
Line Mass per 100m	M_{100m}	M_{100m}/d^2	kg/100m/mm^2	0.055	0.066	0.055	0.062	0.409
Ratio of MBL & W		MBL/M_{100m}	km	100	66.7	31.3	27.7	20
Elongation at break		δ/L	100%	3.5	3.6	13.0	20.0	1.1
Tenacity			N/tex	3.5	2.0	0.8	0.8	0.2

Note: (1) Tenacity = Ultimate breaking strength of the fiber divided by tex;
(2) Tex = Unit of measure for the line mass density of fibers and defined as the mass in grams per 1,000 meters;
(3) HMPE Ropes ($d = \phi 40$mm) is based on Dyneema® SK75 while Steel Wire ($d = \phi 40$mm) is based on 6×36WS+FC;
(4) Polyester & Nylon Ropes are based on a diameter $d = \phi 80$mm;
(5) Aramid Rope is based on a diameter $d = \phi 40$mm.

Pendulous Method to Install Manifolds in Ultra Deep Water

Haitao Zhang[1], Xu Yuan[2], Shichao Lu[2]

[1] Offshore Oil Engineering Co., Ltd., Tianjin, China
[2] Offshore Oil/Gas Research Center, China University of Petroleum, Beijing, China

Abstract This paper focuses on the operations to install subsea manifolds in ultra-deep water. In 2006 and 2007, Petrobras installed Manifold MSGL – RO – 02 (280 tonnes) and Manifold MSGL – RO – 03 (230 tonnes) at the Roncador field on the continental slope at Campos Basin, respectively. Both of them were installed in more than 1,800m water depth and the pendulous method developed by Petrobras was adopted. The overall introduction to pendulous method, the installation planning, the monitoring system and the cable system are all included in this paper.

Key Words Ultra deep water; Subsea manifolds; Pendulous method; Roncador field

INTRODUCTION

With the trend towards deepwater and ultra-deep water in the exploration and development of oil and gas fields, the oceaneering faces the challenge of deploying heavy subsea facilities associated with increasing water depth. The subsea manifold which consists of pipelines, valves, connectors, controllers and structural frame is an important part in subsea production system (see Fig. 1). It is used to transport oil and gas from wells to the main flowline, meanwhile, the water injection, gas injection, chemical injection and the maintainance are also included. So many functions result in the largest structure among the subsea facilities. So the installation of subsea manifolds should be considered firstly. The conventional method of utilization of wire rope and drilling risers has some disadvantages. The allowable tension is consumed by the self-weight (see Fig. 2). In addition, the axial resonance of wire rope is a great problem, the mechanism are as follows:

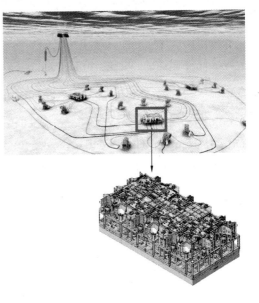

Fig. 1 Subsea production system and manifold

Consider the deploying system as a spring-damper system (see Fig. 3), the kinematic equation based on this one dimentional system is as follows:

$$M_e \ddot{x} + C \dot{x} + K x = K x_0 \qquad (1)$$

$$K = \frac{(EA)_\varepsilon}{L} \tag{2}$$

$$M_e = M + M_a + \frac{2}{5}m \tag{3}$$

Where M_e is the equivalent mass, C is the damping coefficient of wire line and manifold, x_0 is the vertical displacement of crane, x is the vertical displacement of manifold, E is the modulus of elasticity of the wire line, A is the cross-sectional area, L is the length of wire line, K is the equivalent stiffness of wire line, M is the mass of wire line, m is the mass of manifold, M_a is the additional mass of manifold.

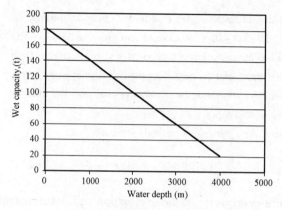

Fig. 2   Allowable tension curve

Steel wire rope diameter 4 in Safety factor 3.1 (DNV)
Weight 44 kg/m Safe working load 214.5 t Weight in water
36.5 kg/m DAF 1.2 Minimum breaking load 665 t

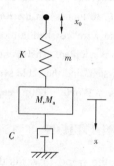

Fig. 3   Spring-damper system

We suppose that the kinematic equation of crane is as follows:

$$x_0 = h\sin(\omega t) \tag{4}$$

According to the solution of second order ordinary differential equation, the amplitude of subsea manifold's motion will be defied in the following equation:

$$\frac{x}{h} = \frac{1}{\left[\left(1 - \frac{\omega^2}{\omega_n^2}\right)^2 + \left(\frac{2\xi\omega}{\omega n}\right)^2\right]^{1/2}} \tag{5}$$

Where $\omega_n = \sqrt{\frac{K}{M_\varepsilon}} = \sqrt{\frac{K}{M + M_a + \frac{2}{5}m}}$ is the natural frequency of undamped system, $\xi = \frac{c}{2M_\varepsilon \omega_n}$ is the damping coefficient.

The maximum value $\left(\frac{x}{h}\right)_{max} = \frac{1}{2\xi\sqrt{1-\xi^2}}$ will be obtained when $\frac{\omega}{\omega_n} = \sqrt{1 - 2\zeta^2}$. In general, the damping coefficient of subsea manifold system ξ is much smaller than 1, so we think that the resonance also occurs when $\omega = \omega_n$. According to the Eq. (1) ~ Eq. (5), the conclusion can be easily got:

The increase of the length of wire line results in the increase of M (the mass of wire line) and the decrease of K (the equivalent stiffness of wire line) and ω_n (the natural frequency of undamped system). Resonance occurs when $\omega = \omega_n$.

Some attempts were made by the engineers in the past years, such as the utilization of special construction vessels associated with handling capacity of heavy loads and the synthetic fiber ropes. Unfortunately, both are infeasible after deep consideration.

Using special construction vessels causes high daily rates which makes the installation costs prohibitive. The issues to be solved in the utilization of synthetic fiber rope are that the heating by bending in the rope during deployment and the axial resonance.

The pendulous method was proposed to overcome the constraints mentioned above. It is an innovative idea which makes the installation of heavy subsea facilities in ultra-deep water conceivable by utilizing conventional vessels.

1 BACKGROUD

Petrobras has installed several subsea facilities in deepwater such as the 240 tonnes manifold in 940 meters water depth at Marlim field, the 175 tonnes manifold MSGL – RO – 01 in Roncador field. They are installed by utilizing drilling risers and sheave method, respectively.

Petrobras is a member of DISH (Deepwater Installation of Subsea Hardware) Joint Industry Project undertaken by BMT Fluid Mechanics Limited and project managed by Offshore Technology Management Limited. The objective of JIP is to gain a common understanding of existing technical limitations and develop feasible, globally acceptable solutions for deploying subsea hardwares in water depth of 2,000 meters and greater. More and more oil fields located in ultra-deep water were found, such as offshore Brazil (2,000m) and offshore Angola (3,000m). In order to meet the installation requirements of a new designed manifold MSGL-RO – 02 (280t, designed for Roncador field P – 52) which would be installed in 1,900 meters water depth, Petrobras developed pendulous method. It is based on the installation of torpedo pile for mooring system and the model test and 1:1 prototype test were undertaken in 2004 and 2005.

2 DESCRIPTION OF THE PENDULOUS METHOD

2.1 Overall Introduction to Pendulous Method

A pendulous motion occurs during the installation where two vessels are employed. The manifold is transported to the installation site on the deck of one vessel equipped with a crane (see Fig. 4).

At the installation site, transportation vessel overboard the manifold and the manifold is lowered submerged in a near sea surface position, the installation cable connects the subsea manifold to the installation vessel (see Fig. 5 and Fig. 6).

The installation cable length is slightly smaller than the water depth and the distance between the vessels is approximately 90% of the cable length. The transportation vessel releases the manifold by using the crane and the system realizes a pendulous motion slowly until the installation cable held by the installation vessel, at this time the installation cable is in vertical position at equilibrium (see Fig. 7).

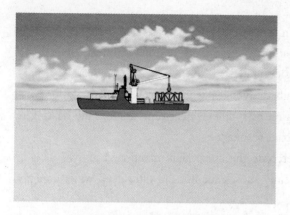

Fig. 4　Transportation vessel

Fig. 5　Manifold over boarding

Fig. 6　Manifold hung off

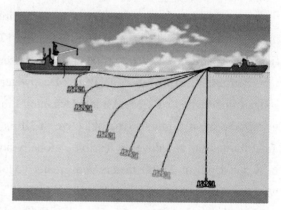

Fig. 7　Pendulous motion

　　The elongation of installation cable is checked to assure the clearance between manifold and seabed. The subsea manifold will be at an appropriate distance from the mud line to be further deployed. It will be landed after having the position adjusted.

　　The installation cable has three main segments, from the manifold towards the installation vessel: one short wire rope segment which equipped with distributed buoyancy modules, one long polyester segment and one short chain segment that ends at the installation vessel deck. Due to drag load on the manifold and installation cable, the system is dumped, so there will not be a reverse pendulum motion, i. e. the system will not swing back and forth.

　　At the end of the pendulous motion, the motion of installation vessel and manifold will be in vertical direction. However, the axial resonance will be prevented because the system is specially designed and the dynamic amplification factor (DAF) is substantially reduced compared with a wire rope system.

2.2 Installation Planning and the Monitoring System

Accurate weather forecasts are important before installation. We should get a favorable three-day weather forecast required for the operation. Current profile should be also measured every six hours by the ROV support vessel. ROV is deployed during the installation to do the visual control of the operation and to do specific tasks.

Due to the length of the BGL-1 crane boom, it is necessary to use another segment of chain to reach the designed water depth for the manifolds when suspended at the side of the installation vessel. When the weather forecast shows a favorable three-day-window and the motions of the barge measured by the IPT monitoring system is in accordance with the results of the previous installation analysis, the operation to install the manifolds begun.

2.3 Cable System

The cable used for installation consists of 3 segments (see Fig. 8).

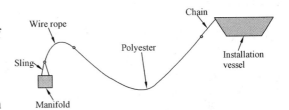

Fig. 8 Cable system

2.4 Detailed Installation Procedure

(1) Assemble the mechanism to restrain rotation on the deck of the transportation vessel;

(2) Connect the four slings, tri-plate, and first segment of chain and link plate from the manifold to the BGL – 1 crane hook;

(3) Connect installation vessel to tri-plate by using installation cable;

(4) Connect the anti-rotation mechanism to the tri-plate;

(5) Prepare the rigging of BGL – 1 crane;

(6) Cut the fastening of the manifold;

(7) Lift the manifold above deck and move to the position showed in Fig. 1;

(8) Lower the manifold and connect the first link plate on the support at the board;

(9) Disconnect the crane hook of the link plate;

(10) Connect a second segment of chain and link plate to the first link plate;

(11) Tension the chains and disconnect the first link plate from the support;

(12) Lower the manifold and connect the second link plate on the support at the board;

(13) Use ROV to survey the motion of manifold;

(14) Connect the work wire between orientation vessel and the orientation hook in the manifold;

(15) Position the orientation vessel in order to orient the manifold properly;

(16) Use the SPI to adjust the position of the manifold;

(17) ROV disconnects the work wire to the orientation hook in the manifold;

(18) ROV cuts the slings to release installation cable from the manifold.

REFERENCES

Christian Cermelli, Denby Morrison, Hugo Corvalan San Martin, et al. 2003. Progression of ultradeep subsea deployment systems [C]. Offshore Technology Conference. OTC15147: 1 - 11.

Kuppens M L. Special Workshop on the Pendulous Installation Method OMAE 2006 Roncador Field Subsea Manifold - A Risk Analysis Approach to Verify the New Installation Procedure.

Machado R D, Stock P F K, Roveri F E, et al. 2005. The Utilization of the Pendulous Motion for Deployment Subsea Hardware in Ultra-deepwater[C]. Proceedings of the 2005 Deep Offshore Technology: 1 - 11.

Maxwell B de Cerqueira, Francisco E Roveri, Luiz Eduardo Peclat, et al. 2006. The need for the pendulous installation method [C]. Proceedings of OMAE06 25nd International Conference on Offshore Mechanics and Arctic Engineering: OMAE 92654.

Rveri F E, Sangrilo L V, Lima E P C, et al. 2003. Comparing Measured and Calculated Forced of a Manifold Deployment in 940 meters Water Deep[C]. Proceedings of OMAE03 22nd International Conference on Offshore Mechanics and Arctic Engineering, OMAE2003 - 37114.

Stock P F K, M B de Cerqueira, F E Roveri. 2002. A new method for deploying subsea hardware in deep water [C]. Proceeding of the 2002 Deep Offshore Technology.

Sverre Rye Torben, Per Ingeberg, yvindBunes, et al. 2007. Fiber Rope Deployment System for Ultradeepwater Installations [C]. Offshore Technology Conference. OTC 18932:1 - 11.

Optimization Research of Jetting Parameters for Conductor Installation in Deepwater

Jin Yang[1], Bo Zhou[1], De Yan[1], Ruirui Tian[1],
Shujie Liu[2], Jianliang Zhou[2], Renjun Xie[2], Guoxian Xu[2]

[1] Petroleum Engineering Laboratory of the Ministry of Education,
China University of Petroleum, Beijing, China;
[2] CNOOC Research Center, Beijing, China

Abstract Jetting was widely used in deepwater conductor operation. Reasonable selection of drilling parameters is very important for conductor jetting in deepwater. Relational model between jetting ROP(penetration of rate) and the drilling fluid flowrate and WOB(weight on bit) parameters were established based on the lab experiment, field test and analysis of field application data, results show that flowrate and WOB parameters are the key factors for ROP of conductor jetting and the soaking time. The research achievement has been successfully applied in several deepwater wells in South China Sea, and it will provide scientific basis and technical guidance for deepwater well design and site operation.

Key Words Deep water; Conductor; Jetting; WOB; Drilling Fluid flowrate; Field test

INTRODUCTION

Jetting was often used to adapt to unconsolidated characteristics of shallow soil and avoid the effects of shallow geologic hazard, and this method could form the borehole by jetting to scour soil mass, jetting drilling while running conductor, which has the feature of saving drilling time, without well cementation, one-trip string for wellbore drilling in second section. The main parameters of conductor jetting technology is the drilling fluid flowrate and WOB, the reasonable drilling fluid flowrate and WOB can ensure that the conductor efficiently works and meet the stability of subsea wellhead the security of operations as required. At the moment, the researches of conductor jetting technology focused mainly on the vertical and horizontal bearing capacity of conductor and field application reports, while the researches on jetting drilling parameters are fewer, so the jetting drilling parameters were determined by the past experience in engineering operations, which is blindness. This paper analyses the technical principle of conductor jetting and indoor and outdoor simulated test.

1 TECHNICAL PRINCIPLE AND ADVANTAGE

The basic principle of conductor jetting technology is by using drill bit to inject water jet and the gravity of conductors to trip in conductor and subsea wellhead with drill bit at target depth. After jetting in position, conductors are firm because of the adsorbing force between conductors and surrounding soil, then the drilling string and conductor are released, continue to wellbore drilling

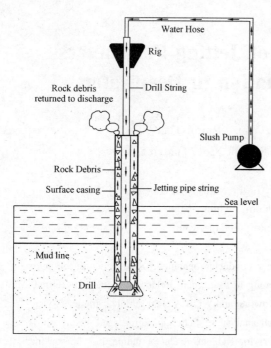

Fig. 1 Cartogram of conductor jetting

in second section, after that, picking up the drilling string and conductor running tool to drill platform, the surface drilling operations in deepwater is finished. As shown in Fig. 1.

The advantages of conductor jetting technology in deep water:

(1) Running conductor while drilling, which avoided the process that perforating then running conductor for finding the wellbore, resolved the problem that the conductor is not easy to find the well head after perforating, even the problem of abandoned wells in deepwater.

(2) This technology significantly reduced the operation time (up to 24 hours or more) by assembling conventional drilling, running conductor, well cementation and wellbore drilling in second section as one. It is very economical for drilling in deepwater that the cost per day is millions of dollars.

(3) It did not need the cement milk to cement after the operation of conductor, which could avoid the phenomenon of the leakage of cement milk because of the higher density of cement milk break down formation in conventional well cementation, the accident of wellhead subsidence even lose effectiveness because of the low temperature in deep water or other factors affecting the cement job quality. At the same time, it is environmental by avoiding the pollution.

(4) This technology needs fewer tools, equipments and personnel staff. It is efficient and convenient to operate in field, which has well application prospect.

2 THE SIMULATED EXPERIMENT OF JETTING DRILLING

In order to know the technological process of conductor jetting, analyze how different drilling fluid flowrate and WOB affects ROP, optimize drilling parameters of conductor jetting, our research group carried out the simulated experiment of jet drilling in Bohai Tianjin.

2.1 Experimental Condition

This experiment was carried out in fishing port of Tanggu center in Tianjin. Test depth of water was 1m. The sea bed soil mainly was bury. Choosing the casing with 339.73mm diameter and 8.50mm wall as experimental running string, the mean length was 11.50m. We chose homemade three-winged drag bit based on soil character in the experiment, the diameter was 269.88 mm, the entire length was 400 mm, the bank angle of bit port was 60°, nozzle diameter was 17mm. In order to simulate the running method of jetting and realize the casing and internal tools synchronous running, we also used home-made top running pup joint, the dimensions installation matched with casing threads. Meanwhile, the framing scaffold with 15 m height was built to better simulate actual drilling with jetting in deep water, the length and width respectively was 40m × 10m. The experimental drilling rig was placed on the framing scaffold. The casing joined with drilling rig through drill pipe and top running pup joint and ran in subbottom through hydraulic jetting of three-winged drag bit. The experimental process was as shown in Fig. 2.

Fig. 2 Experimental picture of jet drilling

2.2 Test Data of Jetting Flowrate and Results Interpretation

WOB set as the constant value to research how the jetting flowrate affects ROP. The value of jetting flowrate started from 300L/min, increased 300L/min each time until 6,000L/min. Three groups of test results were as shown in table 1.

Table 1 Test Data of Jet Drilling in Different Drilling Fluid Flowrate

Flowrate (L/min)	Casing running allied depth (m)	The mean running velocity of casing (m/min)	The mean running velocity of casing (m/min)	The mean running velocity of casing (m/min)
		Test data of group 1	Test data of group 2	Test data of group 3
300	1.5	0.04	0.02	0.06
600	2.0	0.07	0.05	0.10
900	2.5	0.10	0.08	0.13
1,200	3.0	0.13	0.10	0.15
1,500	3.5	0.18	0.15	0.20
1,800	4.0	0.23	0.19	0.26
2,100	4.5	0.38	0.34	0.41
2,400	5.0	0.55	0.50	0.58
2,700	5.5	0.66	0.60	0.69
3,000	6.0	0.71	0.67	0.74
3,300	6.5	0.75	0.73	0.78
3,600	7.0	0.77	0.75	0.80
3,900	7.5	0.79	0.77	0.82
4,200	8.0	0.81	0.78	0.84
4,500	8.5	0.82	0.79	0.85
4,800	9.0	0.82	0.80	0.85
5,100	9.5	0.83	0.81	0.86
5,400	10.0	0.83	0.81	0.86
5,700	10.5	0.84	0.82	0.87
6,000	11.0	0.84	0.82	0.87

By analyzing the test data, it shows the mean running velocity of casing in different drilling fluid flowrate, just as shown in Fig. 3. When the value of drilling fluid flowrate below 1,500L/min, the mean running velocity of casing is less than 0.2m/min, the increase amplitude is less; When the value of drilling fluid flowrate reached 1,800L/min, the increase amplitude of the mean running velocity of casing is obvious; When the value of drilling fluid flowrate reached 4,200L/min, the mean running velocity of casing is 0.8m/min and no longer significantly increased with the flowrate further increased. In the actual operation, if drilling fluid flowrate is too large, the soil mass around the drill bit would be scoured too much and big wellbore would be formed, which led to soil mass around conductor backfill difficultly. This reduces the adsorptive capacity between casing and formation, increases the risk of conductor sinking after jetting and subsea wellhead instability.

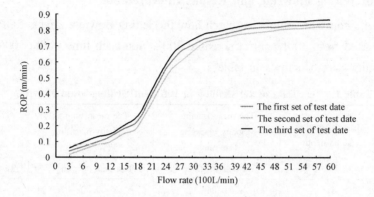

Fig. 3 Cartogram of the mean running velocity of casing in different flowrate

Hence, the values of drilling fluid flowrate range from 2,700L/min to 4,200L/min by analyzing test data and security of conductor jetting in deep water. In addition, because the particularities of extremely soft shallow soil in deep water, in order to avoid over large drilling fluid flowrate scours sea bed too much, the strength of soil in well head is failure. In the first running, the small flowrate is needed (less than 1,000L/min), as the depth increases, the strength of soil increased, so did the jetting flowrate, until the value reaches above reasonable interval. When the conductor is close to the target depth, flowrate should be properly lowed to reduce disturbance for soil mass and guarantee the stability of soil in pipe shoe to enhance bearing capacity of conductor.

2.3 Test Data of WOB and Results Interpretation

Jetting flowrate set as the constant value to research how WOB affects ROP. Choosing running depth of each group of casing respectively was 3m, 5m, 7m and 9m as plotting points, then recording ROP in different running depth. A mark should be made per 0.50m before running casing to conveniently observe. Fig. 4 shows the relationship between WOB and ROP in different running depth.

Fig. 4 have shown that the corresponding relationship between WOB and ROP is well with other constant construction parameters, that is, when casing reached the identical running depth, ROP increased with WOB increases. However, when WOB increased to some value, ROP decreased. The all 4 curves presented the above features. The curves that meet the features can be

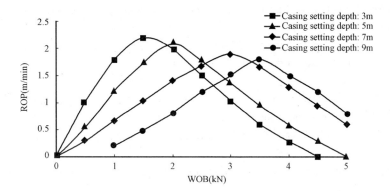

Fig. 4 Cartogram of relationship between WOB and the mean running velocity of casing

used for regression fitting with quadratic function, which means the relationship between WOB and ROP approximately meet the following quadratic function:

$$v = AW^2 + BW + C$$

Where v was ROP, m/h; W was WOB, kN; A, B, C were regression coefficient, which was in connection with running depth of string, formation strength, types of drill bit, rotate speed, hydraulics parameters and other factors. In different areas, the technological value is generally different.

3 FIELD APPLICATIONS

Optimized research results of drilling parameters for conductor jetting in deepwater have been successfully applied in Liwan, Liuhua, Baiyun of South China Sea, Equatorial Guinea of West Africa and several other drilling operations in deepwater. There have 10 deep-sea wells with jetting drilling operations, every well would save operation time for 24.77h and cost for 12.3864 million dollars, which made obvious effect and economic benefit.

CONCLUSIONS

In the technology of conductor jetting in deepwater, reasonable selection of drilling parameters has a significant impact on time efficiency of drilling operations and stability of subsea wellhead. This paper based on the principle of conductor jetting technology and the research of simulated experiment, presenting optimized model of drilling parameters for conductor jetting in deepwater.

Optimized model of drilling parameters for conductor jetting in deepwater have been successfully applied in Liwan, Liuhua, Baiyun of South China Sea and several other deep-sea wells in shallow operations. Every well would save operation time for 24.77h and cost for 12.3864 million dollars, which improved time efficiency of operations and met the safety requirements.

REFERENCES

Akers T J. 2009. Improving Hole Quality and Casing-Running Performance in Riser less Top Holes: Deepwater Angola[C], SPE - 112630.

Beck R D. 1991. Reliable Deepwater Structural Casing Installation Using Controlled Jetting[C]. Society of Petroleum Engineering, SPE - 22542.

Chen Tinggen. 2000. Theory and Technology of Drilling Engineering [M]. Beijing: Press of University of Petroleum. Hu Hailiang. 2008. Deepwater drilling techniques in Baiyun 6 - 1 - 1 well [J]. Drilling & Production Technology, 30(6): 25 - 28.

Liu Shujie. 2011. Relationship between weight-on-bit and drilling rate during jetting drilling in sub-bottom deepwater[J]. Oil Drilling & Production Technology, 33(1): 12 - 15.

Luke F Eaton. 2005. Deepwater Batchset Operations through the Magnolia Shallow Water Flow Sand[C]. Society of Petroleum Engineering, SPE - 92289.

Philippe Jeanjean. 2002. Innovative Design Method for Deepwater Surface Casings[C]. Society of Petroleum Engineering, SPE - 77357.

Quiros G W. 2003. Deepwater Soil Properties and Their Impact on the Geotechnical Program[C]. Offshore Technology Conference OTC.

Su Kanhua, 2008. Determination method of conductor setting depth using jetting drilling in deepwater[J]. Journal of China University of Petroleum, 32(4): 48 - 50.

Tang Xaixiong, 2011. Method of Design of Conductor Setting Depth for Ultra-deepwater Jetting Drilling in South China Sea[J]. Journal of Oil and Gas Technology, 33(3): 147 - 151.

T J Akers. 2006. Jetting of Structural Casing in Deepwater Environments: Job Design and Operational Practices [C]. Society of Petroleum Engineering, SPE - 102378.

Xu Rongqiang. 2007. The Application of Jetting Technology in Deepwater Drilling[J]. Petroleum Drilling Techniques, 35(3): 19 - 22.

Yang Jin. 2010. Research of Conductor Setting Depth Using Jetting in the Surface of Deepwater[C]. Society of Petroleum Engineers, SPE - 130523.

Zhang Hui, 2010. Tubular mechanics in jetting operation of conductor under deepwater condition[J]. ACTAPETROLEI SINICA, 31(3): 516 - 520.

The Mechanical Analysis of Subsea Manifold Lowered into Deep Water

Zhigang Li[1], Weidong Ruan[2], Yong Bai[3], Xu He[4], Junhua Bai[5]

[1] Offshore Oil Engineering Co., Ltd., Tianjin, China
[2,3] Institute of Structural Engineering, Zhejiang University, Hangzhou, China
[4,5] Offshore Pipelines & Risers (OPR) INC, Hangzhou, China

Abstract This paper aims to study two installation methods (drill pipe installation method and winch installation method) of subsea manifold lowered into deep water and presents key results of manifold installation. Based on the material characteristics of drill pipe (or winch wire), two theoretical models are proposed to deal with the horizontal displacement and axial tension along drill pipe (or winch wire) during the lowering process. A MATLAB programming including the high order matrix equation is developed to calculate the horizontal displacement and axial tension. The key results from the theoretical method show good consistency with the finite element model using OrcaFlex.

Key Words Subsea manifold; Drill pipe; Winch wire; Differential segment; Horizontal displacement; Axial tension; Force; Finite element model.

INTRODUCTION

Subsea manifolds have been widely used in the development of oil and gas fields to simplify the subsea system, reduce the demand of pipelines and optimize the fluid flow in the subsea production system. In order to guarantee the efficiency of the subsea production system and maintain the installation safety of subsea manifold, it is an important project to analyze the motion characteristics of subsea manifold lowered into deep water accurately. However, the installation process of subsea manifold lowered into deep water is a little more complex, which is dominated by environmental factors and installation conditions. For a successful installation, the following attentions should be paid to drill pipe (or winch wire): one is the horizontal displacement along drill pipe (or winch wire), if the horizontal displacement has larger difference with the fact, subsea manifold can't be installed at the specific location, which will lead the failure of the installation; the other is the axial tension along drill pipe (or winch wire), in the installation process, the axial tension must be controlled within its ultimate tension, especially the connections between each drill pipe.

So far, the study of subsea manifold installation nationally and abroad is mainly confined to structural design and software simulation. Francisco Edward Roveri et al. (2003) compared the measured and calculated axial forces of a manifold deployment in a certain water depth. However, as the influence of wave and current was disregarded, the horizontal displacement of manifold due to wave and current wasn't studied. Lin and Xiao et al. (2011) applied the finite difference method to analyze the displacement along drill pipe in running deepwater oil tree installation process with considering the effects of wave and current, which was the same with the installation of subsea manifold. But the horizontal hydrodynamic force on the tree was neglected.

In this paper, two installation methods of subsea manifold lowered into deep water are analyzed: drill pipe installation method and winch installation method. As modulus of elasticity of drill pipe is very large, the installation method by drill pipe should take the bending stiffness of drill pipe into consideration. Then the behavior of drill pipe should be analyzed by finite element discretization. Based on the bending deformation theory of material mechanics, the horizontal displacement of each differential segment can be derived by the second order ordinary differential equation and the high order matrix equation. The key results can be obtained by the sum of each differential segment. However, the second method can neglect the bending stiffness of winch wire due to the material characteristics of winch wire. Thus, the analysis of winch installation is the most simple but practical. The behavior of winch wire should be also analyzed by finite element discretization. Then the horizontal displacement of each differential segment can be derived by bending moment balance equation. The key results can be obtained by the sum of each differential segment, too. The validity and accuracy of the two proposed methods are verified by comparing the results simulated by OrcaFlex.

1 ASSUMED CONDITIONS

In general, the hydrodynamic forces acting on drill pipe (or wire rope) are more complicated while subsea manifold is lowered into deep water. Thus, for the sake of convenience in installation analysis, the following assumptions are made:

(1) In the process of motion and deformation, drill pipe is always in a linear elastic range and the bending stiffness of drill pipe is assumed to be a constant.

(2) A planar problem is considered, which means the motions of wave, current and drill pipe (or wire rope) are all restricted in vertical plane.

(3) Lift force caused by wave and current are disregarded.

(4) Subsea manifold is a perfectly rigid body, which means the deformation of subsea manifold is disregarded.

2 DRILL PIPE INSTALLATION

2.1 Configuration calculation

As illustrated in Fig. 1, a global coordinate system is set up by locating the origin at the position of tensioner. Meanwhile, a local coordinate system is set up by locating the origin at the position of the node $i-1$. The part of drill pipe under the water is divided into n equilong differential segments, the length of each segment is $L = D^-/n$. D^+ and D^- represent the length above the water and under the water of drill pipe (or winch wire), respectively. The $(n+3)$ nodes can be signed as shown in Fig. 1. v_i represents current velocity in the node i.

The external forces acting on each differential segment include the drag force and the inertia force caused by wave and current. Traditionally, current conditions in the ocean have been regar-

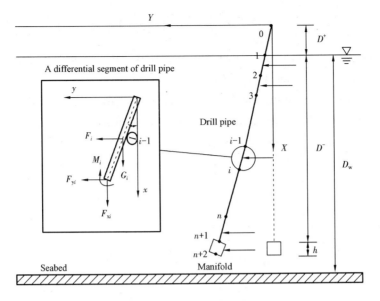

Fig. 1 Mechanics analysis model of drill pipe installation for manifold

ded as being constant (static) in the analysis of marine structure operations. Therefore, the inertia force acting on drill pipe (or winch wire) caused by current can be disregarded. Based on the Morison Formula, the external force acting on each differential segment and subsea manifold can be expressed as:

$$F_i = \frac{\rho}{4} C_d dL (v_i |v_i| + v_{i-1} |v_{i-1}| + u_i |u_i| + u_{i-1} |u_{i-1}|) + \frac{\pi}{8} \rho C_{M1} d^2 L \left(\frac{du_i}{dt} + \frac{du_{i-1}}{dt} \right) \quad (1)$$

$$F_{\text{manifold}} = \frac{1}{2} \rho C_{M2} V \left(\frac{du_{n+1}}{dt} + \frac{du_{n+2}}{dt} \right) + \frac{\rho}{4} C_D A_p (v_{n+1} |v_{n+1}| + v_{n+2} |v_{n+2}|$$

$$+ u_{n+1} |u_{n+1}| + u_{n+2} |u_{n+2}|) \quad (2)$$

where $i = 2, 3, \cdots, n+1$. C_d and C_D represent the drag coefficient of drill pipe (or winch wire) and subsea manifold, respectively. C_{M1} and C_{M2} represent the inertia coefficient of drill pipe (or winch wire) and subsea manifold, respectively. ρ is the density of sea water and d is the diameter of drill pipe (or winch wire). A_p is the drag area of subsea manifold and V is the volume of displaced water of manifold. Water particle velocity u and water particle acceleration du/dt caused by wave can be derived as follows according to Airy theory:

$$u = \frac{\pi H}{T} \frac{\cosh \left[\frac{2\pi (D_W - X)}{\lambda} \right]}{\sinh \left[\frac{2\pi D_W}{\lambda} \right]} \cos \frac{2\pi}{\lambda} \left(Y - \frac{\lambda t}{T} \right) \quad (3)$$

$$\frac{du}{dt} = \frac{2\pi^2 H}{T^2} \frac{\cosh \left[\frac{2\pi (D_W - X)}{\lambda} \right]}{\sinh \left[\frac{2\pi D_W}{\lambda} \right]} \sin \frac{2\pi}{\lambda} \left(Y - \frac{\lambda t}{T} \right) \quad (4)$$

$$\lambda = \frac{gT^2}{2\pi}\tanh\left(\frac{2\pi \cdot D_W}{\lambda}\right) \tag{5}$$

in which H, T and λ represent wave height, wave period and wavelength, respectively. g is the acceleration of gravity. D_W is the water depth.

In the local coordinate system, the (i)th differential segment is located as shown in Fig. 1. The weight of the (i)th differential segment G_i, the vertical force F_{xi}, the horizontal force F_{yi} and the axial tension T_i of the node i can be expressed as:

$$G_i = w_w L \tag{6}$$

$$F_{xi} = F_{\text{total}} + w_w(n+1-i)L \tag{7}$$

$$F_{yi} = \sum_{i+1}^{n+1} F_i + F_{\text{manifold}} \tag{8}$$

$$T_i = \sqrt{F_{xi}^2 + F_{yi}^2} \tag{9}$$

in which, $i = 1, 2, \cdots, n+1$. However, when $i = 1$, $G_1 = w_d D^+$; when $i = n+1$, $F_{y(n+1)} = F_{\text{manifold}}$. w_d and w_w are the dry weight and the submerged weight per unit length of drill pipe (or winch wire), respectively. F_{total} is the submerged weight of manifold.

2.1.1 Boundary Conditions

The top of drill pipe is rigidly restrained with the tensioner:

$$Y_0 = 0 \tag{10}$$

$$\Theta_0 = 0 \tag{11}$$

The end of drill pipe is hinged with subsea manifold by sling:

$$F_{x(n+1)} = F_{\text{total}} \tag{12}$$

$$F_{y(n+1)} = F_{\text{manifold}} \tag{13}$$

$$M_{(n+1)} = 0 \tag{14}$$

In the local coordinate system for the (i)th differential segment:

$$y_{i-1}(0) = 0 \tag{15}$$

$$\theta_{i-1}(0) = \Theta_{i-1} \tag{16}$$

Θ and θ represent the inclination in the global coordinate system and in the local coordinate system, respectively.

2.1.2 Numerical Solution

In the local coordinate system, the external force F_i and the weight of segment G_i cause the bending moment equation of the (i)th differential segment to be too complicated to solve directly. When n gets large enough, which leads $F_i \ll F_{yi}$ and $G_i \ll F_{xi}$, the bending moment caused by F_i and G_i in the bending moment equation can be neglected. Then, the bending moment equation can be obtained which is based on Material Mechanics:

$$M_i(x) = EI\frac{d^2y}{d^2x} = F_{yi}(L - x) + M_i - F_{xi}(y_i(L) - y_i(x)) \tag{17}$$

where M_i is the bending moment of the node i. The bending moment of the node $i-1$ can be expressed as:

$$M_{i-1} = M_i + F_{yi}L - F_{xi}y_i(L) \tag{18}$$

From Eq. (17), we can find that the governing equation is a second order ordinary differential equation. On the basis of boundary conditions of the (i)th differential segment Eq. (15) and Eq. (16), the horizontal displacement of the (i)th differential segment can be derived as:

$$y_i(L) = \left(c_1 e^{\sqrt{\frac{F_{xi}}{EI}}L} + c_2 e^{-\sqrt{\frac{F_{xi}}{EI}}L} - \frac{M_i}{F_{xi}}\right)\mathrm{sech}\left(\sqrt{\frac{F_{xi}}{EI}}L\right) \tag{19}$$

where

$$c_1 = \frac{1}{2}\left[\frac{F_{yi}L + M_i}{F_{xi}} - \left(\frac{F_{yi}}{F_{xi}} - \Theta_{(i-1)}\right)\sqrt{\frac{EI}{F_{xi}}}\right]$$

$$c_2 = \frac{1}{2}\left[\frac{F_{yi}L + M_i}{F_{xi}} + \left(\frac{F_{yi}}{F_{xi}} - \Theta_{(i-1)}\right)\sqrt{\frac{EI}{F_{xi}}}\right]$$

In the local coordinate system, the relationship between Θ_i and Θ_{i-1} can be derived:

$$\Theta_i = \frac{F_{yi}}{F_{xi}} - \left(\frac{F_{yi}}{F_{xi}} - \Theta_{i-1}\right)\mathrm{sech}\left(\sqrt{\frac{F_{xi}}{EI}}L\right) + \sqrt{\frac{F_{xi}}{EI}}\frac{M_i}{F_{xi}}\tanh\left(\sqrt{\frac{F_{xi}}{EI}}L\right) \tag{20}$$

where $i = 1, 2, \cdots, n$. However, when $i = 1$, L should be changed to D^+.

Meanwhile, the relationship between M_i and M_{i-1} can be derived:

$$M_{i-1} = M_i + F_{yi}L - F_{xi}\left(c_1 e^{\sqrt{\frac{F_{xi}}{EI}}L} + c_2 e^{-\sqrt{\frac{F_{xi}}{EI}}L} - \frac{M_i}{F_{xi}}\right)\mathrm{sech}\left(\sqrt{\frac{F_{xi}}{EI}}L\right) \tag{21}$$

in which, $i = 2, 3, \cdots, n+1$.

A $2n$ order sparse matrix can be derived from Eq. (20) and Eq. (21), which contain $2n$ unknowns $\Theta_1, \Theta_2, \cdots, \Theta_n$ and $M_{y1}, M_{y2}, \cdots, M_{yn}$. A MATLAB programming is written to solve the high order sparse matrix, thus $2n$ unknowns, $\Theta_2, \cdots, \Theta_n$ and $M_1, M_2, \cdots, M_n$ can be calculated. y_i can also be calculated from Eq. (19).

In the global coordinate system, the horizontal displacement Y_i can be expressed as the summing form of each segment displacement y_i:

$$Y_i = \sum_1^i y_i \tag{23}$$

At last, the horizontal displacement of manifold bottom can be expressed as:

$$Y_{\mathrm{manifold}} = Y_{n+1} + \frac{F_{\mathrm{manifold}}h}{F_{\mathrm{toatal}}} \tag{24}$$

3 WINCH INSTALLATION

3.1 Configuration Calculation

As illustrated in Fig. 2, the mechanical model of subsea manifold installation by winch wire is similar to the mechanical model of subsea manifold installation by drill pipe. However, as winch wire is flexible, the bending stiffness of wire rope can be neglected. Besides, the connections of both ends of which wire are hinged joints.

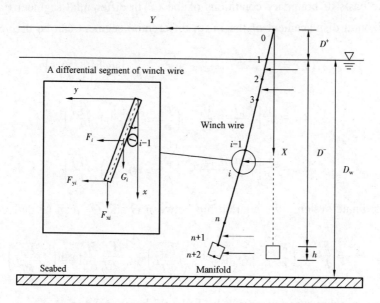

Fig. 2 Mechanics analysis model of winch installation for manifold

3.2 Boundary Conditions

The top of winch wire is hinged with the winch:

$$Y_0 = 0 \tag{25}$$

$$M_0 = 0 \tag{26}$$

The end of winch wire is hinged with subsea manifold by sling:

$$F_{x(n+1)} = F_{\text{total}} \tag{27}$$

$$F_{y(n+1)} = F_{\text{manifold}} \tag{28}$$

$$M_{n+1} = 0 \tag{29}$$

In the local coordinate system for the (i)th differential segment:

$$y_0 = 0 \tag{30}$$

$$M_i = M_{i-1} = 0 \tag{31}$$

3.3 Numerical Solution

In the local coordinate system, the bending moment equilibrium equation of the node $(i-1)$ can be derived as:

$$M_{i-1} = F_{yi}L + \frac{1}{2}F_iL - F_{xi}y_i(L) - \frac{1}{2}G_iy_i(L) = 0 \tag{32}$$

where $i = 2, 3, \cdots, n+1$. When $i = 1, F_{y1}D^* - F_{x1}y_1(D^*) - \frac{1}{2}G_1y_1(D^*) = 0$

For the (i)th differential segment, the increment of horizontal displacement can be derived from Eq. (32):

$$y_i = \frac{\left(F_{yi}L + \frac{1}{2}F_iL\right)}{\left(F_{xi} + \frac{1}{2}G_i\right)} \tag{33}$$

where $i = 2, 3, \cdots, n+1$. While $i = 1$, $y_1 = \dfrac{(F_{y1}D^*)}{\left(F_{x1} + \dfrac{G_1}{2}\right)}$

The horizontal displacement Y_i can be also expressed as the summing form of each segment displacement y_i:

$$Y_i = \sum_1^i y_i \tag{34}$$

The horizontal displacement of manifold bottom can be expressed as:

$$Y_{manifold} = Y_{n+1} + \frac{F_{manifold}h}{F_{toatal}} \tag{35}$$

4 FINITE ELEMENT MODEL

In order to further verify the accuracy and reliability of the proposed methods, general commercial finite element software program OrcaFlex is employed to solve the installation analyses. The main characteristics of the model are as follows:

4.1 Environment

A 3D surface is set up to model the sea surface. The wind load is neglected in the model. The parameters of environmental data are shown in Table 1 ~ Table 3 below.

Table 1 Parameters of sea water

Parameter	Value
Density of sea water ρ (kg/m³)	1,025
Water depth D_w (m)	1,250

Table 2 Data for current profile

Water depth(m)	Current velocity (m/s)
0	1.0
150	0.43
350	0.28
700	0.35
900	0.47
1,200	0.48
1,250	0.48

Table 3 Parameters of wave

Parameter	Value
Wave height H(m)	1
Wave period T(s)	8

4.2 Drill Pipe (or winch wire)

Drill pipe (or winch wire) is modeled by line element because it is particularly suited to model long, slender pipe with both accuracy and high efficiency. In order to simulate the horizontal displacement along drill pipe (or winch wire), it is assumed that the material of line element is homogeneous and invariable in the elastic deformation. The parameters of drill pipe and winch wire are listed in Table 4 as following.

Table 4 Parameters of drill pipe and winch wire

Parameter	Drill pipe	Winch wire
Length above the water D^+ (m)	20	20
Length under the water D^- (m)	1,195.8	1,195.8
Dry weight w_d (N/m)	402.0	329.8
Submerged weight w_w (N/m)	178.5	206.5
Diameter d, (m)	0.1683	0.0889
Drag coefficient C_d	1.32	1.32
Inertia coefficient C_{M1}	2.0	2.0
Modulus of elasticity E(GPa)	210	—

4.3 Subsea Manifold

A 3D buoy is selected to model subsea manifold. The parameters of subsea manifold are shown in Table 5.

Table 5 Parameters of subsea manifold

Parameter	Subsea manifold
Height h(m)	4.2
Drag coefficient C_D	1.32
Inertia coefficient C_{M2}	2.0
Drag area A_P(m^2)	9.627
Submerged weight F_{total}(N)	5×10^5
Volume of displaced water V(m^3)	20

4.4 Boundary Conditions

In drill pipe installation, it is assumed that drill pipe is rigidly linked to the semi-submersible platform at the tensioner. Since the connections between drill pipe and subsea manifold are installation slings, the bending moment of drill pipe end is equal to 0.

In winch installation, manifold is lowered into deep water by winch wire, which means the connection between winch wire and winch is a hinge joint. Meanwhile, winch wire is linked with subsea manifold by rigging.

5 COMPARISION AND DISCUSSION

To discuss the accuracy and reliability of the proposed methods, comparisons are performed with the results from theoretical method and OrcaFlex simulation. The current velocity generally varies irregularly in real sea conditions, but the current velocity of any water depth can be derived by interpolation method from Table 2.

Fig. 3 and Fig. 4 show the comparisons of the horizontal displacement along drill pipe and winch wire; the graphics derived from theoretical method and OrcaFlex simulation are almost the same. As the lifting force caused by wave and current is disregarded, submerged weight of manifold and drill pipe (or winch wire) in OrcaFlex simulation are both slightly less than the theoretical method, which leads to the difference between theoretical method and OrcaFlex simulation. Because both theoretical method and OrcaFlex simulation adopt finite element analysis, different finite element cell number is another reason for the difference between theoretical method and OrcaFlex simulation. Differences of $Y_{manifold}$ and T_0 between the two solutions are listed in Table 6 to Table 7. Within the acceptable error limit, the two theoretical methods can provide a good estimation of the behavior of drill pipe (or winch wire) during installation.

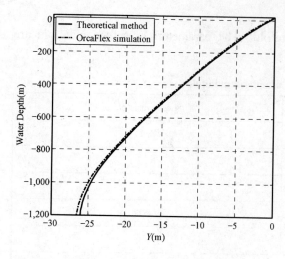

Fig. 3 Comparison of the horizontal displacement along drill pipe

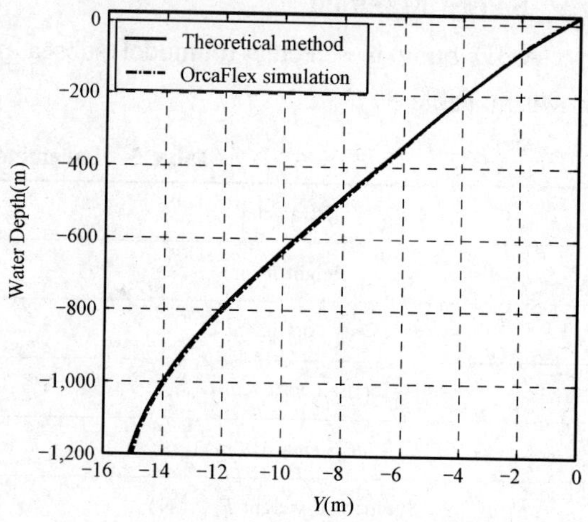

Fig. 4 Comparison of the horizontal displacement along winch wire

Table 6 Comparison of $Y_{manifold}$ and T_0 in drill pipe installation

Item	$Y_{manifold}$ (m)	T_0 ($\times 10^5$ N)
Theoretical method	26.4498	7.2208
OrcaFlex simulation	26.5334	7.1762
difference	0.32%	0.62%

Table 7 Comparison of $Y_{manifold}$ and T_0 in winch installation

Item	$Y_{manifold}$ (m)	T_0 ($\times 10^5$ N)
Theoretical method	15.0387	7.5370
OrcaFlex simulation	15.0146	7.4905
difference	0.16%	0.62%

6 PARAMETRIC STUDY

The horizontal displacement and axial tension along drill pipe (or winch wire) are vital for the security of subsea manifold installation. In this paper, several examples are presented to determine the effects of modulus of elasticity, drag coefficient, submerged weight, current and wave. By comparison, the sensitivity of those variables can be evaluated.

6.1 Effect of Modulus of Elasticity

Compared with winch installation, the modulus of elasticity of drill pipe can't be disregarded. Therefore, modulus of elasticity is an important factor in studying subsea manifold installation. In this paper, to study the influence of modulus of elasticity in drill pipe installation, four different modulus of elasticity ($E_1 = 2,100$ GPa, $E_2 = 210$ GPa, $E_3 = 21$ GPa and $E_4 = 2.1$ GPa) are selected. The horizontal displacements along drill pipe are shown in Fig. 5. The discrepancy among theoretical method and OrcaFlex simulation due to wave and finite element analysis has been previously explained; the overall tendency is the same with the variation of modulus of elasticity: the bottom horizontal displacement $Y_{manifold}$ increases with the decreasing of modulus of elasticity and the maximum axial tension T_0 is nearly keeping a constant. Table 8 lists several key results, and shows that the modulus of elasticity has little effect on the bottom horizontal displacement and the maximum axial tension. In general, modulus of elasticity of rigid body is 210GPa. So in drill pipe installation while lowered into deep water, the bending stiffness of drill pipe can be ignored.

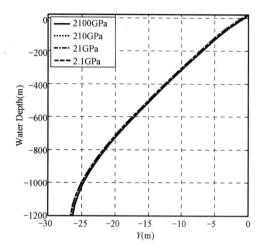

Fig. 5 Horizontal displacement along drill pipe for different modulus of elasticity

Table 8 Comparison of $Y_{manifold}$ and T_0 in drill pipe installation for different modulus of elasticity

Modulus of elasticity $E(GPa)$	Theoretical method		OrcaFlex simulation	
	$Y_{manifold}$ (m)	T_0 ($\times 10^5$ N)	$Y_{manifold}$ (m)	T_0 ($\times 10^5$ N)
2,100	26.1534	7.2208	26.2704	7.1783
210	26.4498	7.2208	26.5334	7.1762
21	26.5447	7.2208	26.5730	7.1757
2.1	26.5747	7.2208	26.5772	7.1757

6.2 Effect of Drag Coefficient

The influence of drag coefficient is very important in practical applications; it is one of the factors directly related to subsea manifold installation. Fig. 6 and Fig. 7 show the change of $Y_{manifold}$ and T_0 with the variation of drag coefficient: the bottom horizontal displacement becomes larger with the increasing of drag coefficient, but the maximum axial tension varied little (shown in Table 9 and Table 10). As the drag coefficient gets larger, the external force acting on each differential segment increases considerably, which leads to the increase of the bottom horizontal displacement. However, the maximum axial tension is derived from the horizontal force and the vertical force of node 0 by mechanical synthesis. As the vertical force is much larger than the horizontal force, the maximum axial tension is mainly dominated by the vertical force. So drag coefficient has little influence on the maximum axial tension.

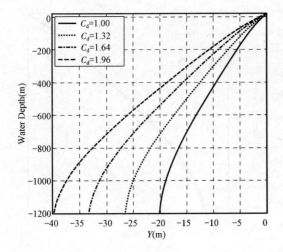

Fig. 6 Horizontal displacement along drill pipe for different drag coefficients

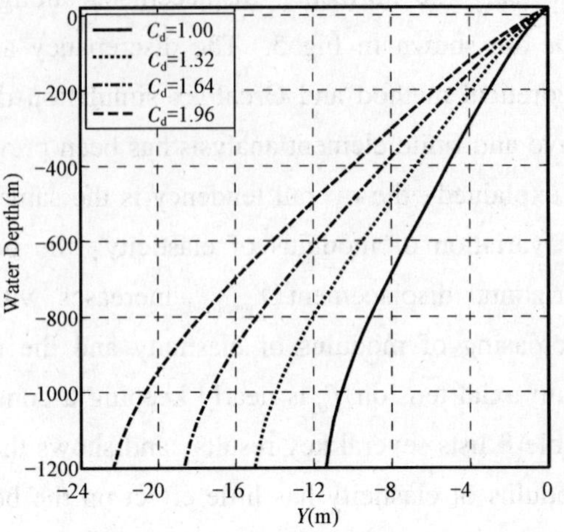

Fig. 7 Horizontal displacement along winch wire for different drag coefficients

Table 9 Comparison of $Y_{manifold}$ and T_0 in drill pipe installation for different drag coefficients

Drag coefficient $C_d = C_D$	Theoretical method		OrcaFlex simulation	
	$Y_{manifold}$ (m)	T_0 ($\times 10^5$ N)	$Y_{manifold}$ (m)	T_0 ($\times 10^5$ N)
1.00	20.0380	7.2183	20.1023	7.1765
1.32	26.4498	7.2208	26.5334	7.1762
1.64	32.8617	7.2240	33.2596	7.1758
1.96	39.2735	7.2279	39.7375	7.1754

Table 10 Comparison of $Y_{manifold}$ and T_0 in winch installation for different drag coefficients

Drag coefficient $C_d = C_D$	Theoretical method		OrcaFlex simulation	
	$Y_{manifold}$ (m)	T_0 ($\times 10^5$ N)	$Y_{manifold}$ (m)	T_0 ($\times 10^5$ N)
1.00	11.3930	7.5363	11.3747	7.4907
1.32	15.0387	7.5370	15.0146	7.4905
1.64	18.6844	7.5380	18.6548	7.4903
1.96	22.3300	7.5391	22.2945	7.4901

6.3 Effect of Submerged Weight

In general, the weight of subsea manifold is between 40t and 200t. Previous studies showed that submerged weight of subsea manifold influences the behavior of drill pipe (or winch wire) significantly. In this paper, to study the influence of submerged weight of subsea manifold, four submerged weights of 5.0×10^5N, 7.5×10^5N, 1.0×10^6N, 1.5×10^6N are chosen for analysis. As shown in Fig. 8 and Fig. 9, the horizontal displacement along drill pipe (or winch wire) decreases with the increase of submerged weight of subsea manifold. Enough submerged weight can help decrease the horizontal displacement along drill pipe (or winch wire) and improve the stability of manifold installation, but it can also bring about higher axial tension (shown in Table 11 and Table 12). Therefore, large submerged weight of manifold should be taken care of in installation process.

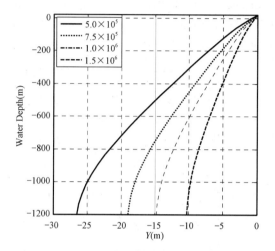

Fig. 8 Horizontal displacement along drill pipe for different submerged weights

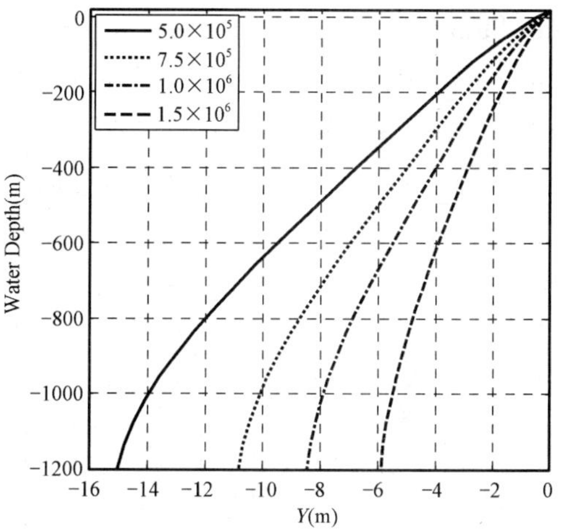

Fig. 9 Horizontal displacement along winch wire for different submerged weights

Table 11 Comparison of $Y_{manifold}$ and T_0 in drill pipe installation for different submerged weights

Submerged weight F_{total} (N)	Theoretical method		OrcaFlex simulation	
	$Y_{manifold}$ (m)	T_0 ($\times 10^5$N)	$Y_{manifold}$ (m)	T_0 ($\times 10^5$N)
5.0×10^5	26.4498	7.2208	26.5334	7.1762
7.5×10^5	18.9235	9.7193	18.9813	9.6668
1.0×10^6	14.7421	12.2184	14.7839	12.1573
1.5×10^6	10.2300	17.2174	10.2556	17.1383

Table 12 Comparison of $Y_{manifold}$ and T_0 in winch installation for different submerged weights

Submerged weight F_{total} (N)	Theoretical method		OrcaFlex simulation	
	$Y_{manifold}$ (m)	T_0 ($\times 10^5$N)	$Y_{manifold}$ (m)	T_0 ($\times 10^5$N)
5.0×10^5	15.0387	7.5370	15.0146	7.4905
7.5×10^5	10.8273	10.0366	10.8092	9.9811
1.0×10^6	8.4653	12.5363	8.4519	12.4716
1.5×10^6	5.8975	17.5360	5.8904	17.4525

6.4 Effect of Current

Current plays a vital role in subsea manifold installation. The horizontal displacement along drill pipe (or winch wire) is mainly dominated by the drag force caused by current; current is one of the most important factors influencing the horizontal displacement in installation process. To study the sensitivity of current to the installation, four different kinds of currents (water depth is invariant, but current velocity is increasing or decreasing at the ratio, the rations are 0.50, 0.75, 1.00, 1.25, respectively) are chosen. As shown in Fig. 10 and Fig. 11 and Table 15 and Table 16, the bottom horizontal displacement becomes larger and larger, but the maximum axial tension varies little from different currents. It can be concluded that current affects the horizontal displacement along drill pipe (or winch wire) significantly and has little influence on the axial tension.

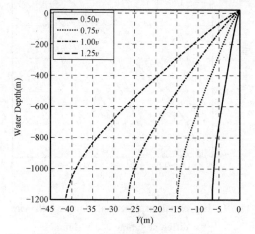

Fig. 10 Horizontal displacement along drill pipe for different currents

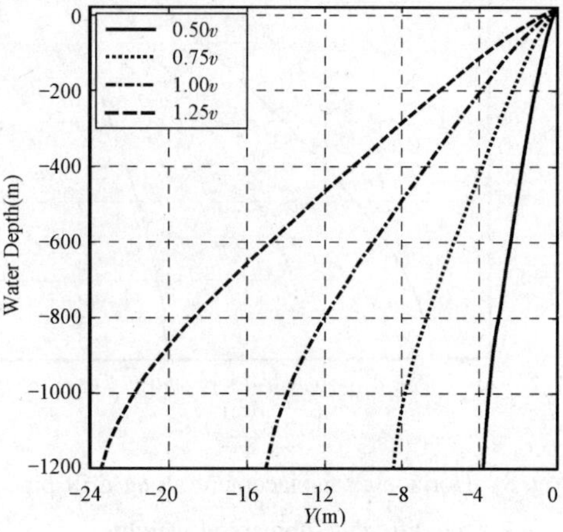

Fig. 11 Horizontal displacement along winch wire for different currents

Table 13 Comparison of $Y_{manifold}$ and T_0 in drill pipe installation for different currents

Water current (m/s)	Theoretical method		OrcaFlex simulation	
	$Y_{manifold}$ (m)	T_0 ($\times 10^5$ N)	$Y_{manifold}$ (m)	T_0 ($\times 10^5$ N)
0.50v	6.6149	7.2153	6.6329	7.1768
0.75v	14.8794	7.2168	14.9257	7.1766
1.00v	26.4498	7.2208	26.5334	7.1762
1.25v	41.3260	7.2291	41.4330	7.1753

Table 14 Comparison of $Y_{manifold}$ and T_0 in winch installation for different currents

Water current (m/s)	Theoretical method		OrcaFlex simulation	
	$Y_{manifold}$ (m)	T_0 ($\times 10^5$ N)	$Y_{manifold}$ (m)	T_0 ($\times 10^5$ N)
0.50v	3.7609	7.5354	3.7697	7.4908
0.75v	8.4600	7.5358	8.4472	7.4907
1.00v	15.0387	7.5370	15.0146	7.4905
1.25v	23.4970	7.5394	23.4588	7.4901

6.5 Effect of Wave

Previous studies showed that the wave has little influence on the horizontal displacement along drill pipe (or winch wire) in deep water, but the effect of wave is obvious in the splash zone. Four different waves are selected to analyze the wave's effect on manifold installation. As shown in Table 17 and Table 18, the bottom horizontal displacement changes little and the difference of the bottom horizontal displacement between the theoretical method and OrcaFlex simulation keeps within 1%. However, the maximum axial tension between the theoretical method and OrcaFlex simulation becomes larger with the larger wave height and smaller wave period. This is because the lifting force caused by wave makes the submerged weight of both subsea manifold and drill pipe (or winch wire) decrease. With larger wave height and smaller wave period, this influence will become more and more obvious.

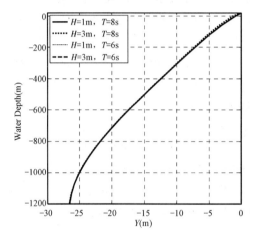

Fig. 12 Horizontal displacement along drill pipe for different waves

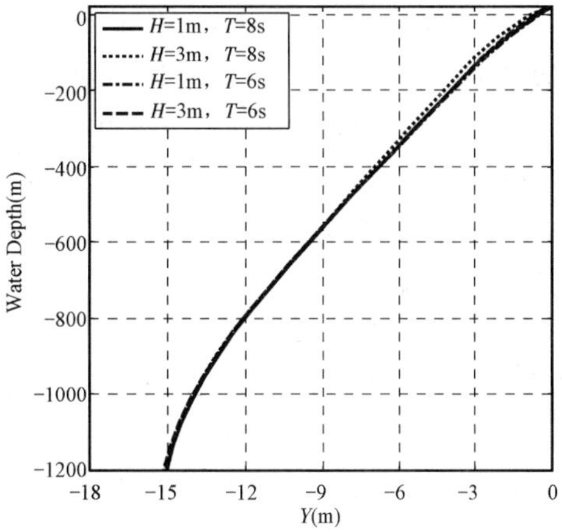

Fig. 13 Horizontal displacement along winch wire for different waves

Table 15 Comparison of $Y_{manifold}$ and T_0 in drill pipe installation for different waves

Height and Period	Theoretical method		OrcaFlex simulation	
	$Y_{manifold}$ (m)	T_0 ($\times 10^5$ N)	$Y_{manifold}$ (m)	T_0 ($\times 10^5$ N)
$H=1m, T=8s$	26.4498	7.2208	26.5334	7.1762
$H=3m, T=8s$	26.4701	7.2210	26.5273	7.0705
$H=1m, T=6s$	26.4485	7.2208	26.4839	7.1025
$H=3m, T=6s$	26.4582	7.2209	26.4173	6.8661

Table 16 Comparison of $Y_{manifold}$ and T_0 in winch installation for different water depths

Height and Period	Theoretical method		OrcaFlex simulation	
	$Y_{manifold}$ (m)	T_0 ($\times 10^5$ N)	$Y_{manifold}$ (m)	T_0 ($\times 10^5$ N)
$H=1m, T=8s$	15.0387	7.5370	15.0146	7.4905
$H=3m, T=8s$	15.0504	7.5370	15.0420	7.4012
$H=1m, T=6s$	15.0379	7.5370	14.9646	7.4288
$H=3m, T=6s$	15.0434	7.5370	14.9326	7.2297

CONCLUSIONS

In this paper, the analyses of drill pipe installation and winch installation of subsea manifold are presented: for drill pipe installation, due to the rigid body characteristics of drill pipe, modulus of elasticity is involved in the governing equations of drill pipe installation, and the high order matrix equation is applied to obtain the horizontal displacement and the axial tension along drill pipe; for winch installation, owing to the flexibility of winch wire, a simple theoretical model is developed to produce the results. Meanwhile, two finite element models are presented to analyze the behavior of drill pipe and winch wire during installation respectively. Results from the two solutions show good agreement, which demonstrates the validity of the theoretical methods. Therefore, the proposed theoretical methods can be used as easy, effective and time-saving methods to study the installation of subsea manifold compared to the finite element method, which will help develop the domestic software platform.

A series of analyses based on the theoretical methods and OrcaFlex simulation are conducted to calculate the horizontal displacement and the axial tension along drill pipe (or winch wire) with changes in several important factors. Based on the comparisons, the following conclusions seem justified.

(1) Modulus of elasticity has a little effect on the manifold installation. When lowered into deep water in drill pipe installation, the bending stiffness of drill pipe can be ignored.

(2) Drag coefficient and current velocity influence the drill pipe (or winch wire) behaviors significantly. While drag coefficient and current velocity are larger, the horizontal displacement and the axial tension are larger.

(3) Large submerged weight of subsea manifold can help decrease the horizontal displacement along drill pipe (or winch wire) and improve the stability of manifold installation, but it can also bring about higher axial tension.

(4) Wave has little influence on the horizontal displacement and axial tension while lowered into deep water. So the effect of wave can be disregarded.

The coincidences of the two solutions further verify the reliability of the theoretical methods. All the works in the paper are meaningful to the installation of subsea manifold. In addition, the theoretical methods of subsea manifold are suitable for some types of underwater installation.

REFERENCES

Bai Y, Bai Q. 2010. Subsea Engineering Handbook[M]. Gulf Professional Publishing:572 – 630.

Berntsen Y, Larsen CM. 2004. On the influence from hydrodynamic forces on speed and footprint for a falling riser [C]. Proc. 23rd International Conference on Offshore Mechanics and Arctic Engineering, OMAE E2004 – 51405.

Buchner B, Bunnik THJ, Honig D, Meskers G. 2003. A New Simulation Method for the Installation of Subsea Structures from the Splash Zone to the Ultra Deep[C]. DOT conference:19 – 21.

Chen S S, Sun B N, Feng Y Q. 2004. Nonlinear Transient Response of Stay Cable with Viscoelasticity Damper in Cable-stayed Bridge[M]. Mathematics and Mechanics,25(6).

Dalmaijer J W, Kuijpers M R L. 2003. Heave Compensation System for Deep Water Installation[R]. Technical report, Gusto Engineering.

Lin X J, Xiao W S, Wang H Y. 2011. Drill String Mechanical Analysis of Running Deepwater Oil Tree[J]. Journal of China University of Petroleum, 35(5).

Morrison D, Cermelli C. 2003. Deployment of Subsea Equipment in Ultradeep Water[C]. Proc. 22nd International Conference on Offshore Mechanics and Arctic Engineering, OMAE E2003 - 37190.

Roveri F E, Sagrilo L V S S, Lima E C P de, Cicilia F B. 2003. Comparing Measured and Calculated Forces of a Manifold Deployment in 940 Meters Water Depth[C]. Proc. 22nd International Conference on Offshore Mechanics and Arctic Engineering, OMAE 2003 - 37114.

Rowe S J, Mackenzie B, Snell R. 2001. Deepwater Installation of Subsea Hardware. Texas Section of the Society of Naval Architects and Marine Engineers.

Skiftesvik P, Sværen J. 2000. Application and Status of New Building Blocks for Subsea Boosting and Processing Systems [C]. 2000 Offshore Technology Conference, Paper 12016, Houston, May 1 - 4.

Tveitnes I, Lindaas O J, Gramnaes J. 2003. Sub Sea Structure Installation on Snohvit Field Using a New Construction Vessel[C]. Deep Offshore Technology Conference.

Yttervik R, Reinholdtsen S A, Larsen C M, Furnes G K. 2003. Marine Operations in Deep Water and a Variable Current Flow Environment[C]. Proc. 3rd Int. Conf. Hydroelasticity Mar.

PIPELINES & RISERS

A Parametric Study Method for J-lay Installation of Deepwater Pipelines

Wenjun Zhong[1], Qingyuan Zhang[2], Menglan Duan[2]

[1] Offshore Oil Engineering Co., Ltd., Tianjin, China
[2] Offshore Oil/Gas Research Center, China University of Petroleum, Beijing, China

Abstract The maximum bending moment or curvature in the neighborhood of the touchdown point(TDP) and the maximum tension at the top are two key quantities that need to be controlled during deepwater J-lay installation, it is essential for ensuring the safety of the pipe-laying operation and the normal operation of the pipeline to control these two quantities effectively. In this paper, the nonlinear governing differential equation during J-lay installation is solved using singular perturbation technique, the asymptotic expression of stiffened catenary is obtained in pipe-laying case, and the theoretical expression of its static geometric configuration as well as axial tension and bending moment has been derived. Validity of this method is verified by comparing with finite element results. The method is simple and fast, the result is reasonable and practicable, it is especially suitable for deepwater J-lay. Then, the influences of the seabed slope, unit weight, flexural stiffness, water depth and the pipe-laying tower angle on the maximum tension and moment of pipeline have been studied by this method, and the expected results with changing individual parameters have been presented.

Key Words Deepwater; Singular perturbation technique; J-lay; Geometric nonlinearity; Stiffened catenary

INTRODUCTION

The commonly used pipeline installation methods are S-lay and J-lay in deep water development (Dang Xuebo, Gong Shunfeng, Jin Weiliang, et al, 2010). As the water depth increases, the suspended length in the traditional S-lay method increases, the tension force imposed by the pipe laying vessel also increases, in the meanwhile, the required length of the stinger increases, and its configuration becomes more complex. To avoid these high demand, the J-lay technique is developed(Chul, 1993), it is the most suitable method for pipe laying in deep and ultra deep water projects.

Although the submarine pipeline is sustained from relatively complex dynamic loads (Tikhonov, Safronnov, Kamyshev, Figarov, 1995), the static analysis is still very important for the design of the entire pipeline, it not only determines the static tension and bending moment of the pipeline under its weight and the tension from the tensioner, but also provides the static equilibrium position for the dynamic analysis(Kouvaros, Sharma, Narayanan, 1994; Kamyshev, Ermolenko, Melnyk, 1992; Takashi Sakamoto, Roger E. Hobbs, 1995; Flavio Torres Lopez da Cruz, Carlos Alberto Nunes Dias, 1997; Suzana Rastelli Sattamini, Agustin J. Ferranti, 1993). Therefore, it not only has higher academic value, but also has very important application to study thoroughly the impact of various parameters on the submarine pipeline laying and control effectively on the key quantities (Seyed, Patel, 1991; Martins, Harada and Costa, Silva, 1999; Martins, Costa, Harada, Silva, 1999; Martins and Higashi, Silva, 2000).

1 MODEL OF J-LAY AND BASIC EQUATION

The schematic diagram of J-lay is shown in Fig. 1. Due to the difference of forces exerted on pipeline during installation, the pipeline is classified into three sections, i.e. flow line, suspended span and incline straight section. The flow line is the section which has been laid on the seabed, the suspended span is the segment from the TDP to the point where the pipeline leaves the framework, and the incline straight section is the segment contact with the framework.

Suppose the weight per unit length of the pipe in the water to be w, the flexural stiffness be EI, the origin of the coordinate system be the TDP. Taken an elementary segment from the deformed pipeline (Fig. 2), according to the equilibrium conditions of forces for the segment, we obtain (Tseng Huang, Qing Liang Kang, 1991; Pesce, Aranha, Martins and Pinto, 1995; Yeong-Soo Bae, Michael M. Bernitsa, 1995).

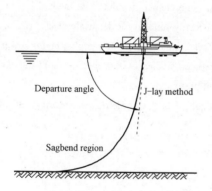

Fig. 1 Schematic of the J-lay installation

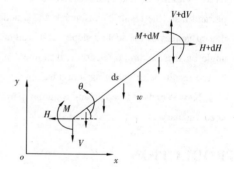

Fig. 2 Forces acting on the elementary segment of the pipeline

$$dH = 0 \tag{1}$$

$$\frac{dV}{dS} = w \tag{2}$$

$$\frac{dM}{dS} + V\cos\theta - H\sin\theta = 0 \tag{3}$$

From Eq. (1) and Eq. (2), we obtain:

$$H = H_0 \tag{4}$$

$$V = V_0 + wS \tag{5}$$

According to the large displacement beam theory, the physical equation is

$$M = EI \frac{d\theta}{dS} \tag{6}$$

Substituting Eq. (4) ~ Eq. (6) into Eq. (3), we have:

$$EI \frac{d^2\theta}{dS^2} + (V_0 + wS)\cos\theta - H_0 \sin\theta = 0 \tag{7}$$

Where, S is the arc length of the pipe, θ is the angle from the horizontal to the tangent at any point on the pipeline, H, V, M are the horizontal, vertical force and moment at the cross-section of the pipe respectively, H_0 and V_0 are the quantities at the origin of the coordinates.

Introducing the dimensionless quantities $s = S/L$, $\omega = wL/H_0$, $\varepsilon = EI/(H_0 L^2)$, in which L is the total suspended arc length of the pipeline. Own to:

$$Q\big|_{S=0} = \frac{dM}{dS}\bigg|_{S=0} = \frac{EI}{L^2}\frac{d^2\theta}{ds^2}\bigg|_{s=0} = H_0 \sin\theta\big|_{s=0} - V_0 \cos\theta\big|_{s=0} \tag{8}$$

Hence

$$\frac{V_0}{H_0} = \tan\theta\big|_{s=0} - \frac{\varepsilon \theta''}{\cos\theta}\bigg|_{s=0} = a + \varepsilon q \tag{9}$$

$$a = \tan\theta\big|_{s=0} = \tan\alpha_0 \quad q = -\frac{\theta''}{\cos\theta}\bigg|_{s=0} \tag{10}$$

Where, the prime stands for the derivation with respect to independent variable. Substituting Eq. (9) into Eq. (7), and expressed by dimensionless quantities, we get after rearrangement:

$$\varepsilon \frac{d^2\theta}{d^2 s} + (\omega s + a + \varepsilon q)\cos\theta - \sin\theta = 0 \tag{11}$$

Eq. (11) is the basic governing equation expressed by the dimensionless quantities and used to determine the static configuration of the pipeline. It is a nonlinear ordinary differential equation with second-order derivative term related with ε. For the normal J-lay problem, ε is a small quantity due to the larger H_0 and L, especially in the case of deep water and high tension, its value is far less than unity. Because the small quantity in the basic equation is related with the coefficient of the highest-order derivative, the solution is sought by the matching expansion in the singular perturbation techniques (Zeng Xiaohui, Liu Chuntu, Xing Jingzhong, 1986; Huang Yuying, Zhu Dashan, 1986; Geir Moe and BjΦrn Larsen, 1997). Given ε, the determination of the solution for Eq. (11) requires two boundary conditions. For the hard seabed, the seabed can be taken as rigid foundation. Because the framework of the J-lay tower is very stiff, it can be regarded as rigid support, therefore, the boundary conditions expressed by dimensionless quantities will be:

$$s = 0 : \frac{d\theta}{ds} = 0 \tag{12}$$

$$s = 1 : \frac{d\theta}{ds} = 0 \tag{13}$$

2 SOLUTION OF THE GOVERNING DIFFERENTIAL EQUATION

The solution by the matching expansion in singular perturbation techniques is composed of external expansion, inner expansion and the correction terms, the inner expansions are valid in the small region near the TDP and near where the pipe leaves the pipe-laying tower, while the external expansion is valid everywhere except the mentioned areas(A H Nayfeh,1990; Wang Yongzheng, 1994; F Guarracino, V Mallardo,1999).

2.1 External Expansion

As long as the function $\theta(s,\varepsilon)$ is analytic for ε, the θ can be expanded as a power series of ε:

$$\theta(s,\varepsilon) = \sum_{n=0}^{\infty} \varepsilon^{\frac{n}{2}} \theta_n(s) \tag{14}$$

q should also be developed as:

$$\left.\begin{array}{l} q = -\sum_{n=0}^{\infty} \varepsilon^{\frac{n}{2}} \dfrac{\theta''_n(0)}{\cos\alpha_0} = q_0 + \varepsilon^{\frac{1}{2}} q_1 + \varepsilon q_2 + \cdots \\[2mm] q_n = -\dfrac{\theta''_n(0)}{\cos\alpha_0} \quad n = 0,1,2,3,\cdots \end{array}\right\} \tag{15}$$

Substituting the expression (14) and (15) into Eq. (11), then equating the coefficients of like powers of ε to zero, we obtain the following set of recurrent equations for $\theta_n(s)$ ($n = 0, 1, 2, \cdots$).

Order ε^0:

$$(\omega s + a)\cos\theta_0 - \sin\theta_0 = 0 \tag{16}$$

Order $\varepsilon^{1/2}$:

$$[(\omega s + a)\sin\theta_0 + \cos\theta_0]\theta_1(s) = 0 \tag{17}$$

Order ε^1:

$$[(\omega s + a)\sin\theta_0 + \cos\theta_0]\theta_2(s) = \theta''_0 + q_0\cos\theta_0 \tag{18}$$

Order $\varepsilon^{3/2}$:

$$[(\omega s + a)\sin\theta_0 + \cos\theta_0]\theta_3(s) = q_1\cos\theta_0 \tag{19}$$

By a simple solution process and substitution, we obtain:

$$\theta_0(s) = \operatorname{arctg}(\omega s + a) \tag{20}$$

$$\theta_1(s) = 0 \tag{21}$$

$$\theta_2(s) = \frac{q_0}{1 + (\omega s + a)^2} - \frac{2\omega^2(\omega s + a)}{[1 + (\omega s + a)^2]^{\frac{5}{2}}} \tag{22}$$

$$\theta_3(s) = \frac{q_1}{1 + (\omega s + a)^2} \tag{23}$$

Substituting Eq. (20) ~ Eq. (23) into Eq. (14), the external expansion can be written as:

$$\theta^0(s,\varepsilon) = \text{arctg}(\omega s + a) + \varepsilon \left[\frac{q_0}{1 + (\omega s + a)^2} - \frac{2\omega^2(\omega s + a)}{[1 + (\omega s + a)^2]^{\frac{5}{2}}} \right]$$

$$+ \varepsilon^{\frac{3}{2}} \frac{q_1}{1 + (\omega s + a)^2} + o(\varepsilon^2) \tag{24}$$

Eq. (24) does not satisfy the boundary conditions Eq. (12) and Eq. (13), it can only be used as the approximate solution of ε^2 order for most region of the pipeline in the middle part.

2.2 Inner Expansion at $S = 0$

At this point, we must amplify the coordinate over the small region near $s = 0$, then expand by means of power series. Hence, we make an independent variable transformation first:

$$\xi = \frac{s}{\sqrt{\varepsilon}}, \frac{d\xi}{ds} = \frac{1}{\sqrt{\varepsilon}} \tag{25}$$

Then Eq. (11) becomes:

$$\frac{d^2\theta}{d^2\xi} + (\omega \sqrt{\varepsilon} \xi + a + \varepsilon q) \cos\theta - \sin\theta = 0 \tag{26}$$

Let

$$\theta(s,\varepsilon) = \phi(\xi,\varepsilon) + \text{arctg} a = \phi(\xi,\varepsilon) + \alpha_0 \tag{27}$$

Substituting Eq. (27) into Eq. (26), and expanding the trigonometric functions, we obtain the following equation after simplifying:

$$\frac{d^2\theta}{d^2\xi} + (\omega \sqrt{\varepsilon} \xi + \varepsilon q) \cos(\phi + \alpha_0) - \beta^2 \sin\phi = 0 \tag{28}$$

where

$$\beta = (1 + a^2)^{\frac{1}{4}} \tag{29}$$

At this point, we can seek the power series expansion of the differential Eq. (28), supposing that the inner expansion of $\theta(s, \varepsilon)$ can be written as:

$$\theta(s,\varepsilon) = \sum_{n=0}^{\infty} \varepsilon^{\frac{n}{2}} \phi_n(\xi) + \alpha_0 \tag{30}$$

Substituting Eq. (15) and Eq. (30) into Eq. (28), then equating the coefficients of like power of ε to zero, we can obtain the recurrent equations for $\phi_n(s)$ $(n = 0, 1, 2, \cdots)$.

Order ε^0:

$$\phi''_0 - \beta^2 \sin\phi_0 = 0 \tag{31}$$

Order $\varepsilon^{1/2}$:

$$\phi''_1 + \omega\xi\cos(\phi_0 + \alpha_0) - \beta^2\cos\phi_0 \cdot \phi_1(\xi) = 0 \tag{32}$$

Order ε^1:

$$\phi''_2 + q_0\cos(\phi_0 + \alpha_0) - \omega\xi\sin(\phi_0 + \alpha_0)\phi_1 - \beta^2\cos\phi_0 \cdot \phi_2 = 0 \tag{33}$$

Order $\varepsilon^{3/2}$:

$$\phi''_3 + \cos(\phi_0 + \alpha_0)\left[q_1 - \frac{\omega\xi}{2}\phi_1^2\right] - \sin(\phi_0 + \alpha_0)[q_0\phi_1 + \omega\xi\phi_2] - \beta^2\cos\phi_0\left[\phi_3 - \frac{\phi_1^3}{6}\right] = 0 \tag{34}$$

Substituting Eq. (30) into the boundary conditions Eq. (12), we obtain:

$$\left.\frac{d\theta}{ds}\right|_{s=0} = \frac{1}{\sqrt{\varepsilon}}\sum_{n=0}^{\infty}\varepsilon^{\frac{n}{2}}\left.\frac{d\phi_n(\xi)}{d\xi}\right|_{\xi=0} = 0 \tag{35}$$

Therefore, the boundary condition of Eq. (31) is written as:

$$\phi'_0(0) = 0 \tag{36}$$

Hence, the solution is:

$$\phi_0(\xi) = 0 \tag{37}$$

The solution satisfies the matching rule obviously.

Substituting Eq. (37) into Eq. (32), the general solution of the resulting differential equation is written as:

$$\phi_1(\xi) = A_1 e^{-\beta\xi} + A_2 e^{\beta\xi} + \frac{\omega\xi}{\beta^4} \tag{38}$$

The equation should satisfy the following boundary condition:

$$\phi'_1(0) = 0 \tag{39}$$

Hence

$$A_1 - A_2 = \frac{\omega}{\beta^5} \tag{40}$$

The other condition can be established by the matching rule between the inner and external expansion. According to Van Dyke's matching rule, i.e., m terms inner expansion within the n items external expansion is equal to the n terms external expansion within the m term inner expansion. The external expansion can be converted into inner variable ($m=2$).

$$\theta^0(s,\varepsilon) = \text{arctg}(\omega s + a) + o(\varepsilon) = \text{arctg}(\omega\sqrt{\varepsilon}\xi + a) + o(\varepsilon) \tag{41}$$

The equation can be expanded in the power form of $\sqrt{\varepsilon}$, we obtain:

$$(\theta^0)^i = \text{arctg}(a) + \frac{\omega\xi}{1+a^2}\sqrt{\varepsilon} + o(\varepsilon) = \alpha_0 + \frac{\omega\xi}{\beta^4}\sqrt{\varepsilon} + o(\varepsilon) \tag{42}$$

The inner expansion can be converted into external variable:

$$(\theta^i)^o = \alpha_0 + \sqrt{\varepsilon}(A_1 e^{\frac{-\beta s}{\sqrt{\varepsilon}}} + A_2 e^{\frac{\beta s}{\sqrt{\varepsilon}}} + \frac{\omega s}{\beta^4 \sqrt{\varepsilon}}) + o(\varepsilon) \quad (43)$$

The first term in the parenthesis approaches zero quickly as the independent variable closes to the scope of the external expansion, therefore:

$$(\theta^i)^o = \alpha_0 + \sqrt{\varepsilon}(A_2 e^{\frac{\beta s}{\sqrt{\varepsilon}}} + \frac{\omega s}{\beta^4 \sqrt{\varepsilon}}) + o(\varepsilon) \quad (44)$$

According to the matching rule:

$$(\theta^i)^o = (\theta^o)^i \quad (45)$$

We obtain:

$$A_2 = 0 \quad (46)$$

So

$$A_1 = \omega/\beta^5 \quad (47)$$

Hence

$$\phi_1(\xi) = \frac{\omega}{\beta^5}(e^{-\beta \xi} + \beta \xi) \quad (48)$$

Substituting Eq. (37) and Eq. (48) into Eq. (33), the general solution of the resulting equation is obtained as:

$$\phi_2(\xi) = A_3 e^{-\beta \xi} + A_4 e^{\beta \xi} + \frac{q_0}{\beta^4} - \frac{a\omega^2}{\beta^8}\xi^2 - \frac{2a\omega^2}{\beta^{10}} - \frac{a\omega^2}{4\beta^8}\xi(\xi + \frac{1}{\beta})e^{-\beta \xi} \quad (49)$$

By the boundary condition:

$$\phi'_2(0) = 0 \quad (50)$$

We obtain:

$$A_4 - A_3 = \frac{a\omega^2}{4\beta^{10}} \quad (51)$$

The external expansion can be converted into the inner variable ($m = n = 3$).

$$\theta^0(s, \varepsilon) = \operatorname{arctg}(\omega s + a) + \varepsilon \left[\frac{q_0}{1 + (\omega s + a)^2} - \frac{2\omega^2(\omega s + a)}{[1 + (\omega s + a)^2]^{\frac{5}{2}}} \right] + o(\varepsilon^{\frac{3}{2}})$$

$$= \operatorname{arctg}(\sqrt{\varepsilon}\omega\xi + a) + \varepsilon \left[\frac{q_0}{1 + (\sqrt{\varepsilon}\omega\xi + a)^2} - \frac{2\omega^2(\sqrt{\varepsilon}\omega\xi + a)}{[1 + (\sqrt{\varepsilon}\omega\xi + a)^2]^{\frac{5}{2}}} \right] + o(\varepsilon^{\frac{3}{2}})$$

$$(52)$$

The equation can be expanded in the power form of $\sqrt{\varepsilon}$, we obtain:

$$(\theta^0)^i = \alpha_0 + \frac{\omega \xi}{\beta^4}\sqrt{\varepsilon} + \varepsilon\left(\frac{q_0}{\beta^4} - \frac{\omega^2 a}{\beta^8}\xi^2 - \frac{2\omega^2 a}{\beta^{10}}\right) + o(\varepsilon^{\frac{3}{2}}) \quad (53)$$

The inner expansion can be transformed into external variable:

$$(\theta^i)^o = \alpha_0 + \sqrt{\varepsilon}\frac{\omega\xi}{\beta^4} + \varepsilon\left(A_4 e^{\beta\xi} + \frac{q_0}{\beta^4} - \frac{\omega^2 a}{\beta^8}\xi^2 - \frac{2\omega^2 a}{\beta^{10}}\right) + o(\varepsilon^{\frac{3}{2}}) \tag{54}$$

By the matching rule, we obtain:

$$A_4 = 0 \tag{55}$$

Hence

$$A_3 = -\frac{a\omega^2}{4\beta^{10}} \tag{56}$$

Therefore

$$\phi_2(\xi,\varepsilon) = -\frac{a\omega^2}{4\beta^8}\left(\xi^2 + \frac{\xi}{\beta} + \frac{1}{\beta^2}\right)e^{-\beta\xi} - \frac{a\omega^2}{\beta^8}\left(\xi^2 + \frac{2}{\beta^2}\right) + \frac{q_0}{\beta^4} \tag{57}$$

Substituting Eq. (37), Eq. (48) and Eq. (57) into Eq. (34), the general solution of the resulting differential equation is obtained as:

$$\phi_3(\xi) = A_5 e^{-\beta\xi} + A_6 e^{\beta\xi} + \frac{q_1}{\beta^4} + \left(\frac{8a^2\omega^3}{\beta^{14}} - \frac{2\omega^3}{\beta^{14}} - \frac{2aq_0\omega}{\beta^8}\right)\xi + \left(\frac{a^2\omega^3}{\beta^{12}} - \frac{\omega^3}{3\beta^{12}}\right)\xi^3$$

$$+ \left[\frac{a^2\omega^3\xi^4}{32\beta^{11}} + \left(\frac{5\omega^3}{48\beta^8} - \frac{3\omega^3}{16\beta^{12}}\right)\xi^3 + \left(\frac{7\omega^3}{32\beta^9} - \frac{11\omega^3}{32\beta^{13}}\right)\xi^2 - \left(\frac{aq_0\omega}{2\beta^8}\right)\right.$$

$$\left. - \frac{7\omega^3}{32\beta^{10}} + \frac{11\omega^3}{32\beta^{14}}\right)\xi\bigg]e^{-\beta\xi} - \frac{\omega^3}{48\beta^{15}}e^{-3\beta\xi} \tag{58}$$

According to the boundary condition:

$$\phi'_3(0) = 0 \tag{59}$$

We obtain:

$$A_5 - A_6 = -\frac{5aq_0\omega}{2\beta^9} + \frac{263\omega^3}{32\beta^{11}} - \frac{329\omega^3}{32\beta^{15}} \tag{60}$$

External expansion can be transformed into the internal variable ($m = n = 4$).

$$\theta^0(s,\varepsilon) = \text{arctg}(\omega s + a) + \varepsilon\left[\frac{q_0}{1 + (\omega s + a)^2} - \frac{2\omega^2(\omega s + a)}{[1 + (\omega s + a)^2]^{\frac{5}{2}}}\right] + \frac{\varepsilon^{3/2}q_1}{1 + (\omega s + a)^2} + o(\varepsilon^2)$$

$$= \text{arctg}(\sqrt{\varepsilon}\omega\xi + a) + \varepsilon\left[\frac{q_0}{1 + (\sqrt{\varepsilon}\omega\xi + a)^2} - \frac{2\omega^2(\sqrt{\varepsilon}\omega\xi + a)}{[1 + (\sqrt{\varepsilon}\omega\xi + a)^2]^{\frac{5}{2}}}\right]$$

$$+ \varepsilon^{\frac{3}{2}}\frac{q_1}{1 + (\sqrt{\varepsilon}\omega\xi + a)^2} + o(\varepsilon^2) \tag{61}$$

The equation can be expanded as the power form of $\sqrt{\varepsilon}$, we obtain:

$$(\theta^0)^i = \text{arctga} + \sqrt{\varepsilon}\frac{\omega\xi}{\beta^4} + \left(\frac{q_0}{\beta^4} - \frac{2\omega^2 a}{\beta^{10}} - \frac{a\omega^2\xi^2}{\beta^8}\right)\varepsilon + \frac{\omega^3\xi^3}{\beta^8}\left[1 - \frac{4}{3\beta^4}\right]\varepsilon^{\frac{3}{2}}$$

$$- \left(\frac{2q_0 a\omega}{\beta^8} - \frac{8\omega^3}{\beta^{10}} + \frac{10\omega^3}{\beta^{14}}\right)\xi\varepsilon^{\frac{3}{2}} + \varepsilon^{\frac{3}{2}}\frac{q_1}{\beta^4} + o(\varepsilon^2) \tag{62}$$

The inner expansion can be transformed into external variable:

$$(\theta^i)^o = \text{arctga} + \sqrt{\varepsilon}\frac{\omega\xi}{\beta^4} + \left(\frac{q_0}{\beta^4} - \frac{2\omega^2 a}{\beta^{10}} - \frac{a\omega^2\xi^2}{\beta^8}\right)\varepsilon + \varepsilon^{\frac{3}{2}}\left[A_6 e^{\beta\xi} + \frac{q_1}{\beta^4}\right.$$

$$\left. + \left(\frac{8a^2\omega^3}{\beta^{14}} - \frac{2\omega^3}{\beta^{14}} - \frac{2q_0 a\omega}{\beta^8}\right)\xi + \left(\frac{a^2\omega^3}{\beta^{12}} - \frac{\omega^3}{3\beta^{12}}\right)\xi^3\right] + o(\varepsilon^2) \tag{63}$$

By the matching rule, we obtain:

$$A_6 = 0 \tag{64}$$

Hence

$$A_5 = -\frac{5aq_0\omega}{2\beta^9} + \frac{263\omega^3}{32\beta^{11}} - \frac{329\omega^3}{32\beta^{15}} \tag{65}$$

Therefore

$$\phi_3(\xi,\varepsilon) = \frac{q_1}{\beta^4} + \left(\frac{8a^2\omega^3}{\beta^{14}} - \frac{2\omega^3}{\beta^{14}} - \frac{2aq_0\omega}{\beta^8}\right)\xi + \left(\frac{a^2\omega^3}{\beta^{12}} - \frac{\omega^3}{3\beta^{12}}\right)\xi^3 + \left[\frac{a^2\omega^3\xi^4}{32\beta^{11}}\right.$$

$$+ \left(\frac{5\omega^3}{48\beta^8} - \frac{3\omega^3}{16\beta^{12}}\right)\xi^3 + \left(\frac{7\omega^3}{32\beta^9} - \frac{11\omega^3}{32\beta^{13}}\right)\xi^2 - \left(\frac{aq_0\omega}{2\beta^8} - \frac{7\omega^3}{32\beta^{10}} + \frac{11\omega^3}{32\beta^{14}}\right)$$

$$\left. \times \xi - \left(\frac{5aq_0\omega}{2\beta^9} - \frac{263\omega^3}{32\beta^{11}} + \frac{329\omega^3}{32\beta^{15}}\right)\right]e^{-\beta\xi} - \frac{\omega^3}{48\beta^{15}}e^{-3\beta\xi} \tag{66}$$

Then substituting Eq. (37), Eq. (48), Eq. (57) and Eq. (66) into Eq. (67), we can get the inner solution in the boundary layer at $s = 0$.

$$\theta^i(\xi,\varepsilon) = \alpha_0 + \varepsilon^{\frac{1}{2}}\frac{\omega}{\beta^5}(e^{-\beta\xi} + \beta\xi) - \varepsilon\left[\frac{a\omega^2}{4\beta^8}(\xi^2 + \frac{\xi}{\beta} + \frac{1}{\beta^2})e^{-\beta\xi} + \frac{a\omega^2}{\beta^8}(\xi^2 + \frac{2}{\beta^2}) - \frac{q_0}{\beta^4}\right]$$

$$+ \varepsilon^{\frac{3}{2}}\left\{\frac{q_1}{\beta^4} + \left(\frac{8a^2\omega^3}{\beta^{14}} - \frac{2\omega^3}{\beta^{14}} - \frac{2aq_0\omega}{\beta^8}\right)\xi + \left(\frac{a^2\omega^3}{\beta^{12}} - \frac{\omega^3}{3\beta^{12}}\right)\xi^3\right.$$

$$+ \left[\frac{a^2\omega^3\xi^4}{32\beta^{11}} + \left(\frac{5\omega^3}{48\beta^8} - \frac{3\omega^3}{16\beta^{12}}\right)\xi^3 + \left(\frac{7\omega^3}{32\beta^9} - \frac{11\omega^3}{32\beta^{13}}\right)\xi^2 - \left(\frac{aq_0\omega}{2\beta^8} - \frac{7\omega^3}{32\beta^{10}} + \frac{11\omega^3}{32\beta^{14}}\right)\xi\right.$$

$$\left.\left. - \left(\frac{5aq_0\omega}{2\beta^9} - \frac{263\omega^3}{32\beta^{11}} + \frac{329\omega^3}{32\beta^{15}}\right)\right]e^{-\beta\xi} - \frac{\omega^3}{48\beta^{15}}e^{-3\beta\xi}\right\} + o(\varepsilon^2) \tag{67}$$

2.3 Inner Expansion at $S=1$

Like the practice of inner solution at $s = 0$, we make an amplification in the vicinity of $s = 1$.

$$\eta = \frac{1-s}{\sqrt{\varepsilon}}, \frac{d\eta}{ds} = -\frac{1}{\sqrt{\varepsilon}} \tag{68}$$

Then Eq. (11) becomes:

$$\frac{d^2\theta}{d^2\eta} + [\omega(1 - \sqrt{\varepsilon}\eta) + a + \varepsilon q]\cos\theta - \sin\theta = 0 \tag{69}$$

Let

$$\theta(s,\varepsilon) = \psi(\eta,\varepsilon) + \text{arctg}(\omega + a) = \psi(\eta,\varepsilon) + \alpha_1 \tag{70}$$

Substituting Eq. (70) into Eq. (69), and using the trigonometric relations, we can obtain the following formulation after simplifying:

$$\frac{d^2\psi}{d^2\eta} + (\varepsilon q - \omega\sqrt{\varepsilon}\eta)\cos(\psi + \alpha_1) - \alpha^2\sin\psi = 0 \tag{71}$$

Where

$$\alpha = [1 + (\omega + a)^2]^{\frac{1}{4}} \tag{72}$$

Let the solution of Eq. (71) be:

$$\theta(s,\varepsilon) = \sum_{n=0}^{\infty} \varepsilon^{\frac{n}{2}} \psi_n(\eta) + \alpha_1 \tag{73}$$

Substituting Eq. (15) and Eq. (73) into Eq. (71), we can also obtain the recurrent equations for $\psi n(\eta)$ after rearrangement:

Order ε^0:

$$\psi''_0 - \alpha^2 \sin\psi_0 = 0 \tag{74}$$

Order $\varepsilon^{1/2}$:

$$\psi''_1 - \omega\eta\cos(\psi_0 + \alpha_1) - \alpha^2\cos\psi_0 \cdot \psi_1(\eta) = 0 \tag{75}$$

Order ε^1:

$$\psi''_2 + q_0\cos(\psi_0 + \alpha_1) + \omega\eta\sin(\psi_0 + \alpha_1)\psi_1 - \alpha^2\cos\psi_0 \cdot \psi_2(\eta) = 0 \tag{76}$$

Order $\varepsilon^{3/2}$:

$$\psi''_3 + \cos(\psi_0 + \alpha_1)[q_1 + \frac{\omega\eta}{2}\psi_1^2] + \sin(\psi_0 + \alpha_1)[-q_0\psi_1 + \omega\eta\psi_2]$$

$$- \alpha^2\cos\psi_0[\psi_3 - \frac{\psi_1^3}{6}] = 0 \tag{77}$$

According to the boundary conditions and the matching conditions between the inner and the external solution, following a similar solution process as described above, we obtain the solution of

Eq. (74) ~ Eq. (77):

$$\psi_0(\eta) = 0 \tag{78}$$

$$\psi_1(\eta) = -\frac{\omega}{\alpha^5}(e^{-\alpha\eta} + \alpha\eta) \tag{79}$$

$$\psi_2(\eta,\varepsilon) = -\frac{\omega^2(\omega+a)}{4\alpha^8}\left(\eta^2 + \frac{\eta}{\alpha} + \frac{1}{\alpha^2}\right)e^{-\alpha\eta} - \frac{\omega^2(\omega+a)}{\alpha^8}\left(\eta^2 + \frac{2}{\alpha^2}\right) + \frac{q_0}{\alpha^4} \tag{80}$$

$$\psi_3(\eta,\varepsilon) = \frac{q_1}{\alpha^4} + \left[\frac{2q_0\omega(\omega+a)}{\alpha^8} - \frac{8\omega^3(\omega+a)^2}{\alpha^{14}} + \frac{2\omega^3}{\alpha^{14}}\right]\eta - \left[\frac{\omega^3(\omega+a)^2}{\alpha^{12}} - \frac{\omega^3}{3\alpha^{12}}\right]$$

$$\times \eta^3 - \left\{\frac{\omega^3(\omega+a)^2\eta^4}{32\alpha^{11}} + \left(\frac{5\omega^3}{48\alpha^8} - \frac{3\omega^3}{16\alpha^{12}}\right)\eta^3 + \left(\frac{7\omega^3}{32\alpha^9} - \frac{11\omega^3}{32\alpha^{13}}\right)\eta^2\right.$$

$$- \left[\frac{q_0\omega(\omega+a)}{2\alpha^8} - \frac{7\omega^3}{32\alpha^{10}} + \frac{11\omega^3}{32\alpha^{14}}\right]\eta - \left[\frac{5q_0\omega(\omega+a)}{2\alpha^9} - \frac{263\omega^3}{32\alpha^{11}} + \frac{329\omega^3}{32\alpha^{15}}\right]\bigg\}e^{-\alpha\eta}$$

$$+ \frac{\omega^3}{48\alpha^{15}}e^{-3\alpha\eta} \tag{81}$$

Substituting Eq. (78) ~ Eq. (81) into Eq. (73), we can get the inner solution of the boundary layer at $s = 1$:

$$\theta^i(\eta,\varepsilon) = \alpha_1 - \varepsilon^{\frac{1}{2}}\frac{\omega}{\alpha^5}(e^{-\alpha\eta} + \alpha\eta) - \varepsilon\bigg[\frac{\omega^2(\omega+a)}{4\alpha^8}\left(\eta^2 + \frac{\eta}{\alpha} + \frac{1}{\alpha^2}\right)e^{-\alpha\eta} + \frac{(\omega+a)}{\alpha^8}$$

$$\times \omega^2\left(\eta^2 + \frac{2}{\alpha^2}\right) - \frac{q_0}{\alpha^4}\bigg] + \varepsilon^{\frac{3}{2}}\bigg\{\frac{q_1}{\alpha^4} - \left[\frac{8\omega^3(\omega+a)^2}{\alpha^{14}} - \frac{2\omega^3}{\alpha^{14}} - \frac{2q_0\omega(\omega+a)}{\alpha^8}\right]\eta$$

$$- \left[\frac{\omega^3(\omega+a)^2}{\alpha^{12}} - \frac{\omega^3}{3\alpha^{12}}\right]\eta^3\bigg\} - \varepsilon^{\frac{3}{2}}\bigg\{\frac{\omega^3(\omega+a)^2\eta^4}{32\alpha^{11}} + \left(\frac{5\omega^3}{48\alpha^8} - \frac{3\omega^3}{16\alpha^{12}}\right)\eta^3$$

$$+ \left(\frac{7\omega^3}{32\alpha^9} - \frac{11\omega^3}{32\alpha^{13}}\right)\eta^2 - \left[\frac{q_0\omega(\omega+a)}{2\alpha^8} - \frac{7\omega^3}{32\alpha^{10}} + \frac{11\omega^3}{32\alpha^{14}}\right]\eta - \left[\frac{5q_0\omega(\omega+a)}{2\alpha^9}\right.$$

$$- \frac{263\omega^3}{32\alpha^{11}} + \frac{329\omega^3}{32\alpha^{15}}\bigg]\bigg\}e^{-\alpha\eta} + \varepsilon^{\frac{3}{2}}\frac{\omega^3}{48\alpha^{15}}e^{-3\alpha\eta} + o(\varepsilon^2) \tag{82}$$

2.4 Synthetic Solution

Eq. (24), Eq. (67) and Eq. (82) are three separate expansions. The first one $\theta_o(s, \varepsilon)$ is valid everywhere except in the small neighborhoods of $s=0$ and $s=1$; the second one $\theta_i(\xi, \varepsilon)$ is valid only in the small neighborhood of $s=0$; the third one $\theta_i(\eta, \varepsilon)$ is valid only in a small neighborhood of $s = 1$. Although $\theta_o(s, \varepsilon)$ has overlapping area with both $\theta_i(\xi, \varepsilon)$ and $\theta_i(\eta, \varepsilon)$, we combine all these three expressions into a composite expansion, to avoid the difficulty of switching from one expansion to another, and get the consistent effective solution throughout the region.

$$\theta^c(s,\varepsilon) = \theta^o(s,\varepsilon) + \theta^i(\xi,\varepsilon) + \theta^i(\eta,\varepsilon) - (\theta^o)_0^i - (\theta^o)_1^i \tag{83}$$

where the subscripts "0" and "1" in the last two terms denote the internal limits of the external solution in the neighborhoods of the boundary layer at $s = 0$ and $s = 1$, respectively. These two limits are:

$$(\theta^o)_0^i = \alpha_0 + \frac{\omega\xi}{\beta^4}\sqrt{\varepsilon} + \left(\frac{q_0}{\beta^4} - \frac{a\omega^2\xi^2}{\beta^8} - \frac{2a\omega^2}{\beta^{10}}\right)\varepsilon + \left[\frac{q_1}{\beta^4} + \left(\frac{8a^2\omega^3}{\beta^{14}} - \frac{2\omega^3}{\beta^{14}} - \frac{2aq_0\omega}{\beta^8}\right)\xi\right.$$

$$\left. + \left(\frac{a^2\omega^3}{\beta^{12}} - \frac{\omega^3}{3\beta^{12}}\right)\xi^3\right]\varepsilon^{\frac{3}{2}} + o(\varepsilon^2) \tag{84}$$

$$(\theta^o)_1^i = \alpha_1 - \frac{\omega\eta}{\alpha^4}\sqrt{\varepsilon} + \left[\frac{q_0}{\alpha^4} - \frac{\omega^2(\omega+a)\eta^2}{\alpha^8} - \frac{2\omega^2(\omega+a)}{\alpha^{10}}\right]\varepsilon + \left\{\frac{q_1}{\alpha^4} + \left[\frac{2q_0\omega}{\alpha^8} \times (\omega+a)\right.\right.$$

$$\left.\left. + \frac{2\omega^3}{\alpha^{14}} - \frac{8\omega^3(\omega+a)^2}{\alpha^{14}}\right]\eta - \left[\frac{\omega^3(\omega+a)^2}{\alpha^{12}} - \frac{\omega^3}{3\alpha^{12}}\eta^3\right]\right\}\varepsilon^{\frac{3}{2}} + o(\varepsilon^2) \tag{85}$$

Substituting Eq. (24), Eq. (67), Eq. (82), Eq. (84) and Eq. (85) into Eq. (83), we can get the approximate solution of Eq. (5) with the boundary conditions Eq. (12) and Eq. (13):

$$\theta^c(s,\varepsilon) = \text{arctg}(\omega s + a) + \varepsilon\left[\frac{q_0}{1+(\omega s+a)^2} - \frac{2\omega^2(\omega s+a)}{[1+(\omega s+a)^2]^{\frac{5}{2}}}\right] + \frac{q_1}{1+(\omega s+a)^2}$$

$$\times \varepsilon^{\frac{3}{2}} + \sqrt{\varepsilon}\frac{\omega}{\beta^5}e^{-\beta\xi} - \varepsilon\frac{a\omega^2}{4\beta^8}\left(\xi^2 + \frac{\xi}{\beta} + \frac{1}{\beta^2}\right)e^{-\beta\xi} + \varepsilon^{\frac{3}{2}}\left\{\left[\frac{a^2\omega^3\xi^4}{32\beta^{11}}\right.\right.$$

$$\left. + \left(\frac{5\omega^3}{48\beta^8} - \frac{3\omega^3}{16\beta^{12}}\right)\xi^3 + \left(\frac{7\omega^3}{32\beta^9} - \frac{11\omega^3}{32\beta^{13}}\right)\xi^2 - \left(\frac{aq_0\omega}{2\beta^8} - \frac{7\omega^3}{32\beta^{10}} + \frac{11\omega^3}{32\beta^{14}}\right)\xi\right.$$

$$\left. - \left(\frac{5aq_0\omega}{2\beta^9} - \frac{263\omega^3}{32\beta^{11}} + \frac{329\omega^3}{32\beta^{15}}\right)\right]e^{-\beta\xi} - \frac{\omega^3}{48\beta^{15}}e^{-3\beta\xi}\right\} - \sqrt{\varepsilon}\frac{\omega}{\alpha^5}e^{-\alpha\eta} - \varepsilon\frac{\omega^2(\omega+a)}{4\alpha^8}$$

$$\left(\eta^2 + \frac{\eta}{\alpha} + \frac{1}{\alpha^2}\right)e^{-\alpha\eta} - \varepsilon^{\frac{3}{2}}\left\{\left[\frac{\omega^3(\omega+a)^2\eta^4}{32\alpha^{11}} + \left(\frac{5\omega^3}{48\alpha^8} - \frac{3\omega^3}{16\alpha^{12}}\right)\eta^3 + \left(\frac{7\omega^3}{32\alpha^9} - \frac{11\omega^3}{32\alpha^{13}}\right)\right.\right.$$

$$\left.\left. \times \eta^2 - \left[\frac{q_0\omega(\omega+a)}{2\alpha^8} - \frac{7\omega^3}{32\alpha^{10}} + \frac{11\omega^3}{32\alpha^{14}}\right]\eta - \left[\frac{5q_0\omega(\omega+a)}{2\alpha^9} - \frac{263\omega^3}{32\alpha^{11}} + \frac{329\omega^3}{32\alpha^{15}}\right]\right]\right.$$

$$\left. \times e^{-\alpha\eta} + \varepsilon^{\frac{3}{2}}\frac{\omega^3}{48\alpha^{15}}e^{-3\alpha\eta} + o(\varepsilon^2) \tag{86}$$

It can be seen from the above asymptotic expansion of synthetic solution that the inner solutions within the two boundary layers are composed of the exponential decay terms, their quantities approach zero quickly as it closes to the external solution zone, so the external solution can be used as the basic solution of the pipe, and the inner solution is only added to describe the level of the dramatic changes in the function θ_c at both ends. Because two unknown H_0 and L are contained in Eq. (86), we need further determine the unknown for calculation. For this purpose, substituting Eq. (86) into Eq. (15), we get:

$$q_0 = -\frac{\theta''_0(0)}{\cos\alpha_0} = -\frac{\frac{d^2}{ds^2}[\arctg(\omega s + a)]_{s=0}}{\cos\alpha_0} = \frac{2\omega^2 a}{\beta^6} \tag{87}$$

$$q_1 = -\frac{\theta''_1(0)}{\cos\alpha_0} = -\frac{1}{\cos\alpha_0} \times \frac{\omega}{\beta^3 \varepsilon} = -\frac{\omega}{\beta\varepsilon} \tag{88}$$

$$q_2 = -\frac{\theta''_2(0)}{\cos\alpha_0} = \frac{\omega^2}{\beta^6}\left[\frac{2q_0(1-3a^2)}{\beta^4} - \frac{10\omega^2 a(3-4a^2)}{\beta^{10}} + \frac{a}{4\varepsilon}\right] \tag{89}$$

Suppose the launching angle of the pipe-laying tower to be β_T, and the water depth to be d. Due to

$$\beta_T = \theta^c(1) = \arctg(\omega + a) \tag{90}$$

so

$$\omega = \frac{wL}{H_0} = \tan\beta_T - a \tag{91}$$

From the view point of geometry:

$$\frac{dx}{dS} = \frac{dx}{Lds} = \cos\theta; \frac{dy}{dS} = \frac{dy}{Lds} = \sin\theta \tag{92}$$

hence

$$L\int_0^1 \sin\theta ds - \int_0^d dy = L\int_0^1 \sin\theta ds - d = 0 \tag{93}$$

Thus, the solution process is: substituting Eq. (87) ~ Eq. (89) and Eq. (91) into Eq. (86), then substituting $\theta_c(s, \varepsilon)$ into Eq. (93), seeking the root of the resulting equation, obtaining the suspended length L, and getting the H_0 by Eq. (91).

The geometrical configuration of the stiffened catenary is determined by:

$$\left.\begin{array}{l} x = L\int_0^s \cos\theta ds \\ \\ y = L\int_0^s \sin\theta ds \end{array}\right\} \tag{94}$$

The tension is determined by:

$$T = H_0\cos\theta + V\sin\theta = H_0[\cos\theta + (\omega s + a + \varepsilon q)\sin\theta] \tag{95}$$

The bending moment is expressed as:

$$M(s) = \frac{EI}{L}\frac{d\theta^c}{ds} = \frac{EI}{L}\left\{\frac{\omega}{1+(\omega s+a)^2} - 2\varepsilon\left[\frac{q_0\omega(\omega s+a)}{[1+(\omega s+a)^2]^2}\right] + \omega^3\frac{1-4(\omega s+a)^2}{[1+(\omega s+a)^2]^{\frac{7}{2}}}\right.$$

$$\left. - \varepsilon^{\frac{3}{2}}\frac{2q_1\omega(\omega s+a)}{[1+(\omega s+a)^2]^2} - \frac{\omega}{\beta^4}e^{-\beta\xi} - \sqrt{\varepsilon}\frac{a\omega^2}{4\beta^8}(\xi-\beta\xi^2)e^{-\beta\xi} - \varepsilon\left[\frac{a^2\omega^3\xi^4}{32\beta^{10}}\right]\right.$$

$$+\left(\frac{5\omega^3}{48\beta^7}-\frac{3\omega^3}{16\beta^{11}}-\frac{a^2\omega^3}{8\beta^{11}}\right)\xi^3-\left(\frac{3\omega^3}{32\beta^8}-\frac{7\omega^3}{32\beta^{12}}\right)\xi^2-\left(\frac{aq_0\omega}{2\beta^7}+\frac{7\omega^3}{32\beta^9}-\frac{11\omega^3}{32\beta^{13}}\right)\xi$$

$$-\left(\frac{2aq_0\omega}{\beta^8}-\frac{8\omega^3}{\beta^{10}}+\frac{159\omega^3}{16\beta^{14}}\right)\Big]e^{-\beta\xi}+\frac{\omega^3\varepsilon}{16\beta^{14}}e^{-3\beta\xi}-\frac{\omega}{\alpha^4}e^{-\alpha\eta}+\sqrt{\varepsilon}\frac{\omega^2(\omega+a)}{4\alpha^8}(\eta-\alpha\eta^2)e^{-\alpha\eta}$$

$$-\varepsilon\Bigg\{\frac{\omega^3(\omega+a)^2\eta^4}{32\alpha^{10}}+\Bigg[\frac{5\omega^3}{48\alpha^7}-\frac{3\omega^3}{16\alpha^{11}}-\frac{\omega^3(\omega+a)^2}{8\alpha^{11}}\Bigg]\eta^3-\left(\frac{3\omega^3}{32\alpha^8}-\frac{7\omega^3}{32\alpha^{12}}\right)\eta^2$$

$$-\Bigg[\frac{q_0\omega}{2\alpha^7}\times(\omega+a)+\frac{7\omega^3}{32\alpha^9}-\frac{11\omega^3}{32\alpha^{13}}\Bigg]\eta-\Bigg[\frac{2q_0\omega(\omega+a)}{\alpha^8}-\frac{8\omega^3}{\alpha^{10}}+\frac{159\omega^3}{16\alpha^{14}}\Bigg]\Bigg\}e^{-\alpha\eta}$$

$$+\frac{\omega^3\varepsilon}{16\alpha^{14}}e^{-3\alpha\eta}+o(\varepsilon^2)\Bigg\} \tag{96}$$

Eq. (95) and Eq. (96) are the tension and bending moment expressions, which are functions of seabed slope, unit weight, flexural stiffness, water depth and the pipe-laying tower angle, from which, we can determine the maximum tension and bending moment, from their derivations with respect to individual parameters, we can obtain their corresponding gradients. Hence, we can change the parameters to control the maximum quantities within permitted scopes. The method is simple and convenient.

3 NUMERICAL VERIFICATION

In order to verify the validity of this method, suppose the flexural stiffness of the pipe to be $156,967,868.8\text{N}\cdot\text{m}^2$, unit weight to be 764.8N/m, seabed slope to be $0°$, water depth to be $1,500\text{m}$, pipe-laying tower angle to be $83.9725°$, the computed results are shown in Fig. 3 ~ Fig. 5. For comparison, the computed results by OCRAFLEX software as well as standard catenary for the same parameters are also shown in the Figures.

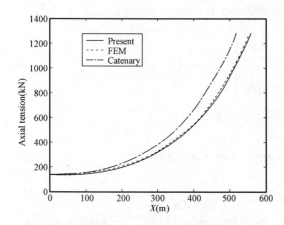

Fig. 3 Axial tension curves by different methods

It is shown in Fig. 3 that results of the present method are in good agree with FEM, their axial tension curves are almost coincident, the relative error of the maximum tension is 0.23%, and both results are lower than that of the standard catenary.

It is shown in Fig. 4 that moment curves of the present result and the FEM are nearly coincident, but their maximum moments are somewhat different, the relative error of them is 3.19%.

The geometrical configuration of the pipe is shown in Fig. 5, it is shown that present solution is in good agree with FEM, and the two curves are practically coincident, both results are lower than that of standard catenary.

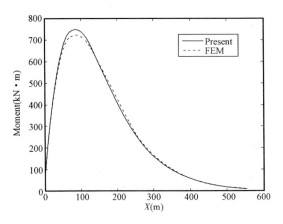

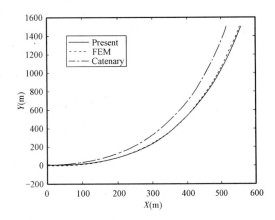

Fig. 4 Bending moment curves by different methods

Fig. 5 Geometric configuration of reinforced catenary

It can be seen by comparison that the results of present method agree well with FEM, the result is reasonable and the validity of the method is verified.

4 APPLICATION AND CONCLUSIONS

Since the maximum tension and bending moment of the pipeline are the key quantities to be controlled during pipe-laying, it is of great significance to study the influence of various parameters on them. So we suppose the unchanging parameters the same as those in Numerical verification, and the changing parameters are indicated in the figures. The computed results are shown in Fig. 6 ~ Fig. 15.

4.1 Effects of the Seabed Slope on the Tension and Bending Moment

It is shown in Fig. 6 that as the seabed slope increases, the distance from the TDP to the pipe laying tower decreases and the axial tension increases with the slope except in the small region near the TDP, where the slope has a lower influence on the axial tension.

It is shown in Fig. 7 that as the slope increases, the distance between the point of the maximum bending moment and the TDP decreases, and the maximum moment is also reduced; the bending moment in the neighborhood of the TDP increases with the slope; for a slope, the bending moment monotonically increases with the distance from the TDP first, as it exceeds the maximum, it decreases gradually, and approaches zero eventually.

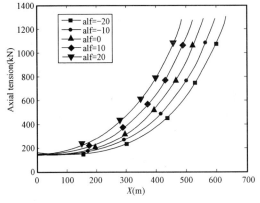

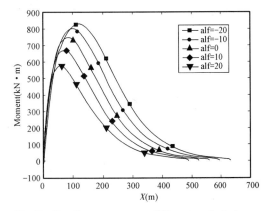

Fig. 6 Axial tension curves at different seabed slopes

Fig. 7 Bending moment at different seabed slopes

4.2 Effects of the Unit Weight on the Tension and Bending Moment

It is shown in Fig. 8 that as the unit weight increases, the distance from the pipe-laying tower to the TDP decreases slightly, and the axial tension increases gradually.

It is shown in Fig. 9 that the impact of the unit weight on the bending moment of the pipe is not significant; for a unit weight, the bending moment increases with the distance from the TDP first, as it reaches the maximum, it begins to decrease, until it approaches zero.

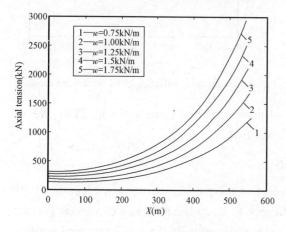
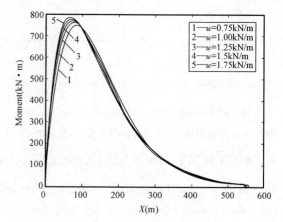

Fig. 8 Axial tension curves at different unit weight   Fig. 9 Bending moment curves at different unit weight

4.3 Effects of the Flexural Rigidity on the Tension and Bending Moment

It is shown in Fig. 10 that as the flexural stiffness increases, the distance from the pipe laying tower to the TDP increases a little, and the axial tension reduces a little, but it does not change significantly in the larger extent near the vicinity of the TDP.

It is shown in Fig. 11 that as the flexural stiffness increases, the bending moment of the pipe increases, and it will bring in significant influence only if the flexural stiffness is larger.

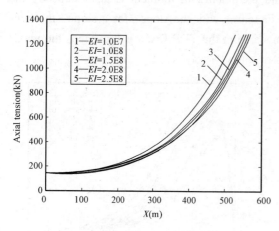
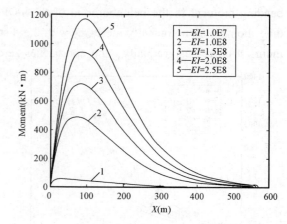

Fig. 10 Axial tension curves at different bending stiffness   Fig. 11 Bending moment curves at different stiffness

4.4 Effects of Water Depth on the Tension and Bending Moment

It is shown in Fig. 12 that as the water depth increases, the distance from the pipe laying towers to the TDP increases significantly, and the axial tension in the vicinity of the TDP also increases. For a water depth, the axial tension increases with the distance from the TDP. For different depth, the maximum axial tension is obviously different, the deeper the water, the greater the maximum axial tension.

It is shown in Fig. 13 that as the water depth increases, the distance from the location of the maximum moment to the TDP decreases slightly, but the maximum bending moment reduces significantly, and the moment curves become increasingly flat. For a water depth, the moment increases with the distance from the TDP first, as it exceeds the maximum, it decreases gradually, finally it approaches zero.

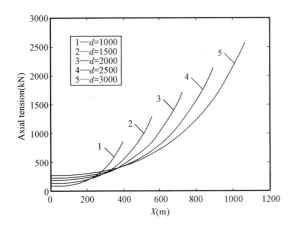

Fig. 12 Axial tension curves at different water depth

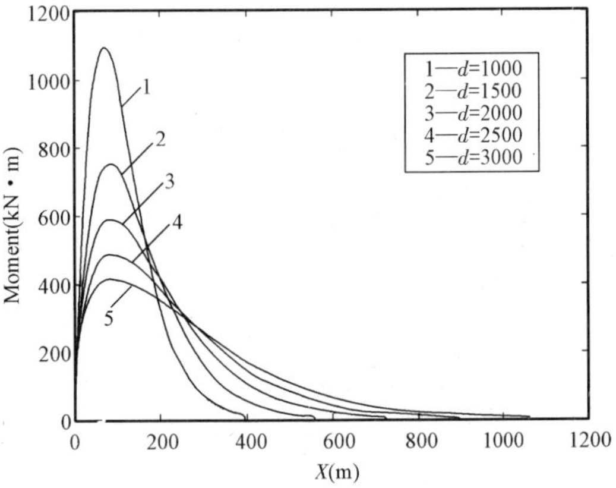

Fig. 13 Bending moment curves at different water depth

4.5 Effects of the Tower Angle on the Tension and Bending Moment

It is shown in Fig. 14 that as the pipe laying tower angle increases, the distance from the pipe-laying tower to the TDP reduces significantly, the greater the pipe laying tower angle, the more dramatic change in the axial tension, but the axial tension changes relatively flat in the vicinity of the TDP, and it decreases with the increase of the tower angle in this region.

It is shown in Fig. 15 that as the pipe laying tower angle increases, the maximum bending moment location away from the TDP increases slightly, but the maximum bending moment increases significantly, and the bending moment curve becomes more pointed; For a water depth, the bending moment increases with the distance from the TDP first, as it exceeds the maximum, it gradually decreases until it approaches zero.

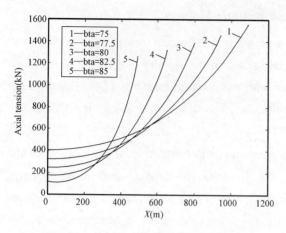

Fig. 14 Axial tension curves at different J-lay angles

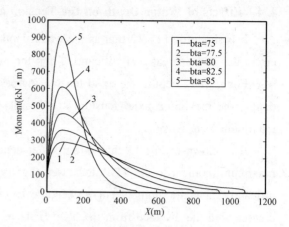

Fig. 15 Bending moment curves at different J-lay angles

CONCLUSIONS

Based on above mentioned analyses, we can conclude:

(1) Changing the slope of the seabed, the maximum tension changes little, while increasing the slope of the seabed, the maximum bending moment or curvature is reduced significantly.

(2) Reducing the unit weight, the maximum tension is significantly decreased, while changing the unit weight, the maximum bending moment or curvature changes little.

(3) Changing the flexural stiffness, the maximum tension changes little, while reducing the flexural stiffness, the maximum moment or curvature is significantly reduced.

(4) Reducing the water depth, the maximum tension is significantly reduced, while increasing the water depth, the maximum moment or curvature is significantly reduced.

(5) Increasing the pipe laying tower angle, the maximum tension is significantly reduced, while decreasing the pipe laying tower angle, the maximum moment or curvature is significantly reduced.

ACKNOWLEDGEMENTS

This paper is financially supported by the National Science and Technology Major Projects of China (Grant No. 2011ZX05027 - 002), the authors would like to express our sincere thanks.

REFERENCES

Chul H Jo. 1993. Limitation and Comparison of S-lay and J-lay Methods[C]. Proceedings of the 3rd International Offshore & Polar Engineering Conference:201 - 206.

Dang Xuebo, Gong Shunfeng, Jin Weiliang, et al. 2010. Research Progress on Submarine Pipeline Laying Technology[C]. China Offshore Platform, 25(5):5 - 10. (in Chinese)

Flavio Torres Lopez da Cruz, Carlos Alberto Nunes Dias. 1997. Structural Analysis of Flexible Pipe Using Finite Element Method[C]. Proceedings of the 7th International Offshore & Polar Engineering Conference:137 - 143.

Geir Moe, Bjørn Larsen. 1997. Dynamics of Deep Water Marine Risers-Asymptotic Solutions[C]. Proceedings of the 7th International Offshore & Polar Engineering Conference: 123 - 130.

Guarracino F, Mallard V 1999. A Refined Analytical Analysis of Submerged Pipelines in Seabed Laying[M]. Applied Ocean Research, 21(1):281 – 293.

Huang Yuying, Zhu Dashan. 1986. Static Analysis of Submarine Pipelines During Installation [J]. The Ocean Engineering, 4(1):32 – 46.

Kamyshev M A, Ermolenko A I and Melnyk L V. 1992. Stress Analysis of Submarine Pipeline During Installation From a Lay-barge[C]. Proceedings of the 2nd International Offshore & Polar Engineering Conference:60 – 67.

Kouvaros D, Sharma C B and Narayanan R. 1994. The Static Response of Submarine Pipelines Under Laying Conditions[C]. Proceedings of the 4th International Offshore & Polar Engineering Conference: 41 – 47.

Martins C A, Costa A B and Harada C A N, Silva R M C. 1999. Parametric Analysis of Steel Catenary Riser: Fatigue Behavior Near the Touchdown Point[C]. Proceedings of the 9th International Offshore & Polar Engineering Conference, 314 – 319.

Martins C A, Harada C A N and Costa A B, Silva R M C. 1999. Parametric Analysis of Steel Catenary Riser Under Extreme Loads [C]. Proceedings of the 9th International Offshore & Polar Engineering Conference: 309 – 313.

Martins C A and Higashi E, Silva R M C. 2000. A Parametric Analysis of Steel Catenary Risers: Fatigue Behavior Near the Top[C]. Proceedings of the 10th International Offshore & Polar Engineering Conference: 54 – 59.

Nayfeh A H 1990. Introduction to Perturbation Techniques[M]. Shanghai:Shanghai Translation & Publishing Co: 304 – 373.

Pesce C P, Aranha J A P, Martins C A and Pinto M O. 1995. Steel Catenary Risers for Deep Water Application [C]. Proceedings of the 5th International Offshore & Polar Engineering Conference:190 – 202.

Seyed F B, Patel M H. 1991. Parametric Studies of Flexible Risers[C]. Proceedings of the 1st International Offshore & Polar Engineering Conference:147 – 156.

Suzana Rastelli Sattamini, Agustin J Ferranti. 1993. Computational Analysis of Deep Water Risers During Installation and Hang off[C]. Proceedings of the 3rd International Offshore & Polar Engineering Conference:301 – 310.

Takashi Sakamoto, Roger E Hobbs. 1995. Nonlinear Static and Dynamic Analysis of Three Dimensional Flexible Risers[C]. Proceedings of the 5th International Offshore & Polar Engineering Conference: 227 – 235.

Tikhonov V, Safronnov A, Kamyshev M, Figarov N. 1995. Numerical Analysis of Pipeline Dynamics in Seabed Laying[C]. Proceedings of the 5th International Offshore & Polar Engineering Conference:53 – 62.

Tseng Huang, Qing Liang Kang. 1991. Three Dimensional Analysis of a Marine Riser With Large Displacements [C]. Proceedings of the 1st International Offshore & Polar Engineering Conference:170 – 177.

Wang Yongzheng. 1994. Foundations to perturbation Technique[M]. Beijing:the Kexue Press:120 – 162.

YeongSoo Bae, Michael M Bernitsa. 1995. Importance of Nonlinearities in Static and Dynamic Analysis of Marine Risers[C]. Proceedings of the 5th International Offshore & Polar Engineering Conference:209 – 218.

Zeng Xiaohui, Liu Chuntu, Xing Jingzhong. 1986. Mechanical Analysis of Pipeline[J]. Mechanics in Engineering, 24(2):19 – 21.

A Theoretical Analysis Method for Sandwich Pipes Under Combined Internal-External Pressure and Thermal Load

Changsheng Xiang[1], Qin Wang[2], Menglan Duan[2]

[1] School of Petroleum Engineering, China University of Petroleum, Qingdao, China
[2] Offshore Oil/Gas Research Center, China University of Petroleum, Beijing, China

Abstract Sandwich pipe is a submarine pipeline for transporting oil and gas resources which can be used in deep and ultra-deep water. As simultaneously sustaining the external pressure which is enormous especially in the deep and ultra-deep water and internal high temperature pressure (HTHP), the pipeline will produce enormous thermal stress when it is in production. The huge internal stresses will lead to buckling that will result in destructive damage to the pipeline both in partial failure and global instability. This paper presents an analytical formulation of the instability of a sandwich pipe system considered for high-temperature products in the oil and gas industry. We analyze the stresses and strains of the pipeline under combined internal-external pressure and thermal load in theory. Through the analysis, a theoretical formula analysis method is reduced which costs effectively and timely and more accurate in the result compared to the analysis from a nonlinear finite element (NLFE).

Key Words Sandwich pipe; Thermal stress; Deepwater; Elasticity solution

Introduction

Nowadays, the global petroleum companies have transferred their attentions to the exploration of deep and ultra-deep water, especially in the North Sea, Gulf of Mexico and subsalt fields off which were discovered recently. Concurrently, the Chinese government and enterprises pay attention to the exploration and development of the South China Sea oil and gas resources. All of these companies need a lot of offshore oil and gas equipments applied to oil exploitation and transportation in deep and ultra-deep water. To the transportation of oil, the pipe should provide enough strength to support external severe stress and well insulation to ensure oil normal transportation. Instead of conventional insulating pipe, new concepts of submarine pipelines and risers have been proposed. Generally speaking, there are two kinds of pipelines which can be applied in the deepwater and ultra-deep water environment, pipe-in-pipe (PIP) and sandwich pipe (SP). Current high-temperature projects widely employ pipe-in-pipe (PIP) systems, which can generally be divided into two categories, compliant and non-complaint systems. Both of the two types of system consist of two concentrically mounted steel pipes and the annular space filled with either circulating hot water or thermal insulated materials. The objective of this type of pipe is to increase the thermal insulation capacity to prevent formatting paraffin and hydrates. The apparent advantages of PIP system are the possibility of using materials with excellent thermal properties, and simultaneously the structural integrity is independently provided by the outer and inner steel layers, Grealish and Roddy.

However, ultra deepwater scenarios, at depths beyond 1,500 meters, require very thick-walled steel pipelines or pipe-in-pipe systems, which are expensive and difficult to install because of their excessive weight. Owing to those disadvantages, a new concept was developed by Segen F. Estefen of the Federal University of Rio de Janeiro and two of his colleagues in the Ocean Engineering Department, Theodoro A. Netto and Ilson P. Pasqualino. This kind of pipe would mitigate some of those disadvantages which composed of two concentric steel pipes separated by and bonded to a polymeric annulus. The design provides a combination of high structural strength with thermal insulation.

The research object of this study is SP with greater structural strength which can ensure the flow assurance. The usual sandwich structures are a particular kind of multi-layered of different materials bonded together, and the single layer property is benefited to the global structure performance. Usually, the sandwich structure contains three layers: two external thin and stiff layers to give enough strength and a central thick and flexible core to ensure the insulation. The external layers are bonded to the core with adhesives to allow the load transfer between the components. Numerical and experimental studies have been carried out by Borselino et al., they analyzed three different woven fibers including class fiber, carbon fiber and Kevlar fiber by performing the flatwised compression test, the edgewise compression test and three point bending test. Special behavior of unidirectional sandwich panels with transversely flexible core has been studied by Sokolinsky et al., where they analyzed the special features response of ordinary and delaminated sandwich panels with a transversely flexible core subjected to external in-plane and vertical statical loads. And their analytical formulation was based on a higher-order theory for sandwich panels with non-rigid bond layers between the face sheets and the core. They consider that their numerical study reveals that the ordinary (fully bonded) sandwich panel behaves as a compound structure in which the local/localized, overall or interactive forms of the response can take place and depend on the geometry, mechanical properties, and boundary conditions of the structure. Although there are a lot of advantages of composite construction, such as the reduction in total life costs, corrosion resistance, high strength and stiffness-to-weight ratios, and improved stealth for military applications, the expenditure for manufacturing this kind of pipe is very expensive now. New higher quality materials with lower costs and new fabrication methods need to be developed before composite materials will be fully accepted for the construction of large ships. Mouring have done this work and his study summarized experimental results of the testing of composite panels stiffened with preformed frames under in-plane uniaxial compressive loads.

Estefen and Netto made a series of small-scale laboratory tests to research the performance of sandwich pipes and developed a finite element model to evaluate the structural performance of such sandwich pipes with two different options of core material. They considered that sandwich pipe systems with either cement or polypropylene cores were feasible options for ultra deepwater applications. And Estefen's work was the basis of this paper. Nevertheless, the paper by Estefen did not give the theoretical analysis method for the structural stability of sandwich pipe while the pipe is concurrently standing external and internal pressure and thermal load. Thermal buckling of high-temperature submarine pipelines is complex and is difficult to control. Besides, it is difficult to maintain the pipe once the pipe appears collapse in deep and ultra-deep water. Therefore, before producing sandwich pipe, the research theoretically and experimentally centering on sandwich pipe is indeed necessary. This paper presents an analytical formulation of the instability of a sandwich pipe system considered for high-temperature products in the oil and gas industry. Through solving the equations of heat transfer and stress equations, the temperature field and stress can be accurately reduced that is the first critical step in the reduction of global analysis of sandwich pipe. Further, this paper also provides the stresses in formula which are verified using a nonlinear finite element (NLFE) model. The accurate prediction of the stresses demands plentiful resources and time through utilizing the advanced nonlinear finite element analysis tools such as ABAQUS. Compared to ABAQUS modeling, the proposed formulations is advantageous and direct if only need a high level analysis in the earlier detailed design stage. It is a pity that due to the complexity and periodicity of the inner pipe temperature range from to production temperature, the processing method in NLFET is different to the theory method in this paper. And it is a difficulty in processing this problem accurately in theory which is not solved in this paper.

1 ANALYSIS PROCEDURE

1.1 Heat Conduction Equation

The key problem in the analysis of multi-layer structure is obtaining the stresses and strains of every layer whose result will be used in the design of multi-layer structure. In the multi-layer structure, due to the coefficient of thermal conductivity of every layer is different, the temperature distribution is dissimilitude. Besides, as the thermal expansion coefficient is also different, thermal stress will be produced in the multi-layer structure. The temperature variation of the structure is big so enough to produce large thermal stress.

The general form of the governing equation of heat conduction in the multi-layer structure as shown in Fig. 1 using cylindrical coordinates is:

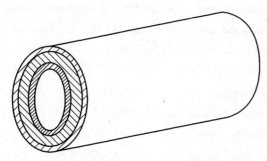

Fig. 1 The sketch plan of the sandwich pipe

$$\frac{\partial^2 T_i}{\partial r^2} + \frac{1}{r}\frac{\partial T_i}{\partial r} + \frac{1}{r^2}\frac{\partial^2 T_i}{\partial \theta^2} + \frac{\partial^2 T_i}{\partial z^2} + \frac{\dot{q}}{k_i} = \frac{1}{\alpha_i}\frac{\partial T_i}{\partial t} \tag{1}$$

Where $\dot{q}$ is the rate of internal energy generation, k_i is the coefficient of heat conduction and α_i is the thermal diffusivity, $i = 1,2,3\cdots,N$.

Consider a pipe of inner and outer radial r_0 and r_a, respectively. Due to long length and axial symmetry and steady-state condition, the temperature distribution in the pipe is a function of radius only, when no heat is generated. The differential equation of heat flow (1) is reduced to:

$$\frac{\partial^2 T_i}{\partial r^2} + \frac{1}{r}\frac{\partial T_i}{\partial r} = 0 \tag{2}$$

Integrating Eq. (2) yields, and to the multi-layer structure:

$$T_i = A_i + B_i \ln r \tag{3}$$

The boundary condition is as below. Both the outer and inner surface are exposed to force convection at fluid temperature T_{os} and the temperature of hot fluid at T_{is} respectively.

$$-k_i \frac{\partial T_i}{\partial r} = \overline{h}_0 (T - T_{is}) \quad r = r_0, i = 1 \tag{4a}$$

$$-k_i \frac{\partial T_i}{\partial r} = \overline{h}_a (T - T_{os}) \quad r = r_a, i = N \tag{4b}$$

$$-k_i \frac{\partial T_i}{\partial r} = -k_{i+1} \frac{\partial T_{i+1}}{\partial r} \tag{5}$$

Where, $\overline{h}_0$ $\overline{h}_a$ are the inner and outer average convective heat transfer coefficients, respectively. Using the above mentioned boundary conditions, we can obtain the constant A_i and B_i.

1.2 Stress and Strain Analysis

Consider a sandwich pipe subjected to axial symmetric thermal and mechanical load and internal-external pressure as shown in Fig. 1. in this work, the inner and outer pipes are all carbon steel pipes, the intermediary layer is polyethylene pipe.

The sandwich pipe has been presented in the cylindrical coordinate system for analysis and modeling purposes, where the cylinder coordinates are: the radial r, the circumferential θ, and the axial coordinates of the cylinder z. Because the thermal load, internal pressure and external pressure are all axial symmetric load. Besides the geometric of sandwich pipe is axial symmetric, the stresses and strains are independent of θ i.e. $\left(\frac{\partial}{\partial \theta}=0\right)$. In addition, the radial and axial displacements depend only on the radial and axial coordinates, respectively. With the above assumption, the field of displacement can be expressed as:

$$u_r = u_r(r) \quad u_\theta = u_\theta(r,z) \quad u_z = u_z(z) \tag{6}$$

Where u_r, u_θ and u_z are radial, hoop and axial displacements, respectively.

Using the cylindrical coordinate system shown in Fig. 2, the stress and strain transformation of the orthotropic material is given by:

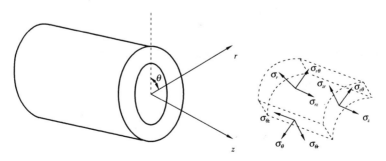

Fig. 2 The sketch map of SP in coordinate system

$$\begin{bmatrix} \sigma_z \\ \sigma_\theta \\ \sigma_r \\ \sigma_{\theta r} \\ \sigma_{zr} \\ \sigma_{z\theta} \end{bmatrix}^{(k)} = \begin{bmatrix} C_{11} & C_{12} & C_{13} & 0 & 0 & C_{16} \\ C_{12} & C_{22} & C_{23} & 0 & 0 & C_{26} \\ C_{13} & C_{23} & C_{33} & 0 & 0 & C_{36} \\ 0 & 0 & 0 & C_{44} & C_{45} & 0 \\ 0 & 0 & 0 & C_{45} & C_{55} & 0 \\ C_{16} & C_{26} & C_{36} & 0 & 0 & C_{66} \end{bmatrix}^{(k)} \begin{bmatrix} \varepsilon_z - \alpha_z \Delta T \\ \varepsilon_\theta - \alpha_\theta \Delta T \\ \varepsilon_r - \alpha_r \Delta T \\ \gamma_{\theta r} \\ \gamma_{zr} \\ \gamma_{z\theta} \end{bmatrix}^{(k)} \quad (7a)$$

This can be written as:

$$\sigma = C\varepsilon \quad (7b)$$

Where C_{ij}^k are the stiffness constants and α_i^k the thermal expansion coefficients. ΔT is the difference in temperature given by:

$$\Delta T = T_i - T_{ref} \quad (8)$$

Where T_{ref} is the reference temperature and T_i is the temperature distribution from Eq. (3).
For the cylinder coordinates, the strain-displacement relations can be described as:

$$\begin{cases} \varepsilon_r^{(k)} = \dfrac{\partial u_r^{(k)}}{\partial r} \\[6pt] \varepsilon_\theta^{(k)} = \dfrac{1}{r}\dfrac{\partial u_\theta^{(k)}}{\partial \theta} + \dfrac{u_r^{(k)}}{r} \\[6pt] \varepsilon_z^{(k)} = \dfrac{\partial u_z^{(k)}}{\partial z} \\[6pt] \gamma_{\theta r}^{(k)} = \dfrac{1}{r}\dfrac{\partial u_r^{(k)}}{\partial \theta} + r\dfrac{\partial}{\partial r}\left(\dfrac{u_\theta^{(k)}}{r}\right) \\[6pt] \gamma_{z\theta}^{(k)} = \dfrac{\partial u_\theta^{(k)}}{\partial z} + \dfrac{1}{r}\dfrac{\partial u_z^{(k)}}{\partial \theta} \\[6pt] \gamma_{zr}^{(k)} = \dfrac{\partial u_z^{(k)}}{\partial r} + \dfrac{\partial u_r^{(k)}}{\partial z} \end{cases} \quad (9)$$

The equilibrium equations in cylindrical coordinates are:

$$\frac{\partial \sigma_r^{(k)}}{\partial r} + \frac{1}{r}\frac{\partial \sigma_{\theta r}^{(k)}}{\partial \theta} + \frac{\partial \sigma_{zr}^{(k)}}{\partial z} + \frac{\sigma_r^{(k)} - \sigma_\theta^{(k)}}{r} = 0 \quad (10a)$$

$$\frac{\partial \sigma_{\theta r}^{(k)}}{\partial r} + \frac{1}{r}\frac{\partial \sigma_\theta^{(k)}}{\partial \theta} + \frac{\partial \sigma_{z\theta}^{(k)}}{\partial z} + \frac{2\sigma_{\theta r}^{(k)}}{r} = 0 \quad (10b)$$

$$\frac{\partial \sigma_{zr}^{(k)}}{\partial r} + \frac{1}{r}\frac{\partial \sigma_{z\theta}^{(k)}}{\partial \theta} + \frac{\partial \sigma_z^{(k)}}{\partial z} + \frac{\sigma_{zr}^{(k)}}{r} = 0 \quad (10c)$$

Substituting Eq. (6) into Eq. (9) and Eq. (10) yields the following simplified forms:
Because the load and the geometry of pipe are axial symmetric, the strain-displacements are rewritten as:

$$\varepsilon_r^{(k)} = \frac{du_r^{(k)}}{dr}$$

$$\varepsilon_\theta^{(k)} = \frac{u_r^{(k)}}{r} \quad (11)$$

$$\varepsilon_z^{(k)} = \frac{du_z^{(k)}}{dz} = 0$$

$$\gamma_{\theta r}^{(k)} = 0 \quad \gamma_{z\theta}^{(k)} = 0 \quad \gamma_{zr}^{(k)} = 0$$

Therefore, the equilibrium equation for the current axial symmetric problem can be expressed as:

$$\begin{cases} \dfrac{d\sigma_r^{(k)}}{dr} + \dfrac{\sigma_r^{(k)} - \sigma_\theta^{(k)}}{r} = 0 \\[2mm] \dfrac{d\sigma_{\theta r}^{(k)}}{dr} + \dfrac{2\sigma_{\theta r}^{(k)}}{r} = 0 \\[2mm] \dfrac{d\sigma_{zr}^{(k)}}{dr} + \dfrac{\sigma_{zr}^{(k)}}{r} = 0 \end{cases} \quad (12)$$

1.3 Three-dimension Laminated-plate Properties

According to the Eq. (7) and the structure features, we can define the sandwich pipe define the three-dimensional sandwich pipe material properties and the material modulus matrix elements C_{ij}, the stress-strain relation including the thermal effects of the compliance matrix is:

$$\begin{bmatrix} \varepsilon_z - \alpha\Delta T \\ \varepsilon_\theta - \alpha\Delta T \\ \varepsilon_r - \alpha\Delta T \\ \gamma_{\theta r} \\ \gamma_{zr} \\ \gamma_{z\theta} \end{bmatrix}^{(1,2,3)} = \begin{bmatrix} S_{11} & S_{12} & S_{13} & 0 & 0 & 0 \\ S_{21} & S_{22} & S_{13} & 0 & 0 & 0 \\ S_{31} & S_{32} & S_{33} & 0 & 0 & 0 \\ 0 & 0 & 0 & S_{44} & 0 & 0 \\ 0 & 0 & 0 & 0 & S_{55} & 0 \\ 0 & 0 & 0 & 0 & 0 & S_{66} \end{bmatrix}^{(1,2,3)} \begin{bmatrix} \sigma_z \\ \sigma_\theta \\ \sigma_r \\ \sigma_{\theta r} \\ \sigma_{zr} \\ \sigma_{z\theta} \end{bmatrix}^{(1,2,3)} \quad (13)$$

This can be written as:

$$\varepsilon = S\sigma$$

Where

$$S = C^{-1}$$

Where the matrix component values can be calculated from the engineering constants, defined by:

$$\begin{cases} S_{11} = \dfrac{1}{E_x} \\ S_{22} = \dfrac{1}{E_y} \\ S_{33} = \dfrac{1}{E_z} \\ S_{12} = \dfrac{-\mu_{xy}}{E_x} \\ S_{13} = \dfrac{-\mu_{xz}}{E_x} \\ S_{23} = \dfrac{-\mu_{yz}}{E_y} \\ S_{44} = \dfrac{1}{G_{yz}} \\ S_{55} = \dfrac{1}{G_{yz}} \\ S_{66} = \dfrac{1}{G_{xy}} \end{cases} \tag{14}$$

For the isotropic material, including the steel and the polypropylene, both of them are isotropic, every side is elastic symmetry plane, every direction is elastic symmetry axis, so we can obtain:

$$E_x = E_y = E_z \quad \mu_x = \mu_y = \mu_z \quad G_{xy} = G_{yz} = G_{xz} \tag{15}$$

And

$$G = \dfrac{E}{2(1+\mu)}$$

For the steel pipe and the polypropylene pipe, The E equals to E_1 and E_2, respectively, the G equals to G_1 and G_2, respectively, and the μ equals to μ_1 and μ_2, respectively.

1.4 The Final Displacement Equation

Substituting the expressions for the stress of Eq. (7b) into Eq. (12), and using Eq. (11), we obtain the following differential equation:

$$\dfrac{d^2 u_r^{(k)}}{dr^2} + \dfrac{1}{r}\dfrac{du_r^{(k)}}{dr} - \dfrac{1}{r^2} u_r^{(k)} = \dfrac{1+\mu^{(k)}}{1-\mu^{(k)}} \dfrac{\alpha^{(k)} B}{r} \tag{16}$$

Integrating Eq. (13) yields:

$$u_r^{(k)} = C^{(k)} r + C^{(k+1)} \dfrac{1}{r} + \dfrac{(1+\mu^{(k)})\alpha^{(k)} B}{2r(1-\mu^{(k)})}\left(r^2 \ln\dfrac{r}{a} - \dfrac{r^2}{2} + \dfrac{a^2}{2}\right) \tag{17}$$

Where $C^{(k)}, C^{(k)}$ are unknown constants of integration, and have to be determined from the boundary conditions and the contact conditions at each interface between layers. B is the heat conduction constant.

1.5 Boundary Conditions

Boundary conditions are imposed by geometric conditions of the structure and the conditions of loading. It is assumed that there are no slips in the interfaces and that there is continuity in stresses and displacements. These boundary conditions allow determining the integration constants.

The traction condition at the inner surface and the outer surface are written as:

$$\left. \begin{array}{l} \sigma_r^{(1)}(r_0) = -p_{is} \\ \sigma_r^{(n)}(r_a) = -p_{es} \end{array} \right\} \tag{18}$$

$$\left. \begin{array}{l} \sigma_{\theta r}^{(1)}(r_0) = \sigma_{zr}^{(1)}(r_0) = 0 \\ \sigma_{\theta r}^{(n)}(r_a) = \sigma_{zr}^{(n)}(r_a) = 0 \end{array} \right\} \tag{19}$$

Continuity conditions for the radial displacements and stresses in the interfaces lead to:

$$\left. \begin{array}{l} u_r^{(k)}(r_k) = u_r^{(k+1)}(r_k) \\ \sigma_r^{(k)}(r_k) = \sigma_r^{(k+1)}(r_k) \\ \sigma_{zr}^{(k)}(r_k) = \sigma_{zr}^{(k+1)}(r_k) \\ \sigma_{\theta r}^{(k)}(r_k) = \sigma_{\theta r}^{(k+1)}(r_k) \end{array} \right\} \tag{20}$$

2 RESULTS AND DISCUSSION

2.1 Basic Date about Property of the Sandwich Pipe

Based on the above theoretical analysis, we analysis a real pipeline, the basic data is shown in table 1.

Table 1 Geometrical physical properties of the sandwich pipe

Inner pipe (flowline)	219.1mm OD × WT
Outer pipe (carrier pipe)	339.7mm OD × WT
Material	API − X65
Insulation system	Polypropylene WT
Design pressure (MPa)	41.4
Heat transfer coefficient (oil-inner pipe) [W/(m² · K)]	1594.6
Heat transfer coefficient (water-outer pipe) [W/(m² · K)]	454.2608
Water depth range (m)	1300 ~ 1800
Steel density (kg/m³)	7850
Polypropylene density (kg/m³)	893
Thermal conductivity of steel [W/(m · K)]	55.6
Thermal conductivity of Polypropylene [W/(m · K)]	0.0415
Modulus of elasticity (steel)	2.1×10^{11}
Modulus of elasticity (Polypropylene)	5.06×10^9
Poisson ratio (steel)	0.3
Poisson ratio (Polypropylene)	0.4

2.2 Temperature Field of the Sandwich Pipe

During the design stage, both the thermal insulation properties and the structural strength of the pipe are very important for the deepwater submarine pipelines. The insulation material in this example is polystyrene (PS) which is a both effective and economical material at present for its heat conduction coefficient is 0.0415. The following Fig. 3 presents the temperature field of the SP when simultaneously supporting internal and external pressures and terminal load caused by great range of temperature in which the blue line stands for the temperature field reduced by the governing heat conduction equation in the multi-layer structure and the date of red dots is from model analysis utilizing the advanced nonlinear finite element analysis tools ABAQUS and two of them fit very well.

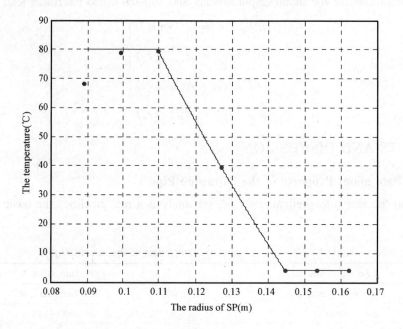

Fig. 3 The temperature field of sandwich pipe

2.3 Thermal Loads and the Internal-External Analysis

2.3.1 The Axial Stress of Sandwich Pipe

Based on the above analysis, we can obtain the axial force and radius stress in theoretically as the following blue line showing (Fig. 4 and Fig. 5). Compared to the result of analysis from ABAQUS, the theoretical solution of the axial stress and radius stress can mainly fit the result from ABAQUS. Nevertheless, due to the complexity and periodicity of the inner pipe temperature range from 4℃ to production temperature, the processing method in NLFET is different to the theory method in this paper. And it is a difficulty in processing this problem accurately in theory which is not solved in this paper.

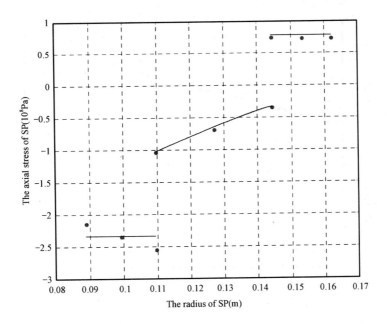

Fig. 4　The axial stress of SP

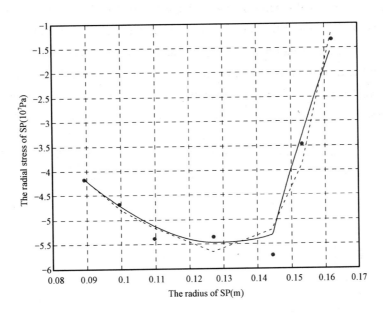

Fig. 5　The radial stress of SP

2.3.2　The Ratio between the Axial Force and the Critical Force for Under Variational Temperature of the Inside Oil Temperature

In Fig. 6, we analyze the ratio between the axial force and critical force of the three layers which is processing under variation temperature of the oil thermal loads. We can make out when the temperature is bigger than , the ratio between the axial force and critical force is bigger than 4.6, so we consider that the middle layer can be buckling easily. That's because, the stiffness of the layer is small comparatively.

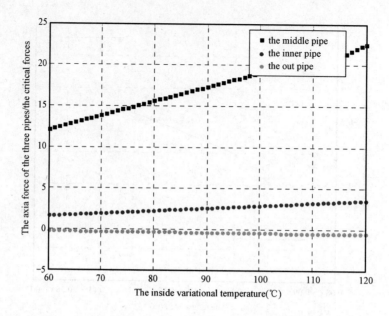

Fig. 6   The three layers axial force for variational temperature inside

2.3.3   THe Ratio between the Axial Force and the Critical Force for Under Variation Depth

In Fig. 7, we analyze the ratio between the axial force and critical force of the three layers which are processed under variation depth of the water. We can make out that the ratio between the axial force and critical force is very great for the insulated layer in the comparison of the inner layer and the outer layer which is negative for the radius stress is opposite direction to the inner and middle layers.

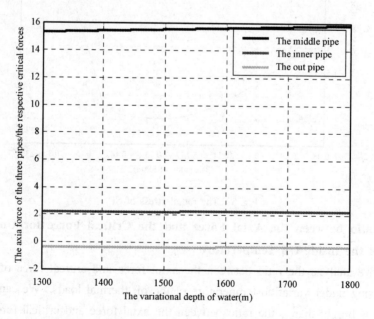

Fig. 7   The three layers axial force for variational depth of water

2.3.4 The Ratio between the Axial Force and the Critical Force for Under Change of the Length

According to Fig 8, we can obtain the ratio is increasing with the increase of the length which is a square relation. It is obvious that the length of the SP is critical for its design in preventing the bulking of SP. Concurrently, the influence of the length of SP to the outer pipe is bigger than inner pipe which can be attained from the following figure. When the length of suspended span pipe is shorter than 3.3m, the inner pipe will not bucking according to the stability theory. Compared to the inner pipe, the outer pipe is more susceptible to the depth than the inner pipe which is visualized by Fig. 8 and Fig. 9, and the outer pipe is secure if the suspended span is shorter than 9m.

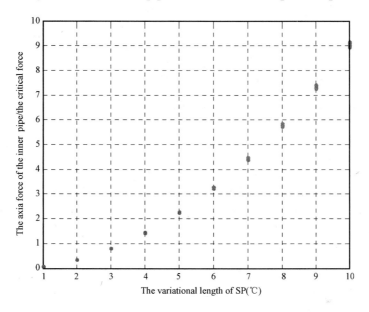

Fig. 8　The inner layer force for variational length of SP

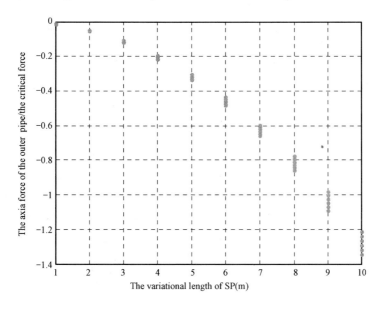

Fig. 9　The outer axial force for variational length of SP

CONCLUSIONS

In this work, we analyze the sandwich pipe under combined internal-external pressure and thermal mechanical loading with thermal variations using theory of elastic mechanics and mechanical of materials, and then we obtain the analytic solution of the radial displacement and the axial force and the redial-axial stresses, The main conclusions are outlined as below:

(1) Through the comparison of Fig. 4 and Fig. 5, the theoretical solution reduced in this paper and the numerical solution by ABAQUS fit very well, in spite of the analysis of inner pipe is a disadvantage that because of the complexity of temperature rise in the beginning of the production. Overall, the theoretical analysis method of sandwich pipe under combined inter-external pressure and thermal load can guide the design of the sandwich pipe accurately and timely.

(2) The middle layer can be buckling easily as the stiffness of the layer is small comparatively, so designers should take measures in designing integral structure especially for the connectors between layers.

(3) For the outer pipe, We can make out that the ratio between the axial force and critical force is very small, and the ratio is negative value, that's because the inner and middle pipe have been buckling, they press the outer pipe which lead to the outer pipe is effect under tensile force, instead of compressive force.

(4) The thickness of the insulating material can not effect the buckling of the inner layer, that's to say, for the inner layer. The thickness of the insulating material is not important to analysis the buckling of the sandwich pipe.

As the depth of water increasing, the outer pipe is most susceptible and the length of the SP is critical for its design in preventing the bulking of SP especially for the outer pipe when the material and size of the outer pipe have been determined.

ACKNOWLEDGEMENTS

This paper is financially supported by the National Natural Science Foundation of China (No. 50979113) and the National Basic Research Program of China (No. 2011CB013702).

REFERENCES

Borselino C, Calabrese L, Valenza A. 2004. Experimental and Numerical Evaluation of Sandwich Composite Structures[J]. Journal of Composites Science and Technology,64:1709 – 15.

Castello X, Estefen S F. 2007. Limit Strength and Reeling Effects of Sandwich Pipes with Bonded Layers[J]. International Journal of Mechanical Sciences ,49 :577 – 588.

Estefen S F, Netto T A, Pasqualino I P. 2005. Strength Analyses of Sandwich Pipes for Ultra Deepwaters[J]. Journal of Applied Mechanics ,72:599 – 608.

Grealish F, Roddy I. 2002. State-of-the-art on Deep Water Thermal Insulation Systems[C]. In: 21st International Conference on Offshore Mechanics and Arctic Engineering, Proceedings of OMAE'02, Oslo,Norway.

Mouring S E. 1999. Buckling and Post Buckling of Composite Ship Panelsstiffened with Preform Frames[J]. Journal of Ocean Engineering ,26:793 – 803.

Sokolinsky V S, Frostig Y, Nutt Y. 2002. Special Behavior of Unidirectional Sandwich Panels with Transversely Flexible Core Under Statical Loading[J]. International Journal of Non-Linear Mechanics ,37:869 – 95.

An Experimental Study of Dynamic Response of a Top Tensioned Riser Model Subject to VIV

Jijun Gu[1], Marcelo Vitola[1], Carlos Levi[1], Jairo Coelho[2], Waldir Pinto[2], Menglan Duan[3]

[1] Ocean Engineering Program, COPPE, Federal University of Rio de Janeiro, Rio de Janeiro, Brazil
[2] School of Engineering, Federal University of Rio Grande, Campus Carreiros, Rio Grande, RS, 96201 - 900, Brazil
[3] Offshore Oil/Gas Research Center, China University of Petroleum, Beijing, China

Abstract Dynamic response of a vertical flexible cylinder vibrating at low mode numbers with combined $x-y$ motion is investigated in this paper. The uniform flow was simulated by towing the flexible cylinder along the tank in still water; therefore the turbulence intensity of the free flow is negligible in order to obtain more reliable results. The dominant frequencies, maximum attainable amplitude, modal analysis and $x-y$ trajectory in cross-flow and in-line directions have been reported and compared with literatures, some good agreements and justifiable discrepancies are obtained. These results could benefit the future experimental and numerical work in this area.

Key Words Model test; Towing tank; Flexible cylinder; Vortex-induced vibration; Integral transform

INTRODUCTION

In deep water, marine risers subject to strong ocean currents, suffer from multi-mode vortex-induced vibrations (VIV), when vortex shedding interacts with the structural properties of the riser, resulting in large amplitude vibrations in both in-line (IL) and cross-flow (CF) directions. When the vortex shedding frequency approaches the natural frequency of the marine riser, the cylinder takes control on the shedding process causing the vortices to be shed at a frequency near to the natural frequency. This phenomenon is called vortex shedding "lock in" or synchronization. Under the "lock in" conditions, large resonant oscillations reduce the fatigue life significantly.

During the last 20 years, its inherent nature has been largely studied by the laboratory model tests using a spring-mounted rigid cylinder in CF motion only or in both IL and CF motion, but some insights are still elusive. Reviews from Sarpkaya (2004) and Williamson and Govardhan (2004) summarized recent fundamental results and discoveries with focus on the importance of mass and damping, the concept of critical mass, the attainable amplitude, and so on. However, the offshore challenge has been extended to the deep water where the risers suffer multi-mode dynamic response.

To understand the multi-mode VIV, numerical simulation and model tests of long flexible risers or pipelines have been conducted under more realistic conditions, where the current flow along the model may vary both in magnitude and direction, thus the structure can respond in several modes in both CF and IL directions. Xie et al. (2011) used finite-volume method to simulate dynamic response of a flexible cylinder at a constant Reynolds number of 1000, Xu et al. (2010) presented a new wake oscillator model and studied dynamics of high aspect ratio(L/D) riser under VIV. As for the experiments, several long flexible cylinder have been performed under towing tank or water channel with pumping system, see Chaplin et al. (2005); Trim et al. (2005); Lie and Kaasen(2006); Huera-Huarte and Bearman(2009a,b). Other tests have been conducted out-door, such as in a fjord Huse et al. (1998) or at sea Vandiver et al. (2006). The CF and IL amplitudes, trajectories and phase synchronization, lift and drag coefficients, dominant frequencies and modal amplitude are reported. As an extended work, the VIV of a tandem and side-by-side arrangement of two flexible cylinders have been studied by Huera-Huarte and Gharib(2011); Huera-Huarte and Bearman(2011). The results showed some interesting findings when they considered different centre-to-centre separations.

Present experiment study is aiming at evaluating the dynamic response of a pinned-pinned flexible cylinder in a towing tank. The uniform flow was simulated by towing the flexible cylinder along the tank in still water, therefore the turbulence intensity of the free flow is negligible in order to be as free as possible from the effects of upstream turbulence, and the results should be more reliable. The maximum vibration amplitude, the resonant frequency and mode shape are reported here to benefit the future experimental and numerical work in this area.

1 EXPERIMENTAL DETAILS

The experiments were carried out in a towing tank which has a test section 0.75m wide, 0.72m high and 16m long, at School of Engineering, Federal University of Rio Grande(FURG), Rio Grande do Sul, Brazil. The uniform flow was simulated by towing the cylinder along the tank in still water; the waiting interval was approximately 10 minutes between runs to avoid the influence of turbulence. The current velocity ranges from 0.10m/s to 0.98m/s, with approximate increments of 0.04m/s. This range of speeds corresponds to Reynolds numbers is from 1950 to 19110.

The sketch of the experimental set-up is shown in Fig. 1. The flexible cylinder had an outer diameter of 19.5mm and was constructed of plastic tube with a 6.4mm steel core as shown in Fig. 1(a). Several nylon pieces were used to clamp the steel core and support the outer plastic tube as shown in Fig. 1(b). The instrumented flexible cylinder was pinned at both ends with two universal joints so the flexible cylinder could oscillate in both IL direction and CF direction. The effective cylinder length was 1.45m, with 0.7m below water surface. The water coverage was 48.3% and therefore drag and lift forces could not be exerted on the full length of the cylinder. However, it has been verified by Chaplin et al(2005) that when a water coverage is higher than a threshold of 30%, the excited modes in the response will not change. Additionally, a similar set-up has been used by Huera-Huarte and Bearman(2009a,b); Huera-Huarte and Gharib (2011); Huera-Huarte and Bearman(2011).

Strain gages were mounted on the flexible cylinder to measure the responses at five locations (identified as S1,S2,S3,S4 and S5), with the relative positions 0.84, 0.67, 0.50, 0.33, and 0.16 in length from the bottom end, respectively. At each location, four strain gages were placed as

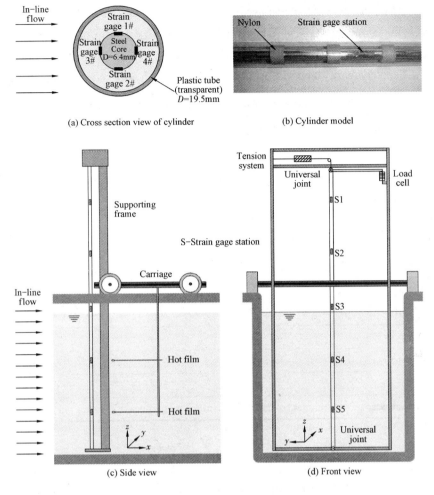

Fig. 1 Sketch of the experiment set-up

shown in Fig. 1(a). The two strain gages in the x-direction were used to measure the IL vibrations, while the other two gages in the y-direction were used to measure the CF vibrations. The signals from strain gages were calibrated with a position sensor to guarantee the accuracy of the response. Two hot films were placed horizontally behind S4 and S5 strain gage stations to measure the wake oscillating frequency. The fundamental natural frequency of the flexible cylinder with variable top tension was measured from decay tests in still water as shown in Table 1. Structural damping ratio was calculated from decay test in the air to be 3.18%.

Table 1 The fundamental natural frequency

Top tension $T(N)$	Symbol	f_n (Hz)
19.6	□	4.5
29.4	○	4.83
49.0	△	5.33
68.6	+	5.67
88.2	*	6.0
107.8	×	6.3

The summary of the experiments showed in Table 2. The mass ratio is defined as $m^* = m/m_a$, m denotes the cylinder model mass, m_a denotes added mass; aspect ratio is defined as L/D; reduced velocity is defined as $u_r = v/f_n D$, v denotes the flow velocity; Reynolds number is defined as $Re = \rho v D/\mu$, ρ and μ denotes density and viscosity of the flow which were corrected for temperature.

Table 2 Summary of main parameters of the experiment

Experiment parameters	Symbol	Unit	Value
Total length	L	m	1.45
Outer diameter	D	m	0.0195
Submerged Length	L_s	m	0.70
Bending stiffness	EI	N·m²	20.4
Mass ratio	m^*	—	1.55
Aspect ratio	Λ	—	74.4
Top tension	T_a	N	19.6 ~ 107.8
Flow speed	u	m/s	0.1 ~ 0.98
Reduced velocity	u_r	—	0 ~ 12
Reynolds number	Re	—	1,950 ~ 19,110
Structural damping	ζ	%	3.18
Mass-damping parameter	$(m^* + C_A)\zeta$	—	0.081

2 RESULTS AND DISCUSSION

The typical CF and IL time histories and the response frequencies are shown in Fig. 2 and Fig. 3, respectively, at T_{top} = 29.4N, v = 0.50m/s, u_r = 4.8, Re = 9,750. The first column graph shows the time history of CF displacement in diameter complete run at each station of strain gage along the cylinder. The second column graph shows the time history in a time interval $t \in [15,17]$ s. The third column graph shows the spectral analysis of each signal without mean. As shown in the first column, the carriage would accelerate for approximately 5s to achieve stable velocity. The second column graph demonstrates a stable instantaneous displacement. As shown in the third column, only one peak of vibration frequency appeared in the CF spectral analysis, whereas two peaks are captured in the IL spectral analysis, which reveal that the IL vibration captured the second mode in addition to the first.

The second mode was not achieved in the IL response for all test cases. Fig. 4 shows the spectral analysis of the midpoint displacement with respect to different towing velocity and applied top tension at midpoint of T_{top} = 29.4N, 68.6N, 88.2N. The right column graph reveals that when the Reynolds number is higher than about 8000, the second mode of IL response appears. However, the spectral power of first mode is larger than that of second mode in most of the cases, which means IL responses are still dominated by the first mode. The reason should be that at such a low aspect ratio (74.4) the cylinder model is more like a beam rather than a string. The bending stiffness plays

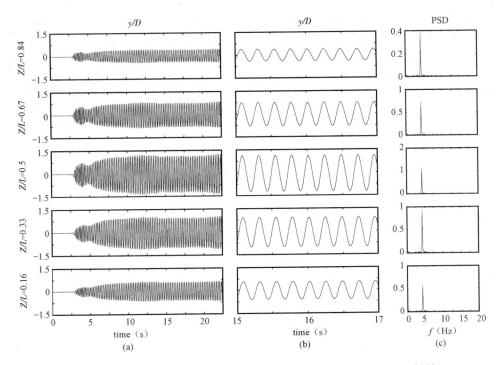

Fig. 2 Time history of test at $T_{top} = 29.4N, v = 0.5m/s, u_r = 4.8, Re = 9750$.
(a) Time history of CF displacement in diameter of the whole run at each strain gage along the cylinder;
(b) Time history in an interval $t \in [15,17]s$; (c) Spectral analysis of each signal without mean

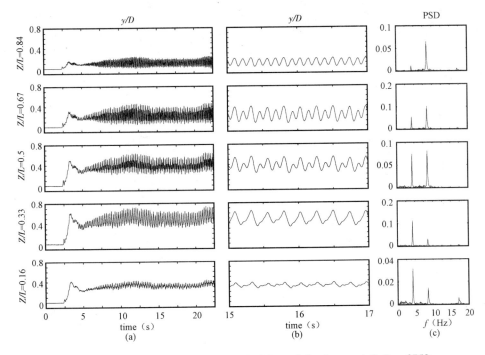

Fig. 3 Time history of test at $T_{top} = 29.4N, v = 0.5m/s, u_r = 4.8, Re = 9750$.
(a) Time history of IL displacement in diameter of the whole run at each strain gage along the cylinder;
(b) Time history in an interval $t \in [15,17]s$; (c) Spectral analysis of each signal without mean

a more important role in this dynamic vibration, which yields a higher natural frequency when compared with Huarte (2006) whose aspect ratio reaches 470. As Huera-Huarte and Bearman (2009a) explained, with lower aspect ratio, the frequency of IL excitation force should be higher enough to reach the natural frequency of the second mode, which happened more easily in some cases with higher flow velocity and lower top tension. This is confirmed by the right column of Fig. 4 which demonstrates the second mode of IL response starts to be excited at higher Reynolds number with the higher top tension.

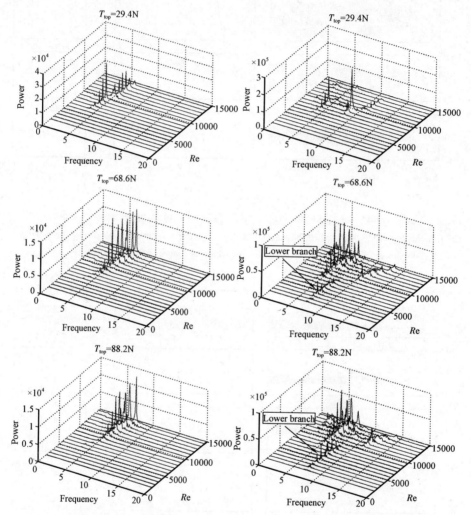

Fig. 4 Spectral analysis of the midpoint displacement, T_{top} = 29.4N, 68.6N, 88.2N, midpoint. Left graphs: CF spectral analysis; Right graphs: IL spectral analysis

The ratios of dominant IL(f_x), CF(f_y) frequency and wake oscillating frequency (f_h) versus natural frequency (f_n) are shown in Fig. 5. The linear fits are implemented in all plots. The slope of the linear fitting of CF ratio (f_y/f_n) is 0.18, as shown in Fig. 5(b); and the slopes of two linear fitting of IL ratio (f_x/f_n) are 0.18 and 0.36, respectively, as shown in Fig. 5(a). These ratios are higher than that published by Huera-Huarte and Bearman (2009a) who observed the slopes 0.16 and 0.32 as shown in Fig. 6.

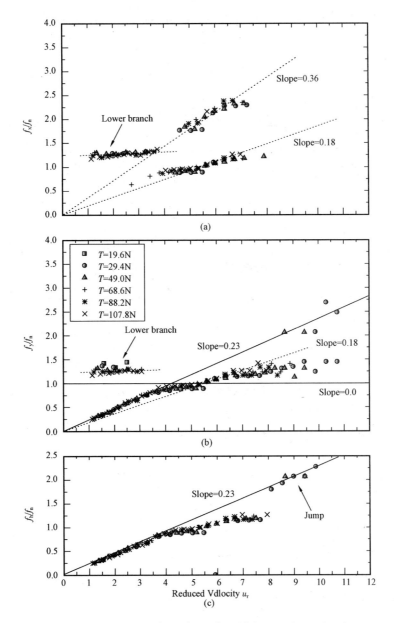

Fig. 5 Normalized IL(f_x/f_n), CF(f_y/f_n) frequencies and wake oscillating frequency(f_h/f_n) as a function of reduced velocity u_r

The CF ratio(f_y/f_n) is 0.23 (solid line with slope = 0.23) when reduced velocity is lower than 4, this value is close to 0.22 found by Khalak and Williamson(1997,1999). When reduced velocity is higher than 5, the CF ratio(f_y/f_n) should be unity based on the conventional "lock in" condition, but it departs from unity (solid line with slope = 0.0) as shown in Fig. 5(b). The reason should be low mass ratio 1.55 in present study. Under low mass ratio condition, added mass would play a more important role, see Sarpkaya(1979). Khalak and Williamson(1997,1999) had a similar phenomenon with the transverse oscillations of an elastically mounted rigid cylinder at low mass ratio. The influence of added mass on the natural frequency was studied by Vikestad et al. (2000) and the results showed that the added mass coefficient decreased from 4.5 to-1 when the reduced

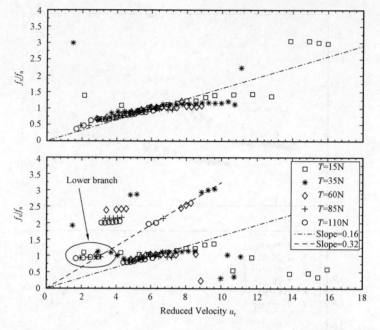

Fig. 6 Normalized IL(f_x/f_n), CF(f_y/f_n) frequencies as a function of reduced velocity u_r (From Huera-Huarte and Bearman, 2009a)

velocity increased from 2 to 14. The definition of reduced velocity was based on the value of the natural frequency measured in still water, which is the same as present study. From the classical beam theory, the actual natural frequency f'_n at the first mode of a pinned-pinned beam is defined as:

$$f'_n = \frac{\pi}{2\pi L^2}\sqrt{\frac{EI}{m + C_A m}} \qquad (1)$$

where C_A denotes added mass coefficient. The natural frequency will increase when added mass coefficient decrease base on Eq. (1). When "lock in" happens, the vortex shedding frequency f_y will be locked on the actual natural frequency f'_n, inducing the departure of f_y/f_n from unity.

From the right column of Fig. 4, it is noticed that a dominant frequency appears more obviously at higher top tension in the IL response when the CF "lock in" has not been widely developed ($u_r < 4.5$), formatting a lower branch as shown in Fig. 5(a). When CF and IL suffer "lock in" ($u_r > 4.5$), these lower branch vibration frequencies disappeared. The ratio of this lower branch is around 1.25. The same lower branch was appeared in Fig. 6 by Huera-Huarte and Bearman (2009a), but he did not comment on it since it was not so remarkable as present study. This lower branch can also be found in CF vibration as shown in Fig. 5(b). According to the experiment at low mass and damping from Khalak and Williamson (1997), the lower branch should be induced by initial "lock in" (where $u_r < 4.5$). Fig. 7 depicts a typical IL and CF time histories of test in an interval $t \in [20, 25]$ s at $T_{top} = 68.6$N, where flow velocity is $v = 0.18$m/s, $v = 0.22$m/s, $v = 0.26$m/s, respectively. The blue lines represent IL vibration and red lines represent CF vibration. The spectral analyses in second and third column reveal that CF vibration is composed of vortex shedding action and initial "lock in" at $u_r < 4.5$, whereas IL vibration only suffers initial "lock in". The first column of Fig. 7 shows vibration amplitudes of IL and CF motions in diameters, these values are so small that the vibrations are not easy to be observed in experiments.

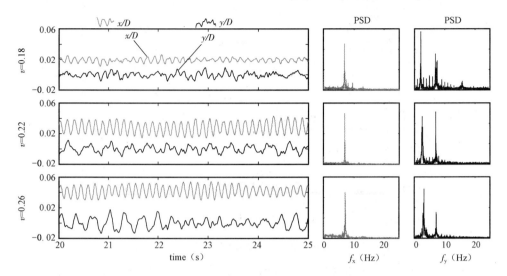

Fig. 7 Time history and spectral analysis of tests at $T_{top} = 68.6$N. First column: Time history of test in an interval $t \in [20,25]$ s; Second column: Spectral analysis of IL vibration; Third column: Spectral analysis of CF vibration. First row: $v = 0.18$m/s, $u_r = 1.63$; Second row: $v = 0.22$m/s, $u_r = 1.99$; Third row: $v = 0.26$m/s, $u_r = 2.35$

By extracting the vibration frequency from the hot films, the ratio (f_h/f_n) between wake oscillating and natural frequency is obtained and plotted as shown in Fig. 5(c). The trend is very similar with CF ratio (f_y/f_n) and the solid line has a same slope of value 0.23. However, the hot films only capture one dominant frequency, hence there is no lower branch observed. When $u_r > 8$, the ratio (f_h/f_n) has a jump to the solid line, which means the vibration of flexible cylinder departs from 'lock in' region. This can be confirmed by Fig. 5(b) where the ratio f_y/f_n becomes scattered when $u_r > 8$.

One of the popular questions in VIV phenomenon is what would be the maximum attainable amplitude? Fig. 8 shows a compilation of the maximum CF amplitude in diameter versus the reduced velocity with various top tensions in the first row, as well as the maximum IL amplitude without mean in the second row. The results from Huera-Huarte and Bearman (2009a); Huarte (2006) have been compiled and compared with present data. The larger aspect ratio (470) experiments were performed by Huarte (2006) in which the CF response could be excited to the second mode in the reduced velocity range $u_r \in [0,12]$. Hence, the maximum CF amplitude increased initially from $u_r \in [0,7.5]$, and then changed to the second mode suddenly. Afterwards, the amplitude increased from $u_r \approx 7.5$ onwards until arrived at the third mode. The largest CF amplitude was 1.02 in the present reduced velocity range $u_r \in [0,12]$ from Huarte (2006) as shown in the figure.

Due to the similar experimental set-up between the present and Huera-Huarte and Bearman (2009a), both of them have the clear initial upper branch and lower branch of maxima amplitudes. For the present cases, the maxima CF amplitudes are distributed in the reduced velocity

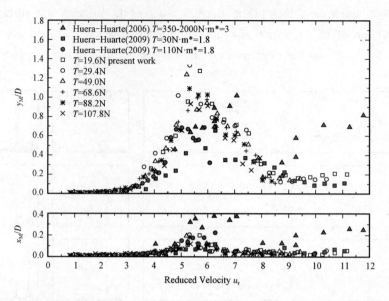

Fig. 8 Maximum of IL(x_M/D) and CF displacements(y_M/D)

range $u_r \in [5,6.5]$, the maximum CF amplitude in diameter is 1.33, lager than 1.02 from Huera-Huarte and Gharib(2011), 0.98 from Huera-Huarte and Bearman(2011) and 0.7 from Huera-Huarte and Bearman(2009a). As for the IL maximum amplitude without mean, the value for present experiment is 0.2 diameters which is less than 0.22 from Huera-Huarte and Bearman(2009a) and much less than 0.37 from Huarte(2006).

If a beam is governed dominantly by top tension, the response behavior is more like a string and the multi-mode response will be clearly shown as the experiment conducted by Chaplin et al. (2005); Trim et al. (2005). The second mode will be more easily reached when the top tension is lower because of the lower natural frequency. In the case of the top tension 19.6N, it is noticed that the maxima amplitudes increased if reduced velocity is above 8.5, in both CF and IL direction. It means that the excited response would approach the second mode slightly. Modal analysis has been used widely to identify dierent modes of vibration. The methodology can be found in the literature Lie and Kaasen(2006). By applying modal analysis here, the mode number up to 3 have been chosen to separate the original response into each mode weight contribution as shown in Fig. 9. The figure of mode weights indicates that in all the cases they have been dominated by the first mode in both CF and IL responses. The second mode only appears weakly in the CF and IL responses if reduced velocity ranges from $u_r \in [5,6]$ as depicted in the second row. The third mode almost has no contribution to the overall response. The present result has a good agreement with Huarte(2006).

Fig. 10 shows the instantaneous x-y trajectories versus reduced velocity at each strain gage station. The case of $T_{top} = 29.4N, u_r = 5.3$ is clearly exhibited a figure – 8 pattern at $z/D = 0.84$, 0.67, 0.5. Fig. 11 indicates the instantaneous deflections in CF and IL directions in a time interval $t \in [8,14]$ s at $T_{top} = 29.4N, u_r = 5.3$m/s corresponding to the case in the third column of Fig. 10. The left column is the instantaneous CF deflection, the middle column is the IL deflection and the third column is the IL deflection without mean. It is noticed that in all cases the IL mean

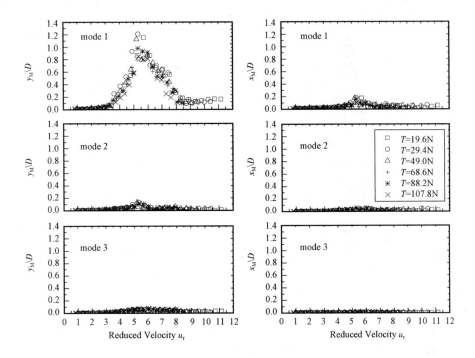

Fig. 9　Modal weights: CF(left) and IL(right) oscillations up to 3rd mode

displacements at $z/D = 0.33$ are larger than the ones at $z/D = 0.84$, this is confirmed by the second column in Fig. 11, due to: (1) the drag force in water is much higher than in air, hence larger drag force exerts on the lower half of the flexible cylinder; (2) lower axial tension force acts on the lower half of flexible cylinder due to its weight.

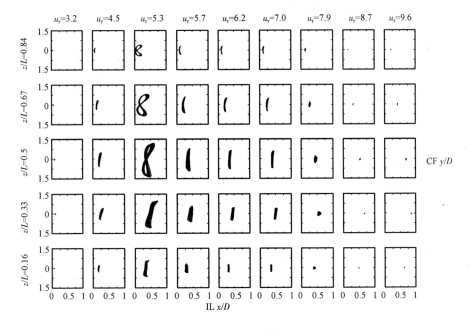

Fig. 10　IL and CF trajectories at diffierent reduced velocity u_r and $T_{top} = 29.4$

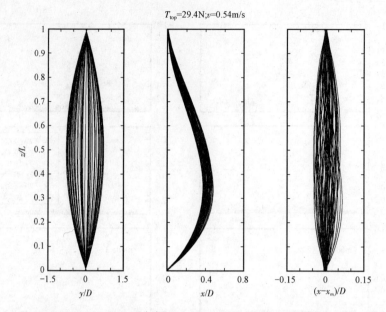

Fig. 11 Instantaneous deflections: (1) Left column graph: the instantaneous CF deflection; (2) Middle column graph: IL deflection; (3) Right column graph: IL deflection without mean. Black lines plotted at every 0.1s in the time interval $t \in [8,14]$ s at $T_{top} = 29.4$ N, $v_r = 0.54$ m/s $u_r = 5.3$

An important question in VIV phenomenon is the attainable amplitude in CF direction. A loglog plot was used by Griffin(1980) for attainable amplitude A^* versus mass damping ($m^*\zeta$) which could collapse peak-amplitude data. Using a large set of experimental data from elastically mounted cylinders, Khalak and Williamson(1999) plotted attainable amplitude A^* versus ($m^* + C_A)\zeta$ and a successful collapse of the data appeared into the upper and the lower branches. For the present data, $(m^* + C_A)\zeta = 0.081$, attainable amplitude is $A^* = 1.33$. A compilation of data including comparison with results from Huera-Huarte and Bearman(2009a); Huera-Huarte and Gharib (2011); Huera-Huarte and Bearman(2011); Huarte(2006); Khalak and Williamson(1999); Griffin(1980); Jauvtis and Williamson(2004) is shown in Fig. 12. From the figure, the attainable am-

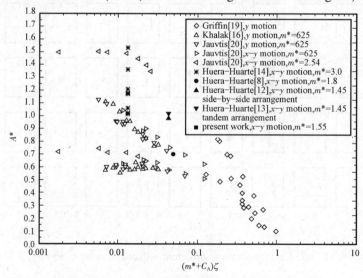

Fig. 12 Peak attainable amplitude A^* versus mass damping($m^* + C_A)\zeta$

plitudes in x-y motion with pinned-pinned flexible cylinders are somewhat larger than that with spring-mounted cylinders. For instance, the attainable amplitude could reach the range $A^* = 1.0 \sim 1.6$ from Huarte (2006), $A^* \approx 1.0$ from Huera-Huarte and Gharib (2011); Huera-Huarte and Bearman(2011), which are larger than most of the cases with spring-mounted cylinders. The present result $A^* = 1.33$ also confirms such a trend.

Conclusions

The experimental results reported in the work are aimed at helping to understand the behavior of vortex-induced vibration of a long flexible cylinder. Comparisons between the present data and results from literature have been shown. The following conclusions can be drawn:

(1) The present set-up could successfully simulate VIV phenomenon and capture some characteristics of VIV. A slope of the linear fit of CF ratio (f_y/f_n) is 0.18. The slopes of two linear line fits of IL ratio (f_x/f_n) are 0.18 and 0.36, respectively. A lower branch is captured both in CF and IL motion; this justifiable discrepancy is caused by initial "lock in".

(2) The maximum attainable amplitude is 1.33 diameters in CF direction, and 0.2 diameters in IL direction. The curve distribution demonstrates clearly the initial upper and lower branches. The modal analysis reveals that CF and IL vibration are dominated by the first mode in most of the cases.

(3) The $x - y$ trajectory plot indicates that the figure – 8 patterns only could be found in a few cases. The "Grinffin" plot has been used to show that the present result $A^* = 1.33$ is higher than the results found in the experiments with spring-mounted cylinder at the same range of $(m^* + C_A)\zeta$.

Due to the more realistic conditions with the set-up of long pinned-pinned flexible cylinders, more work about the dynamic response and wake visualization will be conducted in future to produce a better understanding of the fluid-structure interaction on offshore applications.

ACKNOWLEDGEMENTS

The author would like to thank the Brazilian National Research Council (CNPq) and the National Basic Research Program of China (973 Program) Grant No. 2011CB013702 for the financial support of this research.

REFERENCES

Chaplin J R, Bearman P W, Huera-Huarte F J, Pattenden R J. 2005. Laboratory Measurements of Vortex-Induced Vibrations of a Vertical Tension Riser in a Stepped Current[J]. Journal of Fluids and Structures, 21(1): 3 – 24.

Griffin O M. 1980. Vortex-excited Cross-flow Vibrations of a Single Cylindrical Tube[J], Journal of Pressure Vessel Technology-Transactions of the Asme, 102(2): 158 – 166.

Huarte F J H. 2006. Multi-Mode Vortex-Induced Vibrations of a Flexible Circular Cylinder[D]. Ph. D. thesis, Imperial College London, Prince Consort Road, London SW7 2BY.

Huera-Huarte F J, Bearman P W. 2009a. Wake Structures and Vortex-induced Vibrations of Along Flexible Cylinder-Part 1: Dynamic Response[J]. Journal of Fluids and Structures, 25(6): 969 – 990.

Huera-Huarte F J, Bearman P W. 2009b. Wake Structures and Vortex-induced Vibrations of Along Flexible Cylinder-Part 2: Drag Coefficients and Vortex Modes[J]. Journal of Fluids and Structures, 25(6): 991 – 1006.

Huera-Huarte F J, Bearman P W. 2011. Vortex and Wake-induced Vibrations of a Tandem Arrangement of Two Flexible Circular Cylinders with Near Wake Interference[J]. Journal of Fluids and Structures, 27(2): 193 - 211.

Huera-Huarte F J, Gharib M. 2011. Flow-induced Vibrations of a Side-by-side Arrangement of Two Flexible Circular Cylinders[J]. Journal of Fluids and Structures, 27(3): 354 - 366.

Huse E, Kleiven G, Nielsen F. 1998. Large Scale Model Testing of Deep Sea Risers[J]. vol. 2, Houston, TX, USA, ISSN 01603663: 189 - 197.

Jauvtis N, Williamson C H K. 2004. The Effect of Two Degrees of Freedom on Vortex-Induced Vibration at Low Mass and Damping[J]. Journal of Fluid Mechanics, 509: 23 - 62.

Khalak A, Williamson C H K. 1997. Fluid Forces and Dynamics of a Hydroelastic Structure with Very Low Mass and Damping[J]. Journal of Fluids and Structures, 11(8): 973 - 982.

Khalak A, Williamson C H K. 1999. Motions, Forces and Mode Transitions in Vortex-Induced Vibrations at Low Mass-damping[J]. Journal of Fluids and Structures, 13(7 - 8): 813 - 851.

Lie H, Kaasen K E. 2006. Modal Analysis of Measurements From a Large-Scale VIV Model Test of a Riser in Linearly Sheared flow[J]. Journal of Fluids and Structures, 22(4): 557 - 575.

rpkaya T. 2004. A Critical Review of the Intrinsic Nature of Vortex-Induced Vibrations[J]. Journal of Fluids and Structures, 19(4): 389 - 447.

Sarpkaya T. 1979. Vortex-Induced Oscillations: A Selective Review[J]. Journal of Applied Mechanics, 46(2): 241 - 258.

Trim A D, Braaten H, Lie H, Tognarelli M A. 2005. Experimental Investigation of Vortex-Induced Vibration of Long Marine Risers[J]. Journal of Fluids and Structures, 21(3): 335 - 61.

Vandiver J K, Marcollo H, Swithenbank S B, Jhingran V. 2006. High-Modenumber Vortex-Induced-Vibration Field Experiments[J]. JPT, Journal of Petroleum Technology, 58(2): 6970.

Vikestad K, Vandiver J K, Larsen C M. 2000. Added Mass And Oscillation Frequency for a Circular Cylinder Subjected to Vortex-induced Vibrations and External Disturbance[J]. Journal of Fluids and Structures, 14(7): 1071 - 1088.

Williamson C H K, Govardhan R. 2004. Vortex-Induced vibrations[J]. Annual Review of Fluid Mechanics, 36: 413 - 455.

Xie F F, Deng J, Zheng Y. 2011. Multi-Mode of Vortex-Induced Vibration of a Flexible Circular Cylinder[J]. Journal of Hydrodynamics Ser. B 23(4): 483 - 490.

Xu W H, Wu Y X, Zeng X H, Zhong X F, Yu J X. 2010. A New Wake Oscillator Model for Predicting Vortex Induced Vibration of a Circular Cylinder[J]. Journal of Hydrodynamics, 22(3): 381 - 386.

On Buckling of Subsea Pipe-In-Pipe System

Caiying Yi[1], Qin Wang[2], Menglan Duan[2]

[1] Offshore Oil Engineering Co. , Ltd. , Tianjin, P. R. China
[2] Offshore Oil and Gas Research Center, China University of Petroleum, Beijing, China

Abstract This paper focuses on the methodologies on strength and buckling of typical submarine pipeline(Pipe-in-pipe system) under high temperature and high pressure(HT/HP) conditions. The internal high pressure and high temperature of oil/gas, external hydraulic pressure in deepwater, variational friction force of soil, and the complex structure of the PIP system make the equilibrium equation of the subsea pipeline exhibit strong nonlinear performance in such nonlinear boundary conditions. The numerical methods are summarized for solving the nonlinear equations in this paper. The paper also presents evaluation methods for the lateral and upheaval buckling of two types of pipe-in-pipe systems(compliant and non-compliant) that are widely used in the offshore oil and gas industry.

The post-buckling mode is even more complex, and post-buckling stress and strain are much more difficult to obtain. Researches on post-buckling in both theory and experiment are overviewed in this paper. The advances on thermal stability and safety of compliant and non-compliant pipe-in-pipe systems are also presented which are going to have more application to both design and operation of subsea pipelines in deepwater engineering.

Key Words Pipe-in-pipe system; Buckling; Post-buckling; Sandwich pipe

1 THE INTRODUCTION OF PIPE-IN-PIPE SYSTEM

Nowadays, the global petroleum companies have transferred their attentions to the exploration of deep and ultra-deep water, especially in the North Sea, Gulf of Mexico and subsalt fields off Brazil which were discovered recently. Concurrently, the Chinese government and enterprises pay attention to the exploration and development of the South China Sea oil and gas resources. All of these companies need a lot of offshore oil and gas equipments applied to oil exploitation and transportation in deep and ultra-deep water. To the transportation of oil, the pipe should provide enough strength to support external severe stress and well insulation to ensure normal oil transportation. Instead of conventional insulating pipe, new concepts of submarine pipelines and risers have been proposed. Generally speaking, there are two kinds of pipelines which can be applied in the deepwater and ultra-deep water environment, pipe-in-pipe(PIP) and sandwich pipe(SP). Current high-temperature projects widely employ pipe-in-pipe systems, which can generally be divided into two categories, compliant and non-complaint systems(see Fig. 1 and Fig. 2). Both of the two types of system consist of two concentrically mounted steel pipes and the annular space filled with either circulating hot water or thermal insulated materials. The objective of this type of pipe is to increase the thermal insulation capacity to prevent formatting paraffin and hydrates. The apparent advantages of PIP system are the possibility of using materials with excellent thermal properties, and simultaneously the structural integrity is independently provided by the outer and inner steel layers (Grealish and Roddy, 2002).

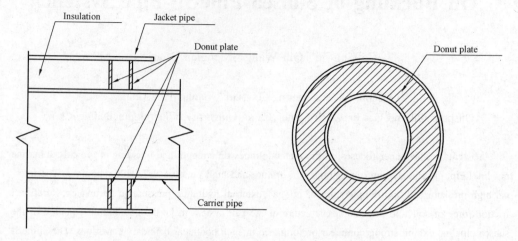

Fig. 1 Typical compliant pipe-in-pipe systems

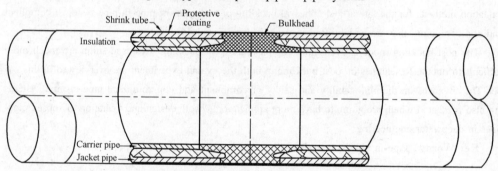

Fig. 2 Typical non-compliant pipe-in-pipe systems

In certain operational or accidental situations, structural elements such as beams, plates or shells may be exposed to large temperature gradients. If displacements are restricted, significant compressive stresses are imparted to the structure, exposing the system to a potential buckling condition. Concomitantly, the structure's mechanical strength may be compromised as the temperature increases. This class of problem is complex because it often exhibits geometrical and physical non-linearities. Buckling of slender structural elements under compression is a considerable failure mode, frequently used in design as a parameter to verify the maximum allowable stresses. Understanding the buckling mechanism may aid in avoiding accidents or severe damage in many engineering. Actually the buckling and post buckling of pipe-in-pipe system caused by high pressure and high temperature is complex, not only for complaint systems but also for non-complaint systems. In complaint system, the carrier pipe and the casing pipe are connected by tulips or donut plates at close intervals which can act as water-stops and allow the transfer of bending moment between the carrier pipe and the casing pipe. The non-complaint system has the connection between the carrier pipe and the casing pipe by bulkheads at an interval of a few kilometers which have high stiffness.

However, ultra deepwater scenarios, at depths beyond 1,500 meters, require very thick-walled steel pipelines or pipe-in-pipe systems, which are expensive and difficult to install because of their excessive weight. Owing to these disadvantages, a new concept was developed by Segen F. Estefen of the Federal University of Rio de Janeiro and two of his colleagues in the Ocean Engineering Department(Theodoro A. Netto and Ilson P. Pasqualino, 2002). This kind of pipe would mitigate some of those disadvantages which composed of two concentric steel pipes separated by and bonded to a polymeric annulus. The design provides a combination of high structural strength with thermal insulation.

The usual sandwich structures are a particular kind of multi-layered of different materials bonded together, and the single layer property is benefited to the global structure performance. Usually, the sandwich structure contains three layers: two external thin and stiff layers to give enough strength and a central thick and flexible core to ensure the insulation. The external layers are bonded to the core with adhesives to allow the load transfer between the components. see Fig. 3.

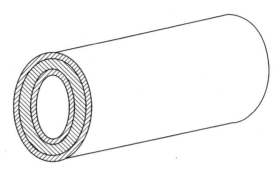

Fig. 3 The sketch plan of the sandwich pipe

2 LATERAL BUCKLING AND VERTICAL BUCKLING OF COMPLAINT AND NON-COMPLAINT PIP

For the HT pipe-in-pipe, the thermal load remains the control load which determines the degree of structure strain. Either the Central Board or Bulkhead which bears the shear between the internal and external of the pipe, the internal pipe will deliver a certain axial load to the external pipe because of thermal expansion. So take all things into consideration, the inner pipe is compressed while the outer pipe is tensioned. When the deliver temperature reaches a certain critical value, the inner pipe may be buckled under the axial force caused by thermal load, but the outer pipe under tension stress will not be buckled. If the PIP system has lateral buckling as a whole, then bending of the internal pipe tube is caused by lateral instability because of the axial force, and the outer tube is due to the load transfer.

In the anchor segment of the flexible connecting PIP system, there will be a large axial load in the middle of the pipe between the adjacent circular plates. When the lateral restraint is weak or the initial deflection is great(for example, the pipe is exposed to sea water because of the ocean current erosion), the overall lateral buckling of PIP system is possible. To the rigid connected PIP systems, because the intervals of the bulkhead are generally wide, the pipe between the adjacent bulkhead will not be vertical buckled as a whole. But the moment of the pipeline getting through the bulkhead contact or the force between the internal and external pipe transferring between the two layers of the pipe is possible. Then, the rigid connected PIP system may lose lateral stability as a whole.

During the transportation of the high temperature liquid in the PIP submarine pipeline, the internal pipe will expand along the axial direction of the pipe under the thermal load. Meanwhile the bulkhead is pushed on and tensile stress in the external pipe causes the pipe axial elongation. But the constraints at both ends of the pipe and the friction between the seabed and pipeline will prevent elongation. Finally an axial balance is established. According to the axial strain, the whole submarine pipeline under the thermal load can be divided into two segments, the central anchor segment and the end slip segment of the pipe. In view of the axial force caused by thermal load will be released at a certain degree in the slip segment because of the expansion of the pipeline. So for the PIP submarine pipeline, the lateral buckling is more likely to happen in the anchor segment of the middle section of the pipe. In the Finite Element Analyses, a section of pipeline is chosen to be modeled to assess the lateral stability of the entire pipeline.

For the HT pipeline, axial force caused by thermal load is always the main internal force and the most important reason which leads to various forms of submarine pipeline buckling. While the effect of the internal and external pressure is much less significant than the temperature. From the actual assessment results of the pipeline buckling finite element analyses, we can completely ignore the axial force which is caused by the transportation pressure load inside the tube and the external environment load undertaken by the external pipe. So actually the thermal buckling analysis of PIP is still the expansion effect analysis of the pipeline under the thermal load.

From the process of the buckling of the pipeline, similar as the Euler beam, when the axial load reaches a certain level, submarine pipeline will produce a sudden lateral deformation, release the axial load, and then establish a new balance. View this as a whole, the new balance is reflected in two forms, namely, lateral buckling deformation and vertical buckling deformation. The buckling location and direction of the submarine pipeline have involved many factors, which mainly include two aspects, internal and external factors.

Beside the basic characteristics of submarine pipeline buckling, the PIP submarine pipeline lateral buckling has its own features. It mainly reflects as the expansion differences between the internal and external pipe under the thermal load and the axial force difference caused by it.

Generally speaking, the researches of offshore pipelines are mostly focused on single pipelines and corresponding practical results have been obtained, which has been included in the specifications (DNV 2007) for structural design against global buckling of high temperature/high pressure submarine single pipelines.

3 THE BUCKLING AND POST-BUCKLING OF SINGLE PIPE

For the single pipeline, according to the material nonlinearity, the quality of the pipe and cross section coefficient, coupled with different boundary conditions, the equation set will be multivariate first order nonlinear differential equations. Methods for solving such equations are the numerical iterative method (shooting method, Rung-Kutta method), progressive method (perturbation method). VAZ has done a lot of research on this non-linear model.

3.1 Double-Hinged with Fixed Ends

VAZ analyzed the elastic buckling and initial post-buckling behavior of slender beams which are assumed to be double-hinged with fixed ends in 2008 (see Fig. 4). When the beams are subjected to uniform heating, the double-hinged with fixed ends prevent thermal expansion that may lead to beam buckling when destabilizing compressive forces arise.

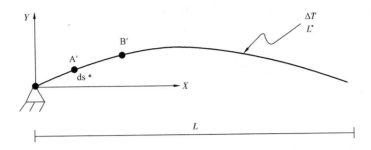

Fig. 4 Beam after buckling(double-hinged with fixed ends)

His work presents significant initial post-buckling results for this model considering the beam material variations in the modulus of elasticity, the thermal strain and the yield stress with temperature. In his work, the perturbation method is employed in reducing the non-linear set of differential equations into a set of sequentially solvable linear equations. Through solving the linear equations, the critical buckling loads, temperatures and modes are obtained from the first order expansion whereas the second order solution provides information on the initial post-buckling behavior including the influence of different initial temperatures on the buckling and post-buckling behavior. Importantly, his results show that it is important to consider the temperature-dependent material properties such as slenderness ratios, especially in engineering design. If the modulus of elasticity of a specific and hypothetical material drops more rapidly with temperature, it is expected that the physically linear and non-linear approaches yield more accentuated divergent results, especially for small slenderness ratios.

3.2 Hinged-Rotational Spring

The post-buckling analysis of slender elastic rods subjected to axial terminal forces was successfully performed by VAZ who assumed that the boundary conditions are hinged and elastically restrained with a rotational spring. This two-point boundary value problem is governed by a set of five first-order non-linear ordinary differential equations. Solutions for buckling, initial post-buckling(perturbation), large loads(asymptotic) and numerical integration are developed. As perturbation methods may be well employed in weakly non-linear problems, they have been extensively used in initial post-buckling analyses. A solution is sought as an expansion in terms of a perturbation parameter, rendering a set of sequentially solvable linear equations. Direct numerical integration is preferable, as higher order expansion improves the solution but it becomes analytically very complex. A two-point boundary value problem is characterized though three boundary conditions which are given at one end while two are specified at the other end. Several techniques have been employed for this class of problem(e. g. finite element methods, finite difference schemes and energy methods). Solutions via the shooting method with direct integration are conveniently employed in linear or non-linear problems when only one parameter is required for interpolation but they become rather complex if two conditions are sought in non-linear systems. Runge-Kutta high order solution may be employed to solve the set of non-linear ordinary differential equations. Vaz chooses a special class of solution for high values of applied loads through an asymptotic expansion. See Fig. 5.

There are several interesting phenomena such as limit load, jump, hysteresis, bifurcation and non-uniqueness as the load-displacement relation is not monotonic and equilibrium configuration may depend on the load-unloading path (i. e. if displacement or load is controlled) which is described by Wang.

Results indicate that the post-buckling response is a rotational spring with stiffness constant independent when both coincide once the bending moment at the rotational spring is zero.

The simple but powerful numerical procedure employed may facilitate further developments in two-point boundary value problems such as post-buckling analysis of slender elastic rods subjected to self-weight and applied terminal moments and is the object of research.

3.3 Double-Hinged with Non-Movable and Movable Ends

Based on the previous research, VAZ gives the solution of the behavior of slender elastic rods subjected to axial terminal forces and self-weight with a double-hinged boundary condition. The numerical and analytical solutions developed in this paper are in sound agreement. A perturbation scheme is developed to provide an analytical solution for the rod initial post-buckling. The rod post-buckling response is obtained from solving a complex two-point boundary value problem governed by a set of six first order non-linear ordinary differential equations.

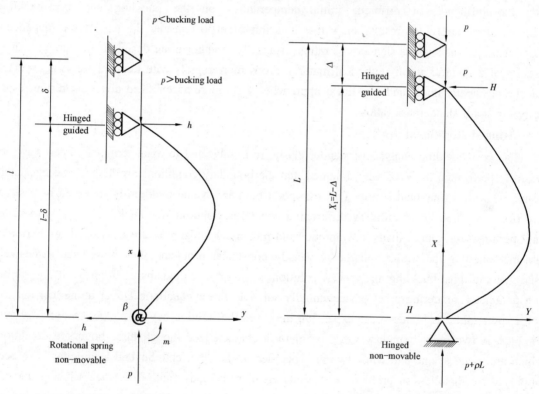

Fig. 5　Beam after buckling
(hinged-rotational spring)

Fig. 6　Beam after buckling
(double-hinged with non-movable and movable ends)

The value of the non-dimensional weight influences the slender elastic rod stability behavior and its post-buckling configuration greatly. The rod is shown to be initially stable up to a critical value of self-weight (for the non-dimensional parameterization employed) and unstable elsewhere. see Fig. 6.

4 THE THEORETICAL ANALYSIS OF PIPE-IN-PIPE SYSTEM

PIP is popular because it enables substantially lower heat loss per unit length than any available external insulation coating for a single pipe flowline. Actually, the buckling and post buckling of the pipe-in-pipe systems are more complex than that of a single pipe, so even following a lot of simplifications, it is difficult to find an effective analytical solution to describe the buckling performance of pipe-in-pipe systems in service (Vaz and Patel, 1999), and there has been no widely accepted method yet published for evaluating the lateral buckling of pipe-in-pipe systems.

Generally, bundled pipe systems consist of one or more internal pipes assembled inside a larger external pipe.

A simplified model of pipe-in-pipe system is shown in Fig. 7.

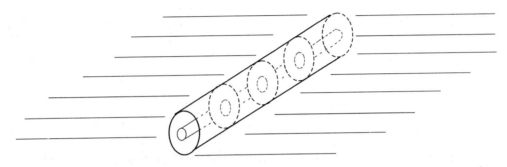

Fig. 7 A simplified model of pipe-in-pipe system

Without considering the presence of sea bed friction, initial imperfections, and by treating the inner and outer pipes as not being constrained axially and rotationally relative to each other at their ends, Vaz presents an analytical formulation of the coupled buckling instability of a pipe-in-pipe system typical of that considered for high-temperature in the oil and gas industry. Post-buckling analysis or simultaneous multi-pipe locking was not addressed in his work.

The formulation divides the problem into two sub-systems, the internal and external pipe, and applies physical restrictions to connect the equations arising from each pipe. These restrictions are based on compatibility of displacements and equilibrium of forces at the spacers. See Fig. 8.

The two sub-systems subjected to an axial compressive force and n concentrated loads are beam-column models. The beam lateral deflection between loads is given by Timoshenko and Gere. Though coupled by imposing force equilibrium and lateral displacement compatibility for both regular and irregularly pitched spacers, the equations of bending for the inner and outer concentric pipes of a pipe-in-pipe system are solved out for the parameter which is the stiffness ratio of internal and external pipe. It defines the extent to which the buckling instability of the inner and outer pipes interacts or one dominates the other. The analysis is useful for considering local buckling effects on short pipe segments and would allow an optimization of pipe dimension and wall thicknesses that would be slightly less conservative than the simple isolated pipe considerations used to date, though there are two drawbacks in this analysis as related above.

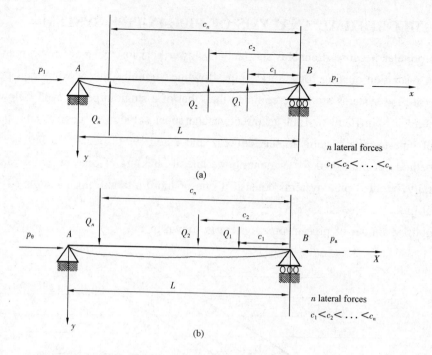

Fig. 8 (a) Model for the internal pipe; (b) Model for the external pipe

5 THE THEORETICAL ANALYSIS OF SANDWICH PIPE

The sandwich pipe is proposed in deepwater engineering because of its greater structural strength which can ensure the flow assurance. Numerical and experimental studies have been carried out by Borselino et al. (2004). They analyzed three different woven fibers including class fiber, carbon fiber and Kevlar fiber by performing the flatwised compression test, the edgewise compression test and three point bending test. Special behavior of unidirectional sandwich panels with transversely flexible core has been studied by Sokolinsky et al. (2002), where they analyzed the special features response of ordinary and delaminated sandwich panels with a transversely flexible core subjected to external in-plane and vertical static loads. And their analytical formulation was based on a higher-order theory for sandwich panels with non-rigid bond layers between the face sheets and the core. Their numerical study reveals that the ordinary (fully bonded) sandwich panel behaves as a compound structure in which the local/localized, overall or interactive forms of the response can take place and depend on the geometry, mechanical properties, and boundary conditions of the structure. Although there are a lot of advantages of composite construction, such as the reduction in total life costs, corrosion resistance, high strength and stiffness-to-weight ratios, and improved stealth for military applications, the expenditure for manufacturing this kind of pipe is very expensive now. New higher quality materials with lower costs and new fabrication methods need to be developed before composite materials will be fully accepted for the construction of large ships. Mouring (1999) did this work and his study summarized experimental results of the testing of composite panels stiffened with preformed frames under in-plane uniaxial compressive loads.

Estefen and Netto (2005) conducted a series of small-scale laboratory tests on the performance of sandwich pipes and developed a finite element model to evaluate the structural performance of such sandwich pipes with two different options of core material. They considered that sandwich pipe systems with either cement or polypropylene cores were feasible options for ultra deepwater applications. Nevertheless, the paper by Estefen did not give the theoretical analysis method for the structural stability of sandwich pipe while the pipe is concurrently standing external and internal pressure and thermal load. Thermal buckling of high-temperature submarine pipelines is complex and is difficult to control. Besides, it is difficult to maintain the pipe once the pipe appears collapse in deep and ultra-deep water. Therefore, before producing sandwich pipe, the research theoretically and experimentally centering on sandwich pipe is indeed necessary. We do same research on an analytical formulation of the instability of a sandwich pipe system considered for high-temperature products in the oil and gas industry. Through solving the equations of heat transfer and stress equations, the temperature field and stress can be accurately reduced that is the first critical step in the reduction of global analysis of sandwich pipe. Our research also provides the stresses in formula which are verified using a nonlinear finite element (NLFE) model. The accurate prediction of the stresses demands plentiful resources and time through utilizing the advanced nonlinear finite element analysis tools such as ABAQUS. Compared to ABAQUS modeling, the proposed formulations is advantageous and direct if only need a high level analysis in the earlier detailed design stage. It is a pity that due to the complexity and periodicity of the inner pipe temperature range from 4℃ to production temperature, the processing method in NLFET is different to the theory method in this paper. And it is a difficulty in processing this problem accurately in theory which is not solved in our research.

6 MEASURES AGAINST BUCKLING IN ENGINEERING

Buckling has always been a troubling problem which harasses the safety of the submarine pipeline. The high/super-high temperature long-distance transportation poses a real challenge to the thermal stability of the pipeline design. Currently, there are two anti-buckling designs, one is to avoid the occurrence of buckling directly, the other is to limit the magnitude of pipe buckling. The former method generally is to reduce the thermal load, increase pipe bending stiffness, or increase the constraints; the latter allows the pipe buckling in a certain range, but it is needed to control the displacement and stress and strain within an allowable range.

The current anti-buckling measures are as follows:

(1) use the heat exchanger to reduce the temperature of the liquid before it enters the pipeline, reduce the heat load of the pipe;

(2) increase the constraints on the horizontal or vertical, such as increase buried depth, increase weight of outsourcing concrete and friction, accumulate stones above the pipeline;

(3) using PIP or tube bundle to increase the flexural rigidity;

(4) pre-tension the pipe through the tow in the installation and the preheated buckling before use;

(5) snake laying.

The above-mentioned measures via the following aspects to solve the problem of HT pipeline buckling:

(1) by lowering the liquid temperature to avoid pipe buckling damage, namely, reduce the thermal load to within the existing range.

(2) and (3) by adding constraints or increasing its bending stiffness to increase the critical buckling load of the pipe, and then avoid buckling.

(4) by using the retained pre-tension to counteract the axial force during transporting the high temperature liquid, so as to achieve the purpose of maintaining the stability of the pipeline.

(5) by providing initial deflection in the predetermined locations, use a certain extent of lateral buckling deformation to release the high-temperature axial force, so as to achieve the purpose of maintaining the stability of the pipeline.

The following table 1 lists the effects of the technical measures, disadvantages, cost comparisons and practical applications, in which the level of project cost is divided into seven levels, 1 means relatively the lowest cost and 7 means the highest cost.

Table 1 Contrast of methods dealing with high temperature

Methods and measures	Function	Disadvantage	Cost	Engineering applications
Per-cold	Reduce the thermal load	need oil cooling device and need to avoid the deposition of paraffin components and the information the hydrate	3	Shell mallard and Erskin-2 oilfield
Increase constrains	Increase the critical buckling load	High cost and have a certain of limit	6	Widely used in North Sea
PIP system	Increase the anti-bending stiffness and limited range of flexion	Increase the cost and construction difficulties	7	Franklin oilfield (160℃)
Per-tension	Reduce the thermal load	Per-heat before the using of the pipe	2	Glamis oilfield
Snake laying	Release the axial force due to the thermal	Need efficiently buckling control and laying is complicate	4	Petrobras oilfield (new pipeline)
The PIP system	Buried laying	Need to confirm the least bury depth	—	Shell mallard oilfield
	Subbed laying	Need efficiently buckling control		Erskin-2 and king oilfield
	Using heat compensator	Installation is difficult and increase the cost		Heron and jade oilfield

In the table, measure (1) doesn't improve the carrying capacity on thermal stresses of pipeline. In most cases, delivery temperature of liquid is the objective technical requirements in terms of flow assurance. Cooling measures are feasible only if the production temperature of crude oil is higher than the required temperature of transportation processes and the later is within the bounds of pipeline's carrying capacity on thermal stresses.

Measure (2) and measure (3) are frequently employed for buckling inhibition in projects. Burying scheme is restricted by the geological condition of seabed and the construction technology of ditching. But the external enclosing concrete layer increases the difficulty of installation operation for pipe-laying vessels.

Besides, it is impractical to pile up stones above the pipe for the sake of inhibit buckling for long distance high-temperature pipeline because of its high cost.

Measure(3) is constantly employed by CNOOC for high temperature transportation at the present stage, and pipe-in-pipe structure is mainly used. The latest research shows that although the structure can inhibit the amplitude of buckling, the post-buckling of PIP structure at high temperature load still needs further research.

Measure(4) and measure(5) is a new method against buckling for offshore pipeline and is more cost-saving than adding constraint or applying PIP structure(or bundle structure). But its design is more complex. There are still a lot of blanks to be filled in relevant design theory. Therefore, from an overall perspective, measure(4) and measure(5) is still at the exploratory stage. But in the long run, in deep water where pipeline installation depth is close to the limit, using the lateral buckling within the control range to release the axial force caused by the high temperature is almost the only way to solve the problem of long distance and high temperature transportation.

Therefore, the lateral buckling(or prebuckling) method is the most potential application and the main research target at present.

REFERENCES

Bernitsas M M, Kokkinis T. 1983. Buckling of Columns with Movable Boundaries [J]. J. Struct. Mech. 11: 351–370.

Borselino C, Calabrese L, Valenza A. 2004. Experimental and Numerical Evaluation of Sandwich Composite Structures[J]. Journal of Composites Science and Technology, 64: 1709–15.

Bru ton D, Carr M, Crawford M, et al. 2005. The Safe Design of Hot Onbottom Pipelines with Lateral Buckling Using the Design Guideline Developed by the SAFEBUCK Joint Industry Project[C]. Deep Offshore Technology Conference, Vitoria, Espirito Santo, Brazil.

Castello X, Estefen S F. 2007. Limit Strength and Reeling Effects of Sandwich Pipes with Bonded Layers[J]. International Journal of Mechanical Sciences, 49: 577–588.

Castello X, Estefen S F. 2008. Sandwich Pipes for Ultra Deepwater Applications, OTC 19704.

Custódio A B, Vaz M A. 2002. A Nonlinear Formulation for the Axisymmetric Response of Umbilical Cables and Flexible Pipes[J]. Appl. Ocean Res. 24: 21–29.

David Bruton, Malcolm Carr. 2005. The Safe of Hot On-Bottom Pipelines with Lateral Buckling using the Design Guideline Developed by the SAFEBUCK Joint industry Project[C]. Deep Offshore Technology Conference, Vitoria Espirito Santo, Brazil.

Estefen SF, Netto TA, Pasqualino IP. 2005. Strength Analyses of Sandwich Pipes for Ultra Deepwaters. Journal of Applied Mechanics, 72: 599–608.

Harrison G E, Brunner M S, Bruton D A S. Thermal Expansion Design and implementation, OTC 15310.

Hobbs R E and Liang F. 1989. Thermal Buckling of Pipelines Close to Restraints[C]. Eighth International Offshore Mechanics and Arctic Engineering Symposium, 5: 121–127. New York, USA.

Hobbs R E. 1984. In-service Buckling of Heated Pipelines, Journal of Transportation Engineering, 110: 175–189.

Hooper J, Maschner E. HT/HP Pipe-in-pipe Snaked Lay Technology-Industry Challenges[C]. OTC 16379.

Huang T, Dareing D W. 1969. Buckling and Frequencies of Long Vertical Pipes[J]. J. Eng. Mech. Div. Proceedings of the ASCE, vol. 95, no. EM1:167 – 181.

Lee B K, Wilson J F, Oh S J. 1983. Elastica of Cantilevered Beams with Variable Cross Sections[S]. Int. J. Non-Linear Mech. 28:579 – 589.

Love A E H. 1944. A Treatise on the Mathematical Theory of Elasticity[M]. New York: Dover Publications, (4).

Mouring SE. 1999. Buckling and Post Buckling of Composite Ship Panelsstiffened with Preform Frames[J]. Journal of Ocean Engineering, 26:793 – 803.

Nayfeh A H. 2000. Nonlinear Interactions[M]. New York: Wiley Interscience, (1).

Nils Øvsthus Kristiansen, Ralf Peek. 2005. Designed Buckling for HP/HT Pipelines. Offshore-October.

Patel M H, Vaz M A. 1996. On the Mechanics of Submerged Vertical Slender Structures Subjected to Varying Axial Tension. Proc. R. Soc. London Ser. A 354:609 – 648.

Riks. 1979. An Incremental Approach to the Solution of Snapping and Buckling Problems[J]. International Journal of Solids and Structures, 15(7):529-551.

Sciammarella CA, Rowland E. 1974. Numerical and Analog Techniques to Retrieve and Process Fringe Information [J]. In: Bartolozzi G, editor. Proceedings of the fifth international conference on experimental stress analysis, Udine, Italy:143 – 152.

Sokolinsky VS, Frostig Y, Nutt Y. 2002. Special Behavior of Unidirectional Sandwich Panels with Transversely Flexible Core Under Statical Loading[J]. International Journal of Non-Linear Mechanics, 37:869 – 95.

Taylor N, Tran V. 1996. Experimental and Theoretical Studies in Subsea Pipeline Buckling[J]. Marine Structures, 9:211 – 257.

Timoshenko S P, Gere J M. 1961. Theory of Elastic Stability, 2nd Edition[M]. McGraw-Hill International Editions, Singapore:76.

Vaz M A, Patel M H. 1995. Analysis of Drill Strings in Vertical and Deviated Holes Using the Galerkin Technique [J]. Engineering Structures, 17(6):437 – 442.

Vaz M A, Patel M H. 1999. Lateral Buckling of Bundled Pipe System[J]. Marine Structure, 12:21 – 40.

Vaz M A, Silva D F C. 2002. Post-buckling Analysis of Slender Elastic Rods Subjected to Terminal Force[J]. Int. J. Non-Linear Mech. 34:483 – 492.

Ver ley R, Lund K M. 1995. A Soil Resistance Model for Pipelines Placed on Clay Soils[J]. Proceedings of International Conference on Offshore Mechanics and Arctic Engineering, 5:225 – 232. Pipeline Technology, Copenhagen, Denmark.

Wang C Y. 1983. Buckling and Post-buckling of a Long-hanging Elastic Column Due To a Bottom Load. ASME J. Appl. Mech. 50:311 – 314.

Wang C Y. 1997. Post-buckling of a Clamped-simply Supported Elastic [J]. Int. J. Non-Linear Mech. 32:1115 – 1122.

Zhao Tianfeng, Duan Menglan. 2010. Lateral Buckling of Non-trench High Temperature Pipelines with Pipe Lay Imperfections[J]. Pet. Sci., 7:123 – 131.

Buckling and Collapse of Sandwich Pipes under External Pressure: A Review

Chen An[1,2], Menglan Duan[2], Segen F. ESTEFEN[1]

[1] Ocean Engineering Program, COPPE,
Universidade Federal do Rio de Janeiro, CP 68508, Rio de Janeiro, 21941-972, Brazil
[2] Offshore Oil/Gas Research Center, China University of Petroleum, Beijing, China

Abstract This article reviews most of the research done in recent years (2002—2012) on the buckling and collapse behavior of sandwich pipes, which emphasizes on the development of theoretical, experimental and numerical methods adopted to analyze such structural behaviour of SP with different core materials. The main mechanical and thermal properties of the previously considered core materials are also given, together with the elastoplastic constitutive model for each material. The experimental and numerical results of collapse pressure of SP under external pressure are summarized. Besides, a general discussion of the mechanical failure modes of SP under external pressure is provided.

Key Words Sandwich pipes; Buckling; Collapse; Hyperbaric Chamber; Failure mode; Adhesion; SHCC

INTRODUCTION

Sandwich pipes (SP), considered to be an effective solution for the ultra-deepwater submarine pipeline combining high structural resistance with thermal insulation capability, have attracted considerable research interest over the last few years. Similar to other sandwich structures, SP possess the advantages of improved strength-to-weight and stiffness-to-weight ratios and tailorable characteristics to satisfy the specific requirements for subsea pipelines, flowlines and risers. To design and develop deepwater pipes, three aspects should be considered: subsea environment, physical properties of crude oil and installation loads, the details of which are described in Table 1.

Table 1 Considerations for design and development of deepwater pipes

Aspects	Details
Subsea environment	High hydrostatic pressure, low temperature, seabed topography, soil condition, etc.
Physical properties of crude oil	High-pressure and high-temperature (HPHT), hydrate plug formation, wax deposition, etc.
Installation loads	High bending moment, high clamping force, high tensile force, etc.

The current most-common industrial solution for deepwater pipes with the requirement of thermal insulation is pipe-in-pipe (PIP) systems, for which, typically, the space between the two pipes is either empty or contains insulation material which provide minimal mechanical support to the system(Kyriakides,2002a;Kyriakides,2002b;Kyriakides and Netto,2004). The carrier pipe and the inner pipe are designed independently to resist the external environmental pressure and the internal hydrocarbon pressure, respectively. Dissimilar to PIP, the design of SP is constrained simultaneously by external collapse pressure, internal bursting pressure and sufficient thermal insulation. In other words, the core layer should withstand the mechanical loads in addition to satisfying the thermal insulation requirements. As shown in Fig. 1, several types of core material have been selected and applied to the laboratory prototypes of SP developed previously by Estefen et al. (2005), Castello and Estefen(2007), Castello(2011) and An(2012a) successively, where load-bearing capacity analysis of SP subjected to external pressure were performed.

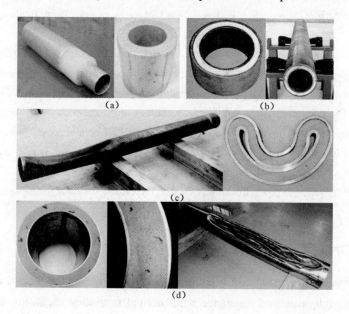

Fig. 1 SP family: (a) SP with polypropylene core(left) and with pure cement mortar(right) (Estefen et al. 2005), (b) SP with polypropylene(Castello and Estefen,2007), (c) SP with polypropylene(Castello,2011) and (d) SP with strain hardening cementitious composites core(An,2012a).

Recently huge subsalt reserves, at a water depth up to 2,400 m, were discovered off the coast of Brazil. Ultra deepwater scenarios, at depths beyond 1,500 m, require very thick-walled steel pipes or PIP systems, which are expensive and difficult to install because of their excessive weight. In this context, SP is being engineered for overcoming the above-mentioned problem and matching all the requirements for deepwater pipes. Therefore, it is necessary to review our concepts and understanding about collapse behaviour of SP under external pressure. The main aim of this review is to collate and share the studies performed on buckling and collapse analysis of SP under external pressure during the last 10 years, thereby providing a broad perspective of the current state-of-art and trends in this field. The remainder of this article is organized as follows: Section "CORE MATERIALS" gives account of the mechanical and thermal properties of the potential core material for SP systems; Subsequently, Section "BUCKLING AND COLLAPSE UNDER EXTERNAL PRESSURE" focuses on the previous investigations on buckling and collapse behaviour of SP under external pressure. Finally, a conclusion remark is presented in Section "CONCLUSIONS".

1 CORE MATERIALS

For sandwich pipes, the core material needs to contribute to external pressure resistance and insulation capabilities, therefore, choosing the right core material is crucial for mechanical and thermal performance of sandwich pipes. With this criterion, solid polypropylene(SPP)(Estefen et al. 2005;Castello and Estefen,2007;de Souza et al,2007;Castello and Estefen,2008), polyetheretherketone(PEEK)(de Souza et al. 2007), polycarbonate(PC)(de Souza et al. 2007), epoxy syntactic foam(ESF)(Castello and Estefen,2008), high density polyimide foam(HDPF)(Castello and Estefen, 2008) were preliminarily selected for evaluation, whose properties are listed in Table 2.

Table 2　Properties of suitable polymeric core materials for SP
(de Souza et al,2007;Castello and Estefen,2008)

Material	Density (kg/m³)	Yield strength (MPa)	Yield strain (%)	Elastic modulus (MPa)	Thermal conductivity [W/(m·K)]	T_{max} [1] (°C)
SPP	900	23	8.0	1000	0.20	145
PEEK	646	68	4.0	2331	0.18	348
PC	679	44	5.0	1599	0.22	188
ESF	720	22	8.5	1580	0.12	177
HDPF	500	26	9.1	521	0.066	300

[1] Maximum service temperature.

With good thermal insulation properties and high compressive strength, SPP was given the most attention as the feasible core material of SP. The elastic behaviour of SPP can be modeled by Ogden model(Ogden,1997) for incompressible isotropic hyperelastic materials, in which the material coefficients should be calibrated by the measured stress-strain data. The plastic behaviour of SPP is usually neglected due to its yield strain value being much higher than the design strain level of pipes considered during installation process, such as the Reel-lay method.

Considering the relatively low thermal insulation capacity of SP with SPP core compared to pipe-in-pipe(PIP), other polymer-based materials with lower thermal conductivity can be employed, such as ESF and HDPF. ESF is composite material synthesized by filling an epoxy resin matrix with hollow glass microspheres, which has the almost same yield strength but only half of the thermal conductivity. Except for the high price, HDPF is an advanced thermoplastic elastomer with excellent mechanical and thermal properties required by SP systems, which is extensively used in the aerospace,automotive and marine industries.

Besides polymeric materials, also cement-based materials can be adopted to fill in the annulus of SP due to high compressive strength, relatively low thermal conductivity, availability and low cost, such as pure cement mortar(Estefen et al. 2005), steel fiber reinforced concrete(SFRC)(An et al. 2012b), strain hardening cementitious composites(SHCC)(An et al. 2012c). The high tensile strain capacity of the latter two materials(SFRC and SHCC), known as specific versions of high-performance fiber-reinforced cementitious composites(HPFRCC), make them particularly suitable for resisting bending loads during deepwater installation of SP.

Pure cement mortar can be modeled by a simplified associative flow rule with isotropic hardening for the plastic regime. The yield surface, which is a function of hydrostatic stress and the von Mises equivalent stress, can be calibrated by the uniaxial compression tests and Brazilian indirect tensile tests(Estefen et al. 2005).

SFRC is proposed as the core material, since the cementitious composite possesses increased tensile strength and improved toughness under flexural loading in comparison to plain concrete, as a result of its superior resistance to cracking and crack propagation(Holschemacher et al. 2010). The Concrete Damaged Plasticity(CDP) model described in ABAQUS(ABAQUS, 2009) is employed to simulate the inelastic behaviour of damaged SFRC including stiffness degradation and crack opening. The material parameters are obtained via uniaxial tensile, uniaxial compression and four-point bending tests.

Different from the tension softening constitutive law and local cracking failure of SFRC under tension, the typical crack pattern of SHCC exhibits strain hardening behaviour and multiple fine cracking characters, which allows for large energy absorption(Li and Xu, 2011). The tensile strain of SHCC can reach to 3% ~7% with the crack width smaller than 0.1mm under uniaxial tensile tests. The typical stress-strain response and crack pattern of SHCC specimens under monotonic tensile loading are shown in Fig. 2. Note that SHCC was formerly named as "Engineered Cementitious Composites(ECC)" by the original developers(Li, 1993). SHCC is heterogeneous composites and have a natural multi-scale behaviour. For the convenience of analysis and design, however, it is often considered as a homogeneous material at the macroscopic scale. Therefore, CDP model can be also used to simulate mechanical properties of SHCC(An et al. 2012c).

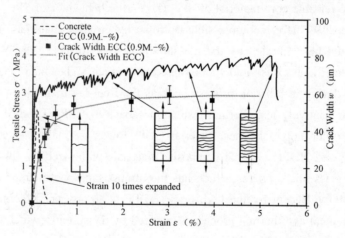

Fig. 2 Typical stress-strain response and crack pattern of ECC specimens under monotonic tensile loading(Weimann and Li, 2003)

2 BUCKLING AND COLLAPSE UNDER EXTERNAL PRESSURE

Before discussing buckling and localized collapse of SP under external pressure, the terms like "bifurcation", "buckling" and "collapse" should be reviewed. Engineering structures are a combination of materials and geometries for supporting loads. For geometrically perfect tubes under external pressure, "bifurcation buckling pressure" refers to "critical buckling pressure", at which the fundamental equilibrium state bifurcates into two equilibrium paths, one unstable path and one stable path. The fundamental equilibrium state is same as the pre-buckling state, and the stable path after bifurcation buckling pressure means the post-buckling path. "Elastic buckling" indicates the stress of the whole material pertaining to the elastic region when buckling happens, while "plastic buckling" means the stress of the partial or whole material pertaining to the plastic region when buckling happens. On the other hand, geometrical imperfection and elastic-plastic materials are usually considered in the real tubes. There is no such bifurcation point in the equilibrium path. As external pressure increases, the stiffness of tubes gradually decrease, and at the "collapse pressure", the negative stiffness can appear, leading to large deformation of the tubes.

It is necessary to perform an analytical investigation to improve understanding of the buckling and collapse behaviour of SP under external pressure. Considering the problem as a two-dimensional plane strain problem, Arjomandi and Taheri (2010) recently presented the mathematical formulation of a long three-layer circular cylindrical shell under external pressure, and obtained analytical solutions by introducing a displacement potential function to simplify the governing equation. Fully bonded interface conditions are expressed by the radial and circumferential displacement continuity, and unbounded interface conditions are described by the radial displacement continuity and zero shear stress conditions. Besides, Kardomateas and Simitses (2005) performed a three-dimensional elasticity buckling study on long sandwich cylindrical shells under external pressure, considering the facings and the core to be orthotropic in cylindrical coordinates. At the face-sheet/core interfaces, the formulations of traction continuity and displacement continuity were employed.

Till now, a series of experimental results on the collapse pressure (p_{co}) tests of SP subjected to hydrostatic pressure were reported (Santos, 2002; Estefen et al, 2005; de Souza, 2008; Castello, 2011; An, 2012a), as shown in Table 3. The core materials adopted for the small-scale SP in Santos (2002) and Estefen et al. (2005) were cement and polypropylene, and inner and outer tubes were made of aluminum, which were assembled in a way that the maximum-diameter directions of the transverse sections with the largest ovality were coincident. The specimens prepared by de Souza (2008) were also small-scale SP with polypropylene core and aluminum tubes. The initial ovality (Δ_0) is defined by:

$$\Delta_0 = \frac{D_{max} - D_{min}}{D_{max} + D_{min}}$$

Table 3 Geometric parameters and experimental for SP (reported by Santos, 2002; Estefen et al. 2005; de Souza, 2008; Castello, 2011; and An, 2012a)

Layer of SP	Tube	Diameter (mm)	Thickness (mm)	Ovality (%)	(MPa)
PIP. M1. G1. I01[①]	Inner	49.15	1.62	0.225	43.35
	Outer	74.9	1.62	0.629	
PIP. M1. G1. I02[①]	Inner	50.36	1.63	0.120	34.09
	Outer	75.92	1.65	0.266	

Continued

Layer of SP	Tube	Diameter (mm)	Thickness (mm)	Ovality (%)	(MPa)
PIP. M1. G2. I01[1]	Inner Outer	50.76 62.16	1.68 1.47	0.205 1.161	10.98
PIP. M1. G2. I02[1]	Inner Outer	50.73 62.28	1.67 1.47	0.260 0.698	12.11
PIP. M2. G1. I02[1]	Inner Outer	49.64 75.40	1.68 1.62	0.456 0.301	37.64
PIP. M2. G1. I03[1]	Inner Outer	49.76 75.19	1.62 1.61	0.186 0.255	31.14
PIP. M2. G2. I01[1]	Inner Outer	49.94 62.10	1.70 1.46	0.364 0.801	20.31
PIP. M2. G2. I02[1]	Inner Outer	50.03 62.40	1.69 1.49	0.547 0.552	17.13
DS. PP. P. 01[2]	Inner Outer	51.40 76.82	1.53 1.65	1.101 0.477	18.2
DS. PP. P. 02[2]	Inner Outer	51.05 76.92	1.49 1.64	0.670 0.569	15.8
DS. PP. P. 03[2]	Inner Outer	51.61 76.90	1.53 1.65	1.117 0.543	17.1
SP1-I[3]	Inner Outer	168.3 219.1	6.35 6.35	0.18 0.40	27.9
SP1-II[3]	Inner Outer	168.3 219.1	6.35 6.35	0.20 0.39	27.2
SP2-I[3]	Inner Outer	168.3 219.1	6.35 6.35	0.28 0.35	21.0
SP2-II[3]	Inner Outer	168.3 219.1	6.35 6.35	0.41 0.34	26.4
SP5-I[3]	Inner Outer	168.3 219.1	6.35 6.35	0.53 0.40	35.7
SP5-II[3]	Inner Outer	168.3 219.1	6.35 6.35	0.43 0.35	29.7
SP1[4]	Inner Outer	152.4 202.7	1.8 2.0	0.32 0.41	30.5
SP2[4]	Inner Outer	152.5 203.0	1.8 2.0	0.22 0.47	30.6
SP3[4]	Inner Outer	152.6 202.9	1.8 2.0	0.23 0.39	29.7

[1] Label of SP used in Estefen et al. (2005), where M1 and M2 indicated the core materials were cement and polypropylene, respectively, and G1 and G2 identified the different geometries.
[2] Label of SP used in de Souza(2008).
[3] Label of SP used in Castello(2011), where only the nominal diameter and thickness were presented.
[4] Label of SP in An(2012a).

And the largest ovality can be determined by the maximum value of at different sections of SP. Tensile tests of the coupons cut off in the longitudinal direction of each tube were performed to obtain the material properties of the tubes. Besides, four strain gages were mounted equidistantly in the hoop direction at the mid-section of the SP, the data of which revealed membrane-strain-dominated status of SP. Shifting to bending-strain-dominated status throughout the collapse tests. The test started from placing the capped SP inside the hyperbaric chamber, then the vessel was sealed and the cavity was completely filled with water, as shown in Fig. 3. A pneumatic control system drove a positive displacement pump that pressurized the chamber gradually. The pressure inside the chamber was monitored at the pump outlet using a pressure transducer, which was connected to a computer for data acquisition. The scheme approximates "volume-controlled" (constant flow at a slow rate) pressurization. Collapse of the SP was recognized by a loudly audible "bong", and a sudden drop in monitored pressure.

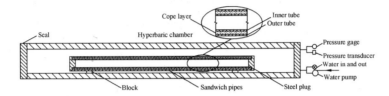

Fig. 3 Sketch of testing facility for hydrostatic pressure collapse of SP

Castello(2011) and An(2012a) performed the large-scale laboratory tests of SP to investigate the collapse pressure subjected to external pressure using hyperbaric chamber. Before testing, the minimum length of SP for was evaluated numerically, as additional rigidity to the ends of SP caused by the metal plugs can influence the of SP. The post-collapse failure modes in flat shape and U-shape of SP with polypropylene core and API X-60 tubes were found in Castello(2011), for which the flat shape was associated with larger initial imperfections and lack of interfacial adhesion, while the U-shape was caused by higher deformation energy and required higher. Fig. 4 shows post-collapse failure modes for SP1 - I and SP5 - I in Castello(2011) by cutting the section from tested SP, where "6 - 24" direction was consistent with the maximum diameter of the outer tube. Three SP with SHCC core and Stainless Steel 304 tubes were tested for the by An(2012a), which had the mean value of 30.3 MPa and the standard deviation of 0.37 MPa. The consistency of experimental results was due to the well-bonded interface between SHCC and tubes and the high ductility of SHCC, which can prevent the localized splitting of core material before the collapse of SP.

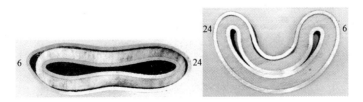

Fig. 4 Post - collapse failure modes for SP1 - I and SP5 - I(Castello, 2011)

Finite element(FE) method was widely used to investigate the collapse behavior of SP under external pressure numerically(Estefen et al. 2005; Castello and Estefen, 2007; de Souza et al. 2007; Arjomandi and Taheri, 2011a; Arjomandi and Taheri, 2011b; Arjomandi and Taheri, 2011c; Castello, 2011; An, 2012a; An et al. 2012b). The types of elements employed by the previous studies and the FE results together with the comparisons with the experimental ones are listed in Table 4.

Table 4 The types of elements in ABAQUS(2009) employed by the previous studies and the related FE results together with the comparisons with the experimental ones

Layer of SP	Types of elements	Model features	(MPa)	(MPa)	(MPa)
PIP. M1. G1. I01[①]	CPE8	Quarter ring	46.23	37.97	43.35
PIP. M1. G1. I02[①]	CPE8	Quarter ring	44.78	38.05	34.09
PIP. M1. G2. I01[①]	CPE8	Quarter ring	24.70	8.11	10.98
PIP. M1. G2. I02[①]	CPE8	Quarter ring	25.74	10.03	12.11
PIP. M2. G1. I02[①]	CPE8	Quarter ring	39.56	12.84	37.64
PIP. M2. G1. I03[①]	CPE8	Quarter ring	38.27	12.52	31.14
PIP. M2. G2. I01[①]	CPE8	Quarter ring	20.84	6.32	20.31
PIP. M2. G2. I02[①]	CPE8	Quarter ring	22.42	6.89	17.13
PIP. M2. G1. I02[②]	C3D20 C3D20H	Half ring	32.7	12.3	37.64
PIP. M2. G1. I03[②]	C3D20 C3D20H	Half ring	32.2	12.0	31.14
PIP. M2. G2. I01[②]	C3D20 C3D20H	Half ring	20.2	6.5	20.31
PIP. M2. G2. I02[②]	C3D20 C3D20H	Half ring	21.5	7.0	17.13
DS. PP. P. 01[②]	C3D20 C3D20H	Half ring		16.4	18.2
DS. PP. P. 02[②]	C3D20 C3D20H	Half ring		16.2	15.8
DS. PP. P. 03[②]	C3D20 C3D20H	Half ring		16.6	17.1
SP1 – I[③]	CPEG4	Full ring	27.6		27.9
SP1 – II[③]	CPEG4	Full ring	27.1		27.2
SP2 – I[③]	CPEG4	Full ring	21.1		21.0
SP2 – II[③]	CPEG4	Full ring	28.8		26.4
SP5 – I[③]	C3D27 C3D27H	Full length	36.5		35.7

Layer of SP	Types of elements	Model features	(MPa)	(MPa)	(MPa)
SP5 – II[3]	C3D27 / C3D27H	Full length	29.6		29.7
SP1[4]	C3D8R	Quarter ring	35.6	31.6	30.5
SP2[4]	C3D8R	Quarter ring	36.1	32.0	30.6
SP3[4]	C3D8R	Quarter ring	35.3	32.1	29.7

① SP simulated by Estefen et al. (2005).
② SP simulated by de Souza(2008), where C3D20 was for tubes, and C3D20H was for polypropylene core.
③ SP simulated by Castello(2011), where C3D27 was for tubes, and C3D27H was for polypropylene core in the full-length model.
④ SP simulated by An(2012a). and mean the numerically predicted collapse pressure of SP with fully bonded and unbonded interface conditions, respectively.

Before starting simulation, the sensitivity to adopted mesh density should be analyzed. Commonly, two limit cases of "perfect-adhesion" (fully bonded) and "no-adhesion" (unbonded) interface conditions were considered in the FE model (Santos, 2002; Segen et al. 2005; de Souza et al. 2007; de Souza, 2008; An, 2012a; An et al. 2012b). Castello and Estefen (2007) investigated the effect of interfacial adhesion degree between polypropylene core and API X-60 tubes, defined by maximum shear stress, on the of SP, and concluded that a maximum shear stress of 14 MPa could be used to simulate the perfect-adhesion interface condition for the specific cases considered in that study. Castello(2011) predicted the of SP with polypropylene core using more realistic interface bonding conditions, which were inspected by ultrasonic technique before testing and examined by cutting through the SP after testing. To formulate the practical equations for evaluating the of SP subjected to external pressure, comprehensive parametric studies were carried out in Arjomandi and Taheri (2011a), Arjomandi and Taheri (2011b) and Arjomandi and Taheri (2011c), where 3840, 3000 and 12000 different SP configurations covering a range of practical design parameters (such as core thickness, core stiffness, tube thickness, steel grades, interface adhesion properties, etc.) were simulated, respectively. In these studies, in-house developed Python scripts, which could generate and run FE models in ABAQUS, were utilized to facilitate and expedite the parametric studies. The effect of plugs, installed at the ends of SP, on the were considered to simulate accurately the hydrostatic pressure tests by de Souza(2008), Castello(2011) and An(2012a), which can decrease the predicted (comparing to the results without considering the effect of plugs) for the SP cored by polypropylene (de Souza, 2008), and increase the predict for the SP cored by SHCC(An, 2012a). It is hard to investigate the failure modes of SP under external pressure through the collapse tests (An, 2012a), therefore, numerical studies are usually performed to understand the mechanism of structural failure. The of SP can decrease significantly due to the increase of initial ovality and wall thickness eccentricity of the tubes and loss of adhesion at the interfaces, which belong to manufacture process-induced failure. The post-collapse failure modes in flat shape and U – shape of SP were well produced by numerical simulation considering the local loss of adhesion at the interfaces in Castello(2011). Some other failure mechanism concerning the collapse of SP under external pressure is design-induced failure, such as compressive crushing of core material (due to not sufficient compressive strength) and collapse of the inner tube(due to structural instability), which were discussed on the case of the SP with SHCC core by introducing the compression damage parameter in the core material constitutive relationship(An,2012a).

CONCLUSIONS

In this study, buckling and collapse behaviour of SP under external pressure are reviewed. Based on the available literature, several conclusions can be drawn as follows: (1) The design and development of more advanced materials for the core layer is essential for SP to satisfy the deepwater operation requirements; (2) A real-time monitoring system should be designed to investigate the failure mechanism of SP under external pressure; (3) FE models further developed should be capable to predict the actual failure modes of SP in addition to the failure loads.

ACKNOWLEDGEMENTS

An Chen and Segen F. Estefen acknowledge gratefully the financial support provided by the Brazilian agencies CNPq, CAPES, FAPERJ and ANP to their research work. The financial support of a group of industrial sponsors including TENARIS/CONFAB and SINOCHEM PETRÓLEO BRASIL LIMITADA is also acknowledged with thanks. This work is also supported by China National 973 Project(No. 2011CB013702) and National Natural Science Foundation of China(No. 50979113). Special thanks to Prof. Dr. Su Jian of Nuclear Engineering Program at COPPE/UFRJ for contributing his expertise. An Chen acknowledges gratefully the financial support provided by the China Scholarship Council.

REFERENCES

ABAQUS(2009). User's and Theory Manuals Version 6.9 – 1. Hibbit, Karlsson and Sorensen, Inc., RI, USA.

An C, Castello X, Duan M L, Toledo Filho, R D, Estefen S F. 2012b. Ultimate Strength Behaviour of Sandwich Pipes Filled with Steel Fiber Reinforced Concrete[J]. Ocean Eng, 55(1): 125135.

An C, Castello X, Oliveira A M, Duan M L, Toledo Filho, RD, Estefen SF. 2012c. Limit Strength of New Sandwich Pipes with Strain Hardening Cementitious Composites(SHCC) Core: Finite Element Modelling [C]. Proc 31st Int C Ocean Offshore Arctic Eng, Rio de Janeiro, Brazil, OMAE 2012 – 83589.

An C. 2012a. Collapse of sandwich pipes with PVA fiber reinforced cementitious composites core under hydrostatic pressure, D. Sc. Thesis, Department of Naval Architecture and Ocean Engineering, the Federal University of Rio de Janeiro, Rio de Janeiro, RJ, Brazil.

Arjomandi K, Taheri F. 2010. Elastic Buckling Capacity of Bonded and Unbonded Sandwich Pipes Under External Hydrostatic Pressure[J]. J Mech Mater Struct, 5(3): 391 – 408.

Arjomandi K, Taheri F. 2011a. A New Look at the External Pressure Capacity of Sandwich Pipes [J]. Mar Struct, 24(1): 23 – 42.

Arjomandi K, Taheri F. 2011b. Stability and Post-buckling Response of Sandwich Pipes Under Hydrostatic External Pressure[J]. Int J Pres Ves Pip, 88(4): 138 – 148.

Arjomandi K, Taheri F. 2011c. The Influence of Intra-layer Adhesion Configuration on the Pressure Capacity and Optimized Configuration of Sandwich Pipes[J]. Ocean Eng, 38(17 – 18): 1869 – 1882.

Castello X, Estefen S F. 2007. Limit Strength and Reeling Effects of Sandwich Pipes with Bonded Layers[J]. Int J Mech Sci, 49(5): 577 – 588.

Castello X, Estefen S F. 2008. Sandwich Pipes for Ultra Deepwater Applications[C]. Proc Offshore Tech C, Houston, Texas, USA, OTC – 19704.

Castello X. 2011. Influence of Adhesion Between Layers on the Collapse Resistance of Sandwich Pipes, D. Sc. Thesis, Department of Naval Architecture and Ocean Engineering, the Federal University of Rio de Janeiro, Rio de Janeiro, RJ, Brazil(in Portuguese).

de Souza A R, Netto T A, Pasqualino I P. 2007. Materials Selection for Sandwich Pipes Under the Combined Effect of Pressure, Bending and Temperature[C]. Proc 28th Int C Ocean Offshore Arctic Eng, San Diego, California, USA, OMAE 2007 – 29128.

de Souza A R. 2008. Structural Strength of Sandwich Pipes Under External Pressure, Longitudinal Bending and Thermal Loading, M. Sc. Thesis, Department of Naval Architecture and Ocean Engineering, the Federal University of Rio de Janeiro, Rio de Janeiro, RJ, Brazil(in Portuguese).

Estefen S F Netto T A, Pasqualino I P. 2005. Strength Analyses of Sandwich Pipes for Ultra Deepwaters[J]. J Appl Mech – T ASME, 72(4):599608.

Holschemacher K, Mueller T, Ribakov Y. 2010. Effect of Steel Fibres on Mechanical Properties of High – strength Concrete[J]. Materials & Design, 31(5):2604 – 2615.

Kardomateas G A, Simitses G J. 2005. Buckling of Long Sandwich Cylindrical Shells Under External Pressure[J]. J Appl Mech-T ASME, 72(4):493 – 499.

Kyriakides S, Netto T A. 2004. On the Dynamic Propagation and Arrest of Buckles in Pipe-in-pipe Systems[J]. Int J Solids Struct, 41(20):5463 – 5482.

Kyriakides S. 2002a. Buckle Propagation in Pipe-in-pipe Systems. Part I. Experiments[J]. Int J Solids Struct, 39(2):351 – 366.

Kyriakides S. 2002b. Buckle Propagation in Pipe-in-pipe Systems. Part II. Analysis[J]. Int J Solids Struct, 39(2):367 – 392.

Li Q, Xu S L. 2011. Experimental Research on Mechanical Performance of Hybrid Fiber Reinforced Cementitious Composites with Polyvinyl Alcohol Short Fiber and Carbon Textile[J]. J Compos Mater, 45(1):528.

Li V C. 1993. From Micormechanics to Structural Engineering-The Design of Cementitious Composites for Civil Engineering Applications[J]. J Struc Mech Earthq Eng-JSCE, 10(2):3748.

Ogden R W. 1997. Non-linear Elastic Deformation. New York: Dover Publication.

Santos JMC. 2002. Sandwich Pipes for Ultra-deep Waters, M. Sc. Thesis, Department of Naval Architecture and Ocean Engineering, the Federal University of Rio de Janeiro, Rio de Janeiro, RJ, Brazil(in Portuguese).

Weimann M B, Li V C. 2003. Hygral Behavior of Engineered Cementitious Composites(ECC)[J] Int J Restor Buildings Monuments, 9(5):513 – 534.

Investigation on Failure Mechanism and Prevention Methods of Steel Catenary Riser

Jing Cao[1], Lei Guo[2], Ying Jiang[3], Lei Yang[2], Ke Tang[2], Menglan Duan[2]

[1] CNOOC Research Institute, Chaoyang, Beijing, PR China
[2] Offshore Oil/Gas Research Center, China University of Petroleum, Beijing, China
[3] Offshore Oil Engineering Co., Ltd., Tianjin, China

Abstract This paper mainly elaborates failure and its causes about steel catenary riser (SCR) as well as preventive measures according to different failure modes. Marine engineering accident is very complex and serious because of its high-tech, high-risk characteristics, especially the complex environment of deep-water, as a result, the emergency treatment to deal with accident of subsea production systems will become a big challenge. So it is very necessary to take prevention in advance of subsea accident. First, we do analysis about failure modes of SCR and give some leading factors about them. Then, we do further study on the cause of formation about corrosion, fracture and fatigue damage of SCR as well as bring forward measures of emergency repair. Especially, a new prevention program in connection with fatigue failure of hanging end of SCR is found, and a new device that can slow down the fatigue of hanging end is designed, too.

Key Words Failure mechanism; Failure; SCR; Anti-fatigue device; Emergency repair

INTRODUCTION

In order to meet the increasingly rapid growth of oil and gas consumption, exploiting offshore oil and gas resources vigorously has become an irresistible trend, and underwater engineering has also gotten a rapid development along with it. However, underwater engineering with a complex environment has the characteristics of high-tech and high-risk, making the accident consequences of subsea production systems very serious, and then the emergency treatment of accidents and failure will be very difficult. According to the survey, the global quantity of petroleum hydrocarbon leaks into the ocean by a variety of routes is about 6 million tons per year, among which a large part is caused because of the accidents during the process of oil exploiting and transporting. As we all know, the explosion of Deepwater Horizon rig in the southern Gulf of Mexico led to a pollution belt over 100km in April 2010, which caused serious damage to the marine ecosystem. So the cost will be very heavy once oil spill occurs and taking early prevention in advance of subsea accidents is very important. Offshore riser as the neck of a subsea production system is a very critical structure. Hence taking study on riser's failure and its prevention based on the preceding analysis is obviously necessary.

There are two typical failure modes of risers, which are corrosion damage and fatigue damage. And scholars have done a lot of scientific research and analysis about riser's fatigue failure, for example, Bybeee and Karen predicted vortex-induced motion (VIM) of the Classic SPAR platform through theoretical analysis, model tests and field measurements and so on in 2005 and concerned VIM fatigue of the anchor chains and risers meanwhile. Halkyard did computational fluid dynamics (CFD) simulation about Truss SPAR models with his team and then compared simulation results with test results during 2005 and 2006, which shows that responses of VIM have a strong relationship with design size and arrangement of the spiral side panels as well as arrangement of the housing attachment outside of the platform. Finnigan and Roddier studied the effect of different spiral side panels through VIM tests about a SPAR in 2007. In addition, some scholars investigated deformed risers with pits which laid the foundation to estimate remaining life about injured risers. Durkin proposed a detailed analytical model for the fitting analysis of pits to determine the ultimate strength of the pitted pipe member with both axial load and bending moment. In recent years, Brooker carried out quasi-static response analysis about a lateral pitted tube with both ends constrained through finite element method based on different wall thicknesses, diameters and length and yield stress level parameters. Since Mattock et al proposed short-term static and cyclic loading pile of $p-y$ curve method of soft soil, many countries explored $p-y$ curve method actively and carried out a number of related numerical simulation as well as experimental studies.

1 TWO TYPICAL FAILURE OF RISERS

1.1 Corrosion Failure

Corrosion damage can be divided into two cases by morphology: general corrosion and localized corrosion.

General corrosion is that corrosion may be distributed across the entire metal surface uniformly or maybe unevenly. For example, corrosion of carbon steel in strong acid is the case. But the danger of uniform corrosion is relatively small, because if we know the rate of corrosion we can infer the life of the material and only need to put this factor into account in the design.

Localized corrosion is defined as that corrosion is mainly concentrated in a certain area of the metal surface, while the other part of the surface is almost unspoiled. Common types of localized corrosion are such as pitting, galvanic corrosion, hydrogen embrittlement, stress corrosion cracking, intergranular corrosion, selective leaching, crevice corrosion, corrosion fatigue, hydrogen induced cracking, and erosion corrosion and so on.

Study found that H_2S and CO_2 are the most dangerous corrosion factors for pipe wall, and especially H_2S could result in sudden sulfide stress cracking (SSC) of the metal material. But, only H_2S and CO_2 dissolved in water can be corrosive. A large number of studies have shown that H_2S and CO_2 tend much stronger corrosive when they are dissolved in the aqueous solution which content salts or residual acid than that of single H_2S and CO_2 solution, and corrosion rate will be several times higher or even hundreds of times.

(1) Process of hydrogen sulfide electrochemical corrosion.

The reaction of H_2S in water can be expressed as:

$$H_2S \Longleftrightarrow H^+ + HS^- \qquad HS^- \Longleftrightarrow H^+ + S^{2-}$$

Electrochemical corrosion process of H_2S solution for steel can be expressed as:

Anodic reaction:

$$Fe \longrightarrow Fe^{2+} + 2e$$

Cathodic reaction:

$$2H^+ + 2e \begin{cases} \longrightarrow H_{ad} + H_{ad} \longrightarrow H_2 \\ \longrightarrow H_{ad} \longrightarrow \text{proliferate in steel} \end{cases}$$

Anode reaction production:

$$Fe^{2+} + S^{2-} \longrightarrow FeS \downarrow$$

Where, Had stands for the hydrogen atom adsorbed on the surface of steel; Hab stands for the hydrogen atom adsorbed in steel which is a factor of hydrogen induced cracking (HIC).

Iron sulfide corrosion products generated by the anodic reaction usually have a defective structure with poor adhesive force which is easy to fall off and to be oxidized. And they constitute active micro batteries with the steel substrate as the cathode, so the steel substrate corrosion continues. The main component of the corrosion products are Fe_9S_8, Fe_3S_4, FeS_2 and FeS.

(2) Corrosion mechanism of carbon dioxide.

Anodic reaction:

$$Fe \longrightarrow Fe^{2+} + 2e$$

Cathodic reaction:

$$H_2O + CO_2 \longrightarrow 2H^+ + CO_3^{2-}$$

$$2H + 2e \longrightarrow H_2 \uparrow$$

Corrosion products:

$$Fe + H_2CO_3 \longrightarrow FeCO_3 + H_2 \uparrow$$

Product $FeCO_3$ is also easy to fall off, which is similar to H_2S corrosion.

1.2 Fatigue Failure

There are mainly three aspects about riser structure fatigue as follows:

(1) The reciprocating movement of operating platform will cause riser's cyclical movement, which maybe cause fatigue failure at the hanging end of the riser especially deepwater risers. Generally speaking, the connecting means of steel catenary riser (SCR) and platform is that flexible connector is linked directly to the platform's support frame. Because the flexible connector is fixed in the slot of the support frame and dozens of tons of deepwater riser's gravity are all focused on the hanging end, so stress concentration and fatigue crack will appear easily near the connecting place of the flexible connector and the riser end. As a result, it will evolve into a sudden rupture or even serious oil spills if the fatigue crack is not found early.

(2) Wave and flow induced riser's reciprocating movement is called vortex induced motion (VIM) which can cause very serious fatigue failure to both hanging end and touch down point (TDP). Scholars have done a lot of researches about riser's TDP fatigue during past years. For

example, Martins et al studied the fatigue behavior of TDP of deepwater SCR, during which they used simplified model, linear approximation, progressive representation and frequency range of solution and so on to do related analysis. They found the progressive formula of TDP's movement using the frequency range solution of the existing linear problem and then got its fatigue life as well as total accumulation damage based on $S - N$ curve which indicates the fatigue behavior of the material.

(3) Riser's vibration formed by the action of water current is called vortex induced vibration (VIV). Because current can act on the entire length of the riser, so water flow caused VIV become the main source of structural fatigue damage, and therefore it is an important significance for deepwater oil field development to take appropriate suppression measures to prevent deepwater riser VIV.

2　PREVENTING MEASURES OF RISER'S FAILURE

2.1　Corrosion Failure Prevention

(1) Adding coat methods.

There are dozens of pipeline anti-corrosion coatings that liquid epoxy coating (H87, HT515, epoxy glass flake and ceramic epoxy, etc.), fusion bonded epoxy powder, cement mortar lining and phenolic resin coating are commonly used. Practice has proved that the most mature and economical coating is undoubtedly epoxy resin-based coating. The American Gas Association (AGA) considers that epoxy resin-based coatings are optimized internal coatings for gas pipelines after research and screen 25 kinds of coatings. But the American Petroleum Institute (API) specification generally recommends using amine-cured epoxy coatings. It has been proved that epoxy and epoxy-polyurethane anti-corrosion coatings with better performance of oil resistance, heat resistance and resistance to oilfield wastewater, so, if the construction quality is reliable it can keep a good protection up to 10 years though tests and practices.

We usually choose extruded polyethylene and coal tar enamel anticorrosive coatings which have long-term water-resistant and chemical resistant ability for offshore oil and gas pipelines. But fusion bonded epoxy or modified polypropylene and other materials will be selected in the case of transmission medium with high temperature.

(2) Cathodic Protection methods.

Cathodic protection not only can slow uniform corrosion of metals in sea water, fresh water, soil and chemical medium but also prevent pitting, intergranular corrosion, stress corrosion cracking, corrosion fatigue, stray current corrosion and biological corrosion from happening well. We can use two specific programs to protect cathode. One way is adding sacrificial anode that an anode is attached to a metal object to inhibit the object's corrosion. The anode is electrolytically decomposed while the object remains free of damage. The other way is adding current that cathode current from an external DC power (rectifier or constant potential instrument) is applied directly to the object structures to make it generate cathodic polarization, so object's corrosion is prevented, too. Therefor an auxiliary anode, a reference electrode, a DC power supply and associated connecting cables will be needed.

2.2 Prevention of Fatigue Failure

Appropriate measures are needed to prevent or suppress VIV in order to reduce the harm to risers, where, there are two methods. On the one hand, adjust the dynamic characteristics of the structure itself to reduce its responses from vortex effect; on the other hand, interfere with or change conditions of vortex occurring or flow pattern to weaken vortex-induced fluid force. The first method is to follow certain guidelines in the structural design, including controlling structure shapes to ensure the reduced speed of the steady flow is not in the scope of causing VIV and increasing effective mass and damping of the structure to suppress or reduce VIV.

The second method is the hydrodynamic method which is to prevent the formation and strengthening of the vortex. So we always use different forms of spoiler device arranging on the structure surface or at the place within the scope of the wake flow to change the position of separation point as well as destruct the required length, location and interactions of a vortex forming. Thereby the formation and bleeding of a vortex is prevented and the vibration of the structure suppressed. But the natural frequency of the structure is possibly reduced if increase mass, so the likelihood of occurrence of VIV will increase. Therefore, we usually add a suppressing device in the outside of the riser to weaken VIV in the actual project.

2.3 Anti-Fatigue Device

There are about three categories of VIV suppressing devices depending on the spoiler way as follows: (1) Surface protrusions, such as lines, helical strakes, hemispherical surface, which enables to separate flow; (2) Devices which can divide flow into many small vortices, for example, porous covering, control rod, axial slat and so on; (3) The near wake stabilizing device, for example, fairings and streamers. And studies show that helical strakes and fairings are two efficient VIV suppression devices (see Fig. 1 and Fig. 2).

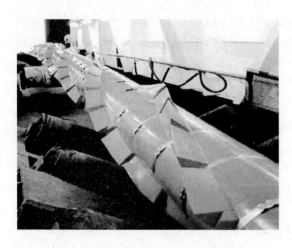

Fig. 1　Helical strakes

Fig. 2　Fairings

The role of fairings is to make flows bypass the riser stably and also slow down the vibration excited by vortex. Fairings on sale are likely as Fig. 2. The better of streamline shape, the more smoothly whirlpools will bypass the riser, and then the vibration generated by Vortex will be much more light. So we design a new kind of fairing based on good streamline shape as Fig. 3.

Scholars have concerned for a long time about the fatigue of TDP caused by VIV, who constructed the pipe-soil interaction model and did many model experiments, too. Yongqiang Dong et al studied riser's movement introduced by semi-submersible platform, single-column platforms and tension leg platforms, and obtained the fatigue life of both top end and TDP through the method that the movement of the platform is applied to the top of the riser based on ABAQUS numerical model during nonlinear analysis. DUAN Menglan et al designed an experimental system which considered the combination of tension, bending, internal and external pressure, pipe-soil interaction and so on, and researched TDP fatigue of SCR through some pipe-soil experiment.

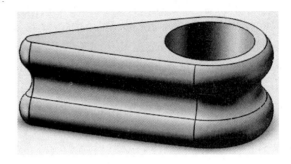

Fig. 3 New kind of Fairings

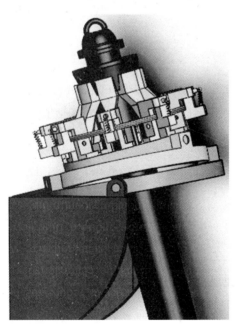

Fig. 4 Anti-fatigue device of hanging end

Perhaps because the distance between platform and the hanging end of risers is very close and the maintenance is relatively easy we concern relatively few about the hanging end, but the fatigue failure of hanging end should not be ignored. It is obvious that the hanging end of the riser is nearly fixed and the motion which is almost rigid passed between platforms and risers continues as time goes on. Therefore, the hanging end suffers strong cyclic bending force and then fatigue failure even fracture will appear if cyclic force keeps for a long time. So we designed an anti-fatigue device based on the principle of universal joint inspired by the gyroscope for riser's hanging end as Fig. 4. The device has three sets of hinges that the angles between the hinge axis are all 60°, which can increase the degree of freedom of the hanging end, and it could eliminate the moment completely that almost any fatigue is prevented here. We can see from the figure that the device's petal-like bracket is composed by eight identical petal structure which can support riser's top end suitably. The process of installing risers on the anti-fatigue device is like this: hang the riser from the down side to the upper side though the ring of the device, and the top end of the riser will extrude the petal components to open, and then they will close under the action of the spring when the riser' stop end gets a upper position. Finally, lower down the riser's top end to the petal-like groove seat and then finish the installation.

CONCLUSION

The complex environment of the ocean bring great inconvenience to offshore oil and gas exploiting and safety monitoring of equipment is relatively complex, too, so the corrosion of the pipe wall is not easy to find in particular. And internal corrosion with strong unpredictability could not be detected yet. In fact, as long as we know the rate of general corrosion, we can predict the general corrosion time, and then the life of the equipment can be designed suitably. Though the area of localized corrosion is generally smaller, corrosion of pits and cracks is usually deep and not easy to find, so the probability of fracture occurs in localized corrosion area is very large if there is cyclic stress in it, too. So, localized corrosion is more dangerous than general corrosion.

We are very concerned about riser's fatigue failure and its prevention after analyzing the mechanism of fatigue failure of deepwater SCR. First, we have designed a new type of Fairings according to the characteristics of the flow line which can be more conducive for currents bypassing the riser stably. As a result, VIV will be much smaller. Second, in order to prevent the fatigue failure of hanging end, we also designed a set of anti-fatigue device in particular at the same when we focus on reducing VIV. Therefore, the fatigue failure of riser's hanging end could be prevented effectively by increasing degrees of freedom of the riser through the anti-fatigue device which can offer free motion between the platform and the riser.

ACKNOWLEDGEMENT

Thank the support of the project of Underwater Emergency Repairs Equipment and Technology(2011ZX05027 -005), thank the co-authors of this paper provide a variety of help.

REFERENCES

Brooker D C. 2004. A Numerical Study on the Lateral Indentation of Continuously Supported Tubesa[J]. Journal of Constructional Steel Research,60:1177 -1192.

Bybee,Karen. 2005. SPAR Vortex Induced Vibration Prediction[J]. Journal of Petroleum Technology,57(2):61 -62.

Dong Yongqiang,Sun Liping. 2008. Comparision Study of Different Floater Effect on Deepwater SCR[C]. Proceedings of the ASME 27th International Conference on Offshore Mechanics and Arctic Engineering OMAE 2008,Portugal.

Durkin S. 1987. An Analytical Method for Predicting the Ultimate Capacity of a Dented Tubular Member[J]. Int. J. Mech. Sci. 29:449 - 467.

Duan Menglan,Hu Zhihui,Cao Jing,Zhao Tianfeng,Jun Fang. 2011. The Design of Fatigue Experimental System for Steel Catenary Riser in Touchdown Zone[J]. Mechanicss and Practice(in Chinese),33(3):15 -19.

Finnigan T,Roddier D. 2007. Spar VIM Model Tests at Supercritical Reynolds Numbers. Proceedings of the 26th International Conference on Offshore Mechanics and Arctic Engineering [C] (OMAE). San Diego, California,USA.

Matlock H. 1970. Correlation for Design of Laterally Loaded Piles in Soft Clay [J]. OTC,25(1):577 -594.

Martins C A,Higashi E,Silva R M C. 2000. A Parametric Analysis of Steel Catenary Risers:Fatigue Behavior Near the Top[C]. Proc of the 10th International Offshore and Polar Engineering Conference,Seattle,USA,II:54 -59.

Numerical Simulation of Submarine Pipeline Upheaval Buckling with Different Imperfection Styles

Guomin[1] Sun, Xiaguang[2] Zeng, Menglan[2] Duan, Xiaoyu[3] Che, Shuangni[2] Tan

[1] Offshore Oil Engineering Co, Ltd. Tianjin, China
[2] Offshore Oil/Gas Research Center, China University of Petroleum, Beijing, China
[3] Department of Mechanics and Engineering Science, Fudan University, Shanghai 200433, China

Abstract Upheaval buckling is likely to damage HT/HP submarine pipeline in service. The simple calculation formulas developed so far cannot give out accurate results and cannot satisfy the more rigid design requirements. Initial imperfection brings the biggest uncertainty on this problem. To investigate its influence, pipeline upheaval buckling has been simulated by numerical calculation technology-FEM in this paper. By a lot of simulation some factors related to initial imperfection are investigated. Simulating results show that pipeline imperfection shape has large influence on upheaval buckling critical load and that's why the analytical methods with simple initial imperfection assumption sometimes cannot give out accurate results.

Key Words Pipeline; Upheaval buckling; Numerical simulation; Influence factor

INTRODUCTION

Buried pipeline may buckle out of the covering soil when it works under high temperature and high pressure (HT/HP) conditions. This phenomenon is called upheaval buckling. It is likely to damage pipeline significantly, so predicting possible upheaval buckling is very important to design a safe buried pipeline.

For submarine pipeline upheaval buckling analysis many researching works have been done in the last three decades. Hobbs examined vertical and lateral buckling of the perfect pipeline without initial out-of-straightness in 1984. Neil Taylor, Aik Ben Gan and Vinh Tran presented a set of analyses of submarine pipeline buckling which incorporate with structural imperfections and deformation-dependent axial friction resistance. P. Terndrup Pedersen and J. Juncher Jensen presented a design criterion for design against gradual upheaval creep of buried marine pipelines subjected to time varying temperature loadings Palmer and Ellinas (1990) developed a semi-empirical design methods based on a numerical analysis MALTBY and CALLADINE (1994) described an experimental study of upheaval buckling of buried pipelines and gave out the theory and analysis about the experimental observations. James G. A. Croll (1997) established a simplified model of upheaval buckling of subsea pipelines and the model provided a direct and simplified analysis of upheaval buckling considering initial imperfect geometries. Barefoot A. J. (1998), Kvalstad T. J. (1999), White D. J., Barefoot A. J. and Bolton M. D. investigated the soil resistance behavior when pipeline upheaval buckling in it. Suzuki and Toyoda(2002) discussed the seismic loading on buried pipelines and deformability of high-strength line pipe. Marius Loen Ommundsen (2009) performed upheaval buckling experiments in 2009 to detect a general behavior of a pipe in different scenarios and simulated these experiments by ANSYS.

The critical axial force is the key physical quantity of pipeline upheaval buckling analysis and different pipeline segment has different critical axial force. Generally speaking, the critical axial force is related to three factors, namely, pipe flexural stiffness, out-of-straightness and downward force on the pipeline. Here the out-of-straightness is initial imperfection of pipeline and it is the most uncertain factor in the analysis of upheaval buckling. So far the problem with these analyzing works mentioned above is that how much influence dose the initial imperfection assumption induce to the critical axial force is not clear. In this paper this influence is researched in detail by numerical simulating some buried pipeline segment with different profiles of out-of straightness.

1 METHOD

To investigate the influence of imperfection on the critical axial force, three types of imperfection profile were established. The first kind imperfection profile is smooth spline through three points, the second is similar to the imperfection assumption mentioned in reference 6, its expression is $h/2 \cdot [\cos(2p_i \cdot x/l) + 1]$ and the third is similar to morlet wavelet function, its expression is $h \cdot \cos(p_i \cdot x/l) \cdot \exp[-(x/10)^2]$. The value of x ranges from $-1/2$ to $1/2$. Where h is the height of imperfection, and l is the length of pipeline segment. In this paper h was assigned five values, namely, 0.1, 0.3, 0.5, 0.7 and 0.9 and l was valued 50. The configurations of those imperfections are shown in Fig. 1 to Fig. 5.

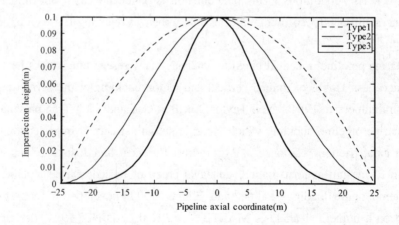

Fig. 1 Three imperfection types with h valued 0.1

15 finite element models were established using ABAQUS. Their parameters are the same except with different imperfection types and height. These data are shown in Table 1, Table 2 and Table 3.

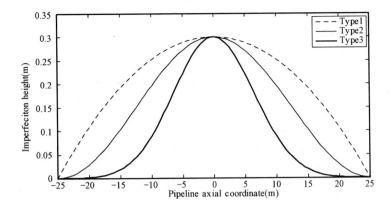

Fig. 2　Three imperfection types with h valued 0.3

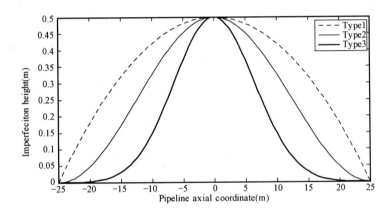

Fig. 3　Three imperfection types with h valued 0.5

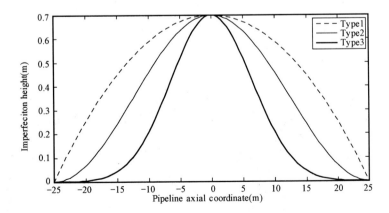

Fig. 4　Three imperfection types with h valued 0.7

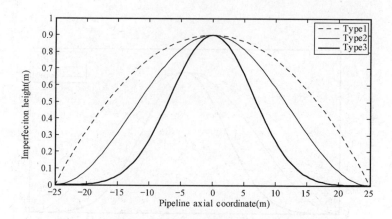

Fig. 5 Three imperfection types with h valued 0.9

Table 1 Parameters of pipeline segment

Diameter (m)	Wall thickness (m)	Steel elastic modulus (Pa)	Passion ratio	Expansion coefficient	Length (m)
0.4	0.01	2.07×10^{11}	0.3	1.17×10^{-5}	50

Table 2 Parameters of seabed

Steel elastic modulus(Pa)	Passion ratio	Depth(m)	Length(m)
2.00×10^{10}	0.2	1.17×10^{-5}	50

Table 3 Other parameters of the models

Friction coefficient	Thermal load(degrees centigrade)	Downward force(N)	Imperfection height(m)
0.3	160	4000	0.1,0.3,0.5,0.7,0.9

Every model is consisted by two parts, pipeline and seabed. The pipeline is modeled by 201 B21 elements and the soil is modeled by 800 CPS4R elements. The interaction between pipeline and seabed are modeled by friction contact with coefficient 0.3. There are two steps in this simulation. The first step is general static analysis applying downward force on pipeline and the second step is Riks static analysis applying thermal predefined field to pipeline from 0 degrees centigrade to 160 degrees centigrade. One of FE models is presented in Fig. 6.

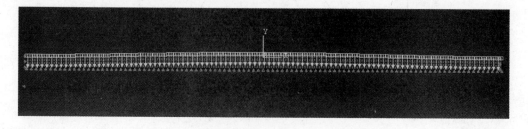

Fig. 6 FE model of pipeline upheaval buckling

2 RESULTS

After calculating the pipeline may buckle up in two forms, as shown in Fig. 7.

Fig. 7 Pipeline upheaval buckling

The critical axial force of every model has been getting. The results are listing in Table 4. Where material yield means pipeline material yields before pipeline upheaval buckling and no critical axial force means there is no snap of axial force during pipeline upheaval lifting from seabed.

Table 4 Results of critical axial force

Imperfection style	$h = 0.1$		
	Type 1	Type 2	Type 3
Critical axial force (N)	material yield	material yield	-3.11623×10^6
Imperfection style	$h = 0.3$		
	Type 1	Type 2	Type 3
Critical axial force (N)	-3.97741×10^6	-2.13412×10^6	-1.63686×10^6
Imperfection style	$h = 0.5$		
	Type 1	Type 2	Type 3
Critical axial force (N)	-2.43069×10^6	-1.43122×10^6	-1.25451×10^6
Imperfection style	$h = 0.7$		
	Type 1	Type 2	Type 3
Critical axial force (N)	-1.75058×10^6	-1.13556×10^6	No critical axial force
Imperfection style	$h = 0.9$		
	Type 1	Type 2	Type 3
Critical axial force (N)	-1.37211×10^6	No critical axial force	No critical axial force

DISCUSSION AND CONCLUSIONS

The pipeline axial shape varied in every pipeline engineering project. The shapes are either symmetrical or unsymmetrical. In this paper three typical symmetrical shapes are studied. The main difference between the three imperfection types can be described by a definite integrate, $\int_{-1/2}^{1/2} f(x) \, dx$. Where $f(x)$ is the expression of imperfection. From the results in Table 2 it can be concluded that:

(1) Imperfection style has big influence on upheaval buckling critical axial force;

(2) It is seems that the value of the definite integrate is big, the critical axial force big;

(3) In different imperfection height the imperfection style has different influence, the higher, the less.

The pipeline axial shapes are very complicated. On the other hand the initial imperfection profile will change during thermal loading. For example, the initial imperfection of the three cases mentioned in reference 5 would change if the pipeline heated by inner fluid. The initial imperfection is described by the assumption suitably before heating, however with the increase of temperature the initial imperfection will change and its assumption may be not suitable. Similarly simple initial imperfection assumptions were used to develop the analytical method, so sometimes they cannot give out accurate results. To predict pipeline upheaval buckling accurately and conveniently, it is suggested that:

(1) Determine actual axial shape of pipeline;

(2) Sort imperfection types of pipeline before analysis;

(3) Establish FE model of imperfection type with minimum value of the definite integrate and calculate.

REFERENCES

Barefoot A J. 1998. Modelling the uplift resistance of buried pipes in a drum centrifuge, MPhil. thesis, Cambridge University Engineering Department.

Hobbs R E. 1984. In-service Buckling of Heated Pipelines[J], Journal of Transportation Engineering, 110(2):175 -189.

Kvalstad T J. 1999. Soil Resistance Against Pipelines in Jetted Trenches[C]. Proceedings of the Twelfth European Conference on Soil Mechanics and Geotechnical Engineering, Amsterdam, Netherlands,2:891-898.

Maltby T C, Calladine C R. 1994. An Investigation into Upheaval Buckling of Buried Pipelines - - II. Theory and Analysis of Experimental Observations:965-983.

Maltby T C, Calladine C R. 1994. An Investigation into Upheaval Buckling of Buried Pipelines - - I. Experimental Apparatus and Some Observations:944-963.

Marius Loen Ommundsen. 2009. Upheaval Buckling of Buried Pipelines[D]. Master thesis at the University of Stavanger.

Neil Taylor, Aik Ben Gan. 1986. Submarine Pipeline Buckling-imperfection Studies[J]. Journal of Thin Walled Structures, N. 4 Elsevier Applied Science Publishers Ltd. ,England:295-323.

Neil Taylor, Vinh Tran. 1996. Experimental and Theoretical Studies in Subsea Pipeline Buckling[J]. Marine Structures, (9):211-257.

Palmer A C, Ellinas C P. 1990. Palmer_Design of submarine pipelines against upheaval buckling. OTC 6335

Pedersen P T, Jensen J J. 1988. Upheaval Creep of Buried Heated Pipelines with Initial Imperfections[J]. Marine Structures, (1):11-22.

Suzuki N, Toyoda M. 2002. Seismic Loadings on Buried Pipelines and Deformability of High Strength Line Pipes [C]. Proceedings of the Pipe Dreamer's Conference, Yokohama, Japan.

Tran V C. 1994. Imperfect Upheaval Subsea Pipeline Buckling [D]. Sheffield Hallam University.

White D J, Barefoot A J, Bolton M D. Centrifuge Modelling of Upheaval Buckling[J]. International Journal of Physical Modelling in Geotechnics, (2):19-28.

White D J, Barfoot A J, Bolton M D. 2001. Centrifuge Modelling of Upheaval Buckling in Sand[J]. Intarnetional Journal of Physical Modelling in Geotechnics, (2):19-28.

Research about the Riser Selection Scheme in the South China Sea Deep Water Oil and Gas Field Development

Ning He[1], Donghong Zhao[2], Menglan Duan[2], Chenggong Sun[2], Wei Yang[1], Wei Liu[1]

[1] Offshore Oil Engineering Co. ,Ltd. ,Tianjin,China
[2] Offshore Oil/Gas Research Center,China University of Petroleum,Beijing,China

Abstract Whatever the form of the floating programs on the development of offshore fields is, it must be required with a riser system, which is a key component of the marine infrastructure. The selection scheme of the riser directly affects the design of the drilling vessel and marine oil exploration program, on the basis of the analysis on the technical characteristics of deep sea riser, combined with the developing situation of the South China Sea and comparative studies on the different programs, this study find that Hybrid Riser has strong applicability in the development of the South China Sea deep water oil and gas field.

Key Words　South China Sea; Deep water oil and gas field development; Riser selection

INTRODUCTION

With the technology, economic and social rapidly developed and the demand of petroleum, natural gas and other energy increasingly needed, offshore oil and gas industry develops in high-speed and more and more oil and gas resources explore and discovery, floating production systems have been widely used. Regardless what kind of development program is, it all requires the use of the riser system, which is a key component of the marine infrastructure. Offshore drilling is the first step, and the marine riser is the key technology and equipment for offshore oil and gas exploration and development.

1　CHARACTER OF DEEP WATER RISER AND ITS COMPARISON

Deep water riser system essentially is the conduit to connect the water floating devices and submarine equipment in sea-bed (such as the wellhead, PLEM, manifold), including risers, support, all the constituent parts of the riser and corrosion protection systems, are important structure in the deep water development, they are mainly used to complete the drilling, producing, well completion, reparation, and external transmission and injection the oil, gas and liquid.

The structure of deep water risers is mainly compliant or tension-type, according to its adaptability, it can be divided into Flexible Riser(FR), Top Tensioned Riser(TTR), Steel Catenary Riser (SCR) and Hybrid Riser(HR), and HR can also be divided into the Free Standing Hybrid Riser (FSHR) and Riser Tower(RT)(Song,2003), as shown in Fig. 1.

Now, this paper will compare the FR, TTR, SCR, and HR riser system from adaptation, installation, cost, and fatigue resistance and other aspect systematically(Wang, Duan, et al. ,2009), as shown in Table 1.

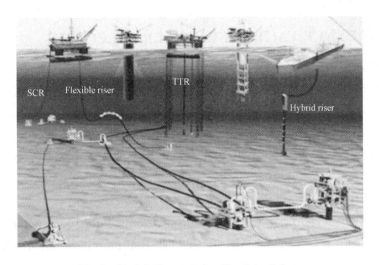

Fig. 1 Description and classification of risers

Table 1 Comparison on characteristics of risers (Petruska, Zimmermann, 2002; Luo, Ye, 2005)

Property	SCR	RT	FSHR	TTR	FR
Adaptation	All platform	TLP, SPAR, FPSO, Semi	TLP, SPAR, FPSO, Semi	TLP, SPAR, FPDSO	All platform
The way reduce the cost	Welded together by a single standard length rigid riser	Make risers bind together	Part of the riser use rigid riser and it can be installed ahead	Independent completion operations, and without a separate platform	Some progress made in production technology
Installation	Easy to install, and can be connected with other pipeline in advance	Advanced installation, and no need for special boats	Advanced installation, and no need for special boats	On-site installation	On-site installation
Resistance for fatigue damage	Touch down point of risers influenced by the eddy current effects, and SCR life influenced by range of motion of platform	Buoyancy tank can make the floating platform movement and the riser motion decouple, and not sensitive to fatigue	Buoyancy tank can make the floating platform movement and the riser motion decouple	Impact by the platform movement	Strength can be improved by the high rigidity of the spiral metal layer
Christmas tree	Wet Christmas tree	Wet Christmas tree	Wet Christmas tree	Dry Christmas tree	Wet Christmas tree
Experience in use	More	Less	Less	More	Many
Applicability in deep water or not	Yes	Yes	Yes	Yes	No
Components made in the local or not	Yes	No	Rigid rises can be	Rigid rises can be	No
Connection with the platform	Flexible joints free-hanging on the platform lateral	Flexible cross over	Flexible cross over	Direct access to platform	

2 MAIN INFLUENCE FACTORS ON RISER SELECTION

Riser selection impacts by many factors, it can be considered mainly from technical and economic feasibility two aspects. When choose the riser, it should be considered adaptation to the water depth, fatigue characteristics, influence and the choice on the body, flow assurance, riser installation and laying and so on from the technical aspects, and considered procurement, manufacturing, installation, engineering and operating costs from economic aspects. Now, riser selection will be analyzed in six aspects as follows.

2.1 Fatigue Characteristics

In the process of oil and gas production in deep water and ultra-deep water, fatigue is an important factor that must be considered. First order load, boat movement, low-frequency second-order boat movement and vortex-induced vibration will affect the riser system for long-term fatigue failure.

Flexible riser can be adapted to large bending deformation and the ensuing tension load, and it also has been shown to apply to shallow water environments. However, the ability to resist fatigue failure has been a great challenge in deep water or ultra-deep water. Flexible risers are generally installed with catenary which is influenced by high tension and low water static load on the ocean floor, while it is opposite in the seabed, and you need to optimize the design. In short, the flexible riser is better than rigid pipe SCR in fatigue resistance, but the cost is higher in deep water.

When using TTR in TLP or SPAR, VIV fatigue damage should be limited, and it is necessary to improve the top tension, which will increase the costs and the load in riser base and wellhead system, and the installation will be very difficult. If TTR is installed on the TLP, TLP will suffer more loads. If TTR is installed on the SPAR, it should be installed more air tank at the bottom of SPAR.

SCR fatigue performance is heavily dependent on environmental load and the boat movement. Generally speaking, SCR is sensitive to fatigue mainly in two aspects: (1) touch down point; (2) Suspension horn.

The biggest advantages of HR in three aspects: it can decouple it from the platform motion, the top tension can be provided by the buoyancy tank, and the rigid riser is away from the sea surface wave zone, all those factors reduce the environmental load and are away from the influence of the waves. However, components in water are low in weight and the tension is small, the influence on riser of VIV will be larger, so it is very important to make a comprehensive analysis.

2.2 Water Depth Applicability

Fundamentally Speaking, four types of riser system can be used for deep water oil and gas development, but depth is still an important factor to affect the riser system's designation and cost.

SCR is basically used in 3000m in deep water, and its diameter is not limited. FR can be used in less than 2,500m, but with water depth increasing, the diameter can be applied decreases rapidly. The top tension of TTR increases when the depth reach 1,500~2,000m, what's more, with the water depth increasing, the weight, and riser diameter at the bottom, as well as the top tension should be increased accordingly, all those lead to increase the costs. HR is a new deep water riser and adapted to the deep water, especially in the ultra-deep water.

2.3 Requirements and Implications on the Floating Structure

At present, the type of floating platform basically includes TLP, SPAR, Semi, FPSO and FPDSO. These platforms are able to adapt to different types of riser (as shown in Table 2), when the oilfield is in a larger scale, it is important to consider the huge payload due to the risers own weight while risers connected to floating structures.

Table 2 Riser selection on boat (Petruska, Zimmermann, 2002; Luo, Ye, 2005)

	TLP	SPAR	Semi	FPSO	FPDSO
SCR	Unlimited	Unlimited	Need good movement performance	Only in a moderate environment	Only in a moderate environment
TTR	Use	Use	No use	No use	Use
FR	Unlimited	Unlimited	Unlimited	Unlimited	Unlimited
FSHR	Unlimited	Unlimited	Unlimited	Unlimited	Unlimited

2.4 Flow Assurance

Flow assurance plays an important role in riser adiabatic designation. The flow in TTR which adopt the dry tree has the shortest distance, and TTR performs better than SCR in flow assurance. Commonly, FSHR designs pipe in pipe, and it can be control by a valve at the bottom of the riser, which improves the operational flexibility. SCR has difficulty in dealing with insulation and flow assurance technology. When using thicker insulation material outside the SCR, it will reduce the stability of the riser and result in fatigue failure at the touch down point. What's more, it is difficult to achieve gas lift in traditional designation. SCR can also adopt pipe in pipe designation, but it will increase its weight and its installation will waste time and energy, so it is not a good choice.

2.5 Influence Factors on Material Selection and Wall Thickness

Generally speaking, the level of material used for riser are API 5L, X60, X65 and X75. Material selection is generally considered as follow:

(1) Weld ability;

(2) Reservoir properties (HP/HT) and corrosive fluid (H_2S, CO_2);

(3) Comparison on installation method, cost and fatigue performance of the riser;

(4) Estimation on upper weight.

The specific design of the riser wall thickness need refer to the current two standard design standards API RP 2RD and DNV OS F201 only as follows (Song, 2003):

(1) API RP 2RD (1998): "Design of Risers for Floating Production Systems and Tension Leg Platforms", First Edition.

(2) DNV OS F201 (2001): "Dynamic Risers".

In addition, some extensions on riser designation standards made by administration have published, such as ABS(2001).

2.6 Riser Installation and Laying

Installation is an important step before riser starting work. Installation costs account a large part of the cost of the riser. When riser installing, welding, transportation and down will influence the strength and fatigue life of the riser, it should be analyzed in advance. Table 3 is the comparison of several main riser installations.

Comparison on riser laying will be shown in Table 4.

Table 3 Comparison of several main riser installations(Wang, Duan, 2009)

	Riser-laying method	Installation boat	Contractors	Weather window	Main advantages and disadvantages
FR	Reel	More in shallow water; less in deep water	Less	Limited by the response to the boat on environment, and can be full-year operated	To large diameter FR, need special design for reel
SCR	S-lay J-lay reel	Sea welding Special ships	More	Limited by the response to the boat when the riser welding	Welding operations on the sea, costs increased
FSHR	MODU J-lay S-lay reel	Many options	Less	Generally full-year operating	Pre-installation before arrival at production platform

Table 4 Advantages and disadvantages on riser laying method(Song, 2003; Hecdo 2004; Baker McClure 2002; Andrei Dumitrescu, Massima Pulici, Marian Trifon, 2003; Choi S, 1999)

Installation method		Feature
Drag laying	Advantage	Pipes or risers made on the shore, good quality of welding acquired in the workshop; Can use cheap tug; Can use varieties of drag method
	Disadvantage	Installation length is limited; Require a higher status on the seabed; Use this method only in shallow water area
Reel laying	Advantage	99.5% of the welding work is completed in a controlled environment in the land; Tension is relatively reduced, efficiency and cost lower; Continuous laying, operational risk is small
	Disadvantage	Need support from shore bases; Steel plastic properties are required higher, and smaller diameter; The typical laying speed of 600 meters per hour

Continued

Installation method		Feature
J-lay	Advantage	Need no stern pipe supporter; Pipe from the angle is very close to vertical, so the tension is smaller
	Disadvantage	Only one welding station, so low in efficiency; All operations are completed in vertical direction, so stability is a problem; Laying is slow, typical laying rate is 1 ~ 1.5 km/d
S-lay	Advantage	Pipe assembled in horizontal direction using single or dual connector, and efficiency; High laying rate, the typical laying rate is 3.5 km/d; The sea conditions adaptability and continuous operational capability is better than any other laying method
	Disadvantage	Must deal with the bigger tension; greater depth = longer pipe Supporter = stability loss; Greater depth = higher tension = greater risk

3 RISER SELECTIONS SCHEME IN THE SOUTH CHINA SEA OIL AND GAS FIELDS

3.1 Challenges on South China Sea Deep Water Oil and Gas Field Development

The South China Sea deep-water area has rich oil and gas resources, because of their complex natural environment and storage conditions, exploration and development in there has high investment, high return, high-tech, high-risk characteristics. In addition to the high cost of investment, technology development problems and maritime disputes, the development in South China Sea faces a series of natural environment difficulties and challenges (Zhang, 2009).

(1) South China Sea is typhoon-prone area, and typhoon will pass through this region at least four times each year.

(2) Flow waves in South China Sea are serious frequent marine natural disasters.

(3) Geological conditions in South China Sea are very complex. Sand and sand ditch is obvious, what's more, sand ridge is moving, its speed can reach about 300 m/a, which is a huge challenge for the production facilities and pipelines on the seabed near the oil and gas fields.

(4) The South China Sea own environment will cause severe corrosion on marine engineering facilities and subsea pipelines.

(5) Deep water is in high static pressure, low temperature environment (usually about 4℃).

In addition, offshore crude oil has the problem such as high temperature, high CO_2 and is full with high viscosity, easy condensate, high content of wax and other feature. Moreover, the South China Sea deep-water area is far from land, because of this oil and gas processing and separation technology is more complex, which also brings many challenges on offshore oil and gas gathering and transportation.

3.2 Riser Selection Scheme on South China Sea Deep-water Oil and Gas Fields

Natural environmental conditions in the South China Sea are very poor. From the riser extensive applicable perspective, HR can decouple from boat movement, and its rigid pipes can avoid the severe storms 150m below the sea level, which has the strong ability to adapt to the development in the South China Sea. From the economic cost perspective, the South China Sea is far away from the world oil and gas development center, there are less alternative professional construction and installation boat, which cost is also much higher than other area. Compared with other risers, FSHR and RT have little demand on installation boat, which can reduce the cost in the South China Sea oil and gas field development effectively. At the same time, because the demand on installation boat is not strict, building or retrofitting boat in the local is a good method to promote the local employment and economic development, what's more, it can reduce the costs compared to adopt the special pipe-laying boat. So, for the South China Sea's development, HR has a considerable advantage and strong applicability.

CONCLUSIONS

In this paper, the practical problems of South China Sea deep water oil and gas field development facing have been analyzed in detail. According to the comparison of the four kinds of deep water risers(FR, TTR, SCR and HR) characteristics and six main impact factors analysis on risers' selection, this paper comes to the conclusion that HR has a strong applicability in the South China Sea development.

ACKNOWLEDGEMENTS

This paper is financially supported by the National Science and Technology Major Projects of China(Grant No. 2011ZX05027 - 002).

REFERENCES

Andrei Dumitrescu, Massima Pulici, Marian Trifon. 2003. Deep Water Seahnes Installation by Using the J-lay Method-The Blue Stream Experience[J]. Thirteenth Intematlonal Offshore and Polar Engineering Conference,3:38 - 43.
Baker B, McClure L. 2002. Reel Method Speeds Lay of Pipe in Pipe[J]. Offshore,62(8):72,156.
Choi H S. 1999. Deep-water Pipe Lay Analysis[J]. Journal of Offshore Technology,7(2):18 - 20.
Hecdo D. 2004. Loading History Effects for Deep Water S - Lay of Pipelines[J]. Journal of Offshore Mechanics and Arctic Engineering,26(2):156 - 163.
Luo Y, Ye W. 2005. Mooring and Riser Design for Gom Fpsos in 10,000ft Water Depth[C]. Offshore Technology Conference.
Petruska D J, Zimmermann C A. 2002. Riser System Selection and Design for a Deep water FSO in the Gulf of Mexico[C]. Offshore Technology Conference, Houston, Texas.
Song Ruxin. 2003. Pipeline and Riser in the Deep Water Development[J]. Technology and Economy Information of Shipbuilding Industry,218:31 - 42.
Wang Yi, Duan Menglan, et al. 2009. Progress of Deep Water Riser Installation[J]. Oil Field Equipment,06.
Zhang Fengjiu. 2009. Outlook for Natural Gas Development in the South China Sea[J]. Natural Gas Industry,29(1):17 - 20.

Simulation of the Scour Around a Submarine Pipeline Under Uniform Flows

Yang He[1], Pan Gao[2], Caiying Yi[1], Jianguo Wang[3], Zhilin Yuan[3] and Menglan Duan[2,3]

[1] Offshore Oil Engineering Co., Ltd., Tianjin, China
[2] Department of Mechanics and Engineering Science, Fudan University, Shanghai, China
[3] Offshore Oil and Gas Research Center, China University of Petroleum, Beijing, China

Abstract The main objective of the present study is to simulate the scour around a pipeline using Fluent. A scour model is incorporated into the Fluent by UDF. Both bed-load transport and suspended transport are considered. The standard $k - \varepsilon$ turbulence model is used to solve turbulence. A sand slide model is proposed to avoid numerical protrusion. The predicted result of bed shear stress agrees well with documented results. And the results of scour simulation are compared with the experimental results in Mao (1986). At early stage, the predicted results are larger than the experiment due to the initial gap. The predicted results go closer to the results given in Mao (1986) in the later stage. The larger the shields parameter is, the faster the scour develops.

Key Words Scour; Marine pipeline; Turbulence; UDF; Sand slide

INTRODUCTION

In offshore oil and gas industry, it has long been concerned that local scour occurs around marine structures, such as pipelines and pile foundations. When a pipeline is laid on the seabed, local scour under the pipeline leads to pipeline suspension in some locations. As the scour propagates along the pipeline, it may experience vortex induced vibration (VIV), which is known to be one of the major threats to pipelines.

In the past several decades, a great amount of work on pipeline scour has been done, both experimentally and numerically. Mainly two kinds of experiments have been done. One kind is to study the equilibrium scour depth. Kjeldsen et al. (1973) firstly conducted a scour experiment under controlled conditions, which is a milestone in this field. Following this experiment, more experiments have been done to supplement Kjeldsen's job (Leeuwestein and Bijker 1985; Mao, 1986; Sumer and Fredsoe, 1990; Chiew, 1991). From all these work, the equilibrium scour depth and how it is influenced by the factors such as flow velocity, pipe diameter and sediment properties have been well understood. What is more, Sumer et al. (1988) and Gao et al. (2006) even conducted two experiments to check how the pipe vibration influenced the scour. The other one kind is to research the propagation of scour along the pipeline, which is known as three dimensional pipeline scour. Wu and Chiew (2010) and Cheng et al. (2009) conducted two experiments focusing on how a 3 - D pipeline scour propagate with respect to different velocities, gap ratios and Sheilds parameters. Their main finding is that the propagation of the pipeline span can be divided into a rapid phase and a slack phase.

Over the last three decades, many numerical models for scour prediction have been proposed. They can be divided into two categories. One includes those based on the potential flow theory, which includes Gao et al(2006), Hansen(1986), Chiew(1991), Li and Cheng(1999). The other one involves those employing turbulent flow models, such as Leeuwestein and Wind(1984), Beek and Wind(1990), Brors(1999), Liang et al. (2005) and Zhao(2010). The potential flow models can make a good prediction of the equilibrium scour depth and the upstream part of the scour hole. However, the downstream part of the scour hole cannot be well predicted with these models. It is most probably because the gentle slope downstream of the pipeline is mainly due to the vortex shedding. And this has been confirmed by the the turbulent flow models. Brors(1999) presented a model using a turbulent flow model, and both the suspended and bed load sediment transport model. It set a large critical Shields parameter of 0.25 for the suspended load, so the scour development studied by the model was mainly attributed to bed load transport. Only a case of clear water scour was predicted, and it agreed well with the lab experiment. Liang et al. (2005) compared the $k - \varepsilon$ and SGS turbulence model. It was found that the $k - \varepsilon$ model can predict the sand accretion behind the pipeline at the initial stage of scour. And it predicted the time dependent scour profile better than the SGS model. The $k - \varepsilon$ model is recommended for the scour model in Liang et al. (2005). Zhao(2010) proposed a model that is very much like the model of Liang et al (2005). It used the $k - \omega$ turbulence model and studied the scour below a vibrating pipeline. It was found that the pipeline vibration enhanced the scour depth below it.

In summary, the experiments about scour below both fixed pipeline and vibrating pipeline have been widely conducted. And the most successful numerical model now is one that combines the turbulence model, suspended and bed load sediment transport and pipeline vibration together. A $k - \varepsilon$ turbulence model is confirmed appropriate to be used in a numerical model for scour prediction. However, the propagation of the pipeline span has not been numerically studied. All the numerical models are programs self-developed without pre-process and post-process, which is very inconvenient for pipeline engineers to use it. The goal of this paper is to develop a scour model with the rather stable and friendly commercial CFD software Fluent, and make it easily accessible to pipeline engineers. The user defined functions are used to fulfill the scour model. A sand slide model is introduced into it to avoid numerical protrusions on the seabed. It is found that Fluent with UDF can predict the scour depth and time dependent scour profile precisely. And it can further be used to study the interaction between a vibrating pipeline and an erodible bed.

1 MATHEMATICAL MODELS

1.1 Governing Equations for Flow Field

Reynolds-averaged Navier-Stokes(RANS) equations are used to solve the flow field. RANS equations for incompressible flow contain the following equations:

$$\frac{\partial u_i}{\partial x_i} = 0 \tag{1}$$

$$\frac{\partial u_i}{\partial t} + u_j \frac{\partial u_i}{\partial x_j} = -\frac{\partial p}{\partial x_i} + \frac{\partial}{\partial x_j}(-\overline{u_i' u_j'}) + \frac{\partial}{\partial x_j}\left[\mu\left(\frac{\partial u_i}{\partial x_j} + \frac{\partial u_j}{\partial x_i} - \frac{2}{3}\delta_{ij}\frac{\partial u_l}{\partial x_l}\right)\right] \quad (2)$$

where is the Reynolds stresses. It can be modeled with the Boussinesq hypothesis:

$$-\overline{u_i' u_j'} = \nu_t\left(\frac{\partial u_i}{\partial x_j} + \frac{\partial u_j}{\partial x_i}\right) - \frac{2}{3}\left(k + \nu_t \frac{u_l}{x_l}\right)\delta_{ij} \quad (3)$$

In Eq. (3), the Reynold stresses are approximated by mean velocities, the turbulence kinetic energy k and turbulent viscosity ν_t. Several models can be used to compute k and ν_t. In the present study, the standard $k-\varepsilon$ model is adopted. The transport equations of turbulence kinetic energy k and its rate of dissipation ε are as follows:

$$\frac{\partial k}{\partial t} + \frac{\partial}{\partial x_i}(ku_i) = \frac{\partial}{\partial x_j}\left[\left(\nu + \frac{\nu_t}{\sigma_k}\right)\frac{\partial k}{\partial x_j}\right] + G_k + G_b - Y_M - \varepsilon + S_k \quad (4)$$

$$\frac{\partial \varepsilon}{\partial t} + \frac{\partial}{\partial x_i}(\varepsilon u_i) = \frac{\partial}{\partial x_j}\left[\left(\nu + \frac{\nu_t}{\sigma_\varepsilon}\right)\frac{\partial \varepsilon}{\partial x_j}\right] + C_{1\varepsilon}\frac{\varepsilon}{k}(G_k + C_{3\varepsilon}G_b) - C_{2\varepsilon}\frac{\varepsilon^2}{k} + S_\varepsilon \quad (5)$$

and

$$\nu_t = C_\mu \frac{k^2}{\varepsilon} \quad (6)$$

In the $k-\varepsilon$ turbulence model, it solves turbulence kinetic energy k and its dissipation rate ε first and then the eddy viscosity ν_t. The model coefficients are listed in Table 2.1. Along with the standard $k-\varepsilon$ turbulence model, the standard wall function is used. It makes it affordable to calibrate the high Reynolds-number flows relatively well. This scheme has been widely used in engineering and researches for more than three decades.

Table 1　Model coefficients for the standard $k-\varepsilon$ model

C_μ	σ_k	σ_ε	$C_{1\varepsilon}$	$C_{2\varepsilon}$
0.09	1.0	1.3	1.44	1.92

1.2　Sediment Transport Equations

Sediment can be transported in both bed-load form and suspended-load form. Sediment transported in bed-load form moves in continuous contact with the bed. Its transport rate is the function of the sediment properties and the bed shear stress. There are a number of bed load transport formulas that can be used. Among them, the one proposed by Van Rijn(1987) is adopted in this research.

$$q_b = \begin{cases} 0.053 \sqrt{(s-1)g} \dfrac{d_{50}^{1.5} T^{2.1}}{D_*^{0.3}} & T > 0 \\ 0 & T \leq 0 \end{cases} \quad (7)$$

In Eq. (7), D_* is the dimensionless grain size defined as Eq. (8); q_b is the volumetric bed load transport rate per unit width; s is the specific weight of sediment; g is the gravity acceleration;

d_{50} is the median grain diameter; T is the transport stage parameter which is defined in Eq. (9); u_* is the friction velocity induced by water flow; u_{*cr} is the threshold bed friction velocity for motion of sediment.

$$D_* = \left[\frac{g(s-1)}{\nu^2}\right]^{1/3} d_{50} \tag{8}$$

$$T = \frac{u_*^2 - u_{*cr}^2}{u_{*cr}^2} \tag{9}$$

According to the definition of friction velocity, u_* can be expressed as:

$$u_* = \sqrt{\frac{\tau_b}{\rho}} \tag{10}$$

Shields parameter which is first proposed in 1930's can be used to calculate the threshold friction velocity u_{*cr}. Soulsby and Whitehouse (1997) proposed a Shield diagram which is expressed as:

$$\theta_{cr0} = \frac{0.3}{1 + 1.2D_*} + 0.055[1 - \exp(-0.02D_*)] \tag{11}$$

θ_{cr0} is only right if it is on horizontal bed. For a sloping bed, it should be modified to:

$$\theta_{cr} = \theta_{cr0}\left[\cos\alpha + \frac{\sin\alpha}{\tan\varphi}\right] \tag{12}$$

where α is the slope angle and φ is the angle of repose of sediment. u_{*cr} can be solved by

$$u_{*cr} = \sqrt{\theta_{cr}g(s-1)d_{50}} \tag{13}$$

Sediment transported in suspended form moves along with the water flow. The suspended sediment transport is governed by a convection diffusion equation.

$$\frac{\partial c}{\partial t} + u_1\frac{\partial c}{\partial x_1} + (u_2 - \omega_s)\frac{\partial c}{\partial x_2} = \frac{\partial}{\partial x_j}\left(\frac{\nu_t}{\sigma_c}\frac{\partial c}{\partial x_j}\right) \tag{14}$$

In Eq. (14), c is the volumetric sediment concentration in the water column; t is time; x_1 and x_2 are the two coordinates, u_1 and u_2 are the corresponding velocities; ω_s is the settling velocity of in water; σ_c is the turbulent Schmidt number, which is set 0.8 in this research. The formula proposed by Soulsby is used to determine the settling velocity of natural sands in clear water.

$$\omega_s = \frac{\nu}{d_{50}}(\sqrt{10.36^2 + 1.049D_*^3} - 10.36) \tag{15}$$

If the sediment concentration is large enough, the interaction among sediment particles will decrease the settling velocity. In the present study, the effect of sediment concentration is neglected.

It should be noted that the governing equations of flow field and that of sediment transport are solved independently. It is because that the concentration of sediment is not high enough to have a significant influence on the flow field. After the sediment concentration is solved, he suspended load transport rate can be easily achieved integrating the concentration and flow velocity:

$$q_s = \int_{y_b+\Delta_b}^{y_b+H} uc\,dy \tag{16}$$

1.3 Scour Model

Based on mass balance of sediment over the whole water column, the change of bed elevation can be computed.

$$\frac{\partial y_b}{\partial t} = \frac{1}{1-p_o}\left[\frac{\partial q_T}{\partial x} - \frac{\partial}{\partial t}\left(\int_{y_b+\Delta_b}^{y_b+H} c\,dy\right)\right] \tag{17}$$

q_T is the total sediment transport rate

$$q_T = q_b + q_s \tag{18}$$

For a bed consisting of sands, if its slope angle is greater than a certain value, the sediment on the top will slide to the valley due to gravity. In other words, the bed slope will never be greater than a certain value, which is called the repose angle of sediment and symbolized as phi. During the iterations, the angle of slope some-where on the bed may exceed the repose angle, while which cannot exist in the real environment.

In order to avoid this kind of difference between the simulation and the real situation, a sand slide model is introduced into the scour model.

$$dy = \begin{cases} \dfrac{x_i - x_{i-1}}{2}(\tan\alpha - \tan\varphi) & y_{bi}^{old} > y_{bi-1}^{old} \\[2mm] -\dfrac{x_i - x_{i-1}}{2}(\tan\alpha - \tan\varphi) & y_{bi}^{old} < y_{bi-1}^{old} \end{cases} \tag{19}$$

As shown in Fig. 1, if the width of the cell on either side of the present cell is the same as that of the present cell, the y_{bi-1}^{old} will move to y_{bi-1}^{temp} by plus dy and y_{bi}^{old} to y_{bi}^{temp} by subtracting dy. If not, doing this will not guarantee the conservation of sediment. Thus the following scheme of sand slide is proposed in the present study.

$$y_{bi-1}^{new} = y_{bi-1}^{old} + 2\xi\,dy \tag{20}$$

$$y_{bi}^{new} = y_{bi}^{old} + (2 - 2\xi)\,dy \tag{21}$$

In which

$$\xi = \frac{1 + \xi_+}{2 + \xi_- + \xi_+}$$

$$\xi_- = \frac{x_{i-1} - x_{i-2}}{x_i - x_{i-1}}$$

$$\xi_+ = \frac{x_{i+1} - x_i}{x_i - x_{i-1}}$$

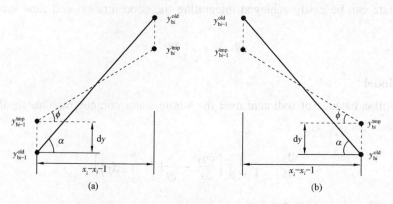

Fig. 1   Illustration sketch of the present sand slide model
(a) Upward slope; (b) Downward slope

The sand slide model can significantly improve the quality of the profile of the sand bed. As shown in Fig. 2, the sand slide model significantly prevents the seabed change from numerical protrusions.

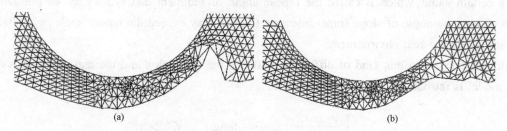

Fig. 2   Comparison of dynamic meshes with and without sand slide model
(a) Without sand slide model; (b) With sand slide model

2 NUMERICAL METHOD

The governing equations for flow field and the suspended sediment transport equation are solved separately. It is because that the suspended sediment concentration in the flow column is not high enough to have a slight influence on the flow field. The governing equations for flow field are solved with the pressure – based solver, in which the continuity equation and momentum equation are solved separately. There is a Reynolds stress item in the momentum equation, so the $k - \varepsilon$ equations have to be solved prior to it. The velocity and pressure are coupled using the SIMPLE method. In each iteration, the velocity field and pressure field are solved alternately. The iterations stop when the pressure and velocity reach convergence. Then the governing equation for suspended sediment transport is solved with the flow velocity already achieved. With all the governing equations solved, the bed shear stress and sediment concentration are used to calculate the bed profile change. This morphological update is completed by a user defined function (UDF) in Fluent. Eq. (17) is solved in the UDF and then the mesh is updated by Fluent. Simply discretized in time and space, Eq. (17) has a form of

$$\frac{\Delta y_{b,i}}{\Delta t} = \frac{1}{1 - p_o} \frac{q_{T,i+1} - q_{T,i}}{\Delta x} + \frac{[\sum_j (c_j \Delta y)]^{n+1} - [\sum_j (c_j \Delta y)]^n}{(1 - p_o) \Delta t} \tag{22}$$

It is preferred to carry out the flow field calculation and morphological update simultaneously. However, the flow field changes much faster than the morphological change due to scour. For example, the vortex in the wake of a circular cylinder has a dimensionless period of 5 (assuming the Strouhal number to be 0.2). If the scour under a pipe reach the equilibrium profile in 200 min, the corresponding dimensionless time would be $(200 \times 60)/(100/400) = 48000$ (assuming a pipe with 100mm in diameter and an average velocity of 0.4m/s). However, each step of morphological update costs as much time as or more than a time step of flow field calculation. It would be very uneconomical to calibrate the flow field change and morphological change simultaneously. Experimental results (Mao, 1986) showed that the morphology changed in a different speed in different stage. In the early stage it changed rather rapidly, then slower and slower.

Based on these facts, a new time marching scheme is proposed to reduce the overall time for simulation. The central idea of the scheme is to use a different time steps for flow and scour calculations. A larger morphological update time step is used to speed up the simulation. As Shown in Fig. 3, within one time step of morphological update, several time step of flow calculation is completed. At the time step of scour calculation, the latest flow field is used to calculate the bed change.

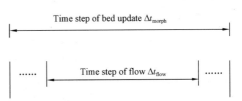

Fig. 3 Illustration of time marching scheme for flow and morphological updates

At the early stage of scour, as suggested in Liang et al. (2005), ten steps of flow field calculation are implemented between two morphological change. At the early stage of scour, twenty steps of flow field calculation are implemented.

The solution procedure in the present model can be summarized as follows:

(1) Calculate the velocity, pressure, turbulence quantities and suspended sediment concentration for a number of time steps (10 for the early stage and 20 for the near equilibrium stage).

(2) Calculate the morphological change and update using UDF, and move the nodes on the bed.

(3) Apply the the sand slide model to make sure that the new bed slope is not larger than the repose angle of sediment.

(4) Remesh the grids near the bed and update the mesh.

(5) Return to step(1) and repeat the above steps until a specified time has been reached.

3 SCOUR CALCULATIONS

3.1 Basic Parameters

The scour model is validated against the experimental data in Mao (1986), the numerical model calculates two cases which are the same as those in the experiment. The numerical model setup is shown in Fig. 4. The computational domain extends $10D$ upstream the pipeline to $20D$ downstream the pipeline. The vertical scale (water depth) is set $5D$. The computational mesh of the entire domain has 180×60 grids. The local mesh near the pipe is shown in Fig. 5. There are three stages in the scour process.

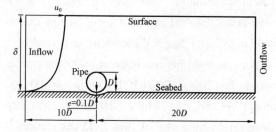
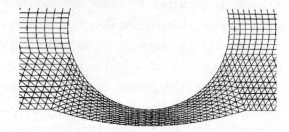

Fig. 4　Illustration of setup of the numerical model　　　Fig. 5　Local mesh near the pipe

When a pipeline is laid on the seabed, first, there would be a pressure difference between the upstream and down-stream pipeline due to the flow. The pressure difference would cause piping, which is known as the onset of scour. The second stage is the tunnel scour. At this stage, there is a small gap between the pipeline and seabed. As the water flows through the gap, it leads to an amplified shear stress which will accelerate the scour. When the scour hole is enlarged to its final state, it comes into the final stage, the so-called lee-wake scour. The lee-wake scour is responsible to the scour downstream the pipeline. In the present numerical model, the first stage of scour is avoided by pre-set a scour hole under the pipeline. However, it is believed this will only have a little influence on the scour hole at the very early stage of the scour. The initial scour depth is taken to be 0:1D There are three kinds of boundaries in this model which are the inflow boundary, the outflow boundary and three wall boundaries including the free surface, the pipe and the seabed. The rigid lid assumption is applied to approximate the free surface. Flow velocity, turbulence parameters and the value of sediment concentration are specified on the inflow boundary. Three profiles of the fully developed velocity, the turbulence kinetic energy profile and its dissipation in Eq. (23) to Eq. (25) are applied to the inflow boundary. The velocity profile has a form as shown in Fig. 4.

$$u(y) = \min\left\{\frac{u_*}{\kappa}\ln\left(\frac{y}{y_0}\right), u_0\right\}; v(y) = 0 \tag{23}$$

$$k(y) = \max\left\{C_\mu^{-1/2}\left(1 - \frac{y}{\delta}\right)^2 u_*^2, 0.0005 u_0^2\right\} \tag{24}$$

$$\varepsilon(y) = \frac{C_\mu^{3/4} k(y)^{3/2}}{l(y)} \tag{25}$$

$$l(y) = \min\{\kappa y(1 + 1.5y/\delta)^{-1}, C_\mu \delta\} \tag{26}$$

The concentration at the inflow boundary is set as Eq. (27):

$$\frac{c}{c_b} = \left(\frac{\Delta b}{H - \Delta b}\frac{H - y}{y}\right)^Z$$

where

$$Z = \frac{\sigma_c \omega_s}{\kappa u_*} \tag{27}$$

$$c_b = 0.015 \frac{d_{50} T^{1.5}}{\Delta b D_*^{0.3}} \tag{28}$$

The concentration at the reference level near the bed c_b is set according to van Rijn (1986). The fluxes of sediment concentration at the free surface, outflow and pipe surface boundary are set zero.

Two cases are carried out to make a comparison with Mao's experiment. The mean flow velocity is 0.35m/s for the $\theta_\infty = 0.048$ case, and 0.4m/s for the $\theta_\infty = 0.065$ case. They represent a clear water scour case and a live bed scour case respectively. The sands on the bed have representative diameters of $d_{50} = 0.36$mm and $d_{90} = 0.48$mm. The pipeline diameter is 100mm. The roughness height k_s is set to be $2.5d_{50}$, and the reference level Δb is set to be $3d_{90}$.

3.2 Predicted Results

The bed shear stress is compared to the numerical results given in Liang et al. (2005), because it is of great concern in the scour prediction. Only if the bed stress is accurate can the sediment transport rate be close to the real situation. As shown in Fig. 6, the bed shear stresses upstream the pipeline of the two models are very close. The bed stress near the inflow boundary is

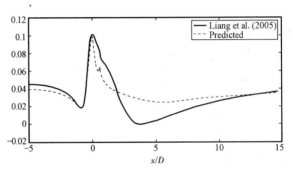

Fig. 6 Comparison of the present predicted friction velocity and that in Liang et al. (2005)

smaller and the bed stress downstream the pipeline calculated by the present model is a little larger than that in Liang's model. This may be ascribed to that the velocity profile is not exactly the same as a fully developed flow field. However, this does not seem to have an influence on the scour because the stress upstream the pipeline is well predicted. For the live bed case, the undisturbed Shields parameter predicted using the present model is 0.06, which is very close to the value of 0.065 reported in Mao (1986).

Two cases of numerical simulation are conducted and compared to the experiment result. For the $\theta_\infty = 0.048$ case, profiles at three different time are present in Fig. 7. The following points are observed from it.

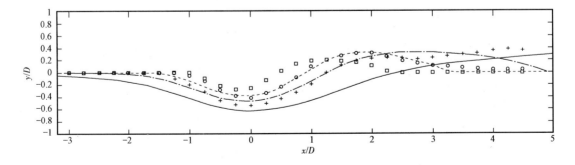

Fig. 7 Development of bed profile, clear water case ($\theta_\infty = 0.048$)

(1) The profile does not change much from 10min to 200min. This is mainly because that the lee – wake scour is much slower than the tunnel scour.

(2) The numerical model over-predicts the sand accretion downstream the pipe at early stage ($t<30$min). Later the sand accretion disappears due to the vortex shedding.

(3) The numerical simulation seems to be much faster than the experiment. The numerical results are a-head of the experiment results. It is expected that the initial gap between the pipe and bed has a larger effect on the slow process. Because the Shields parameter is small ($\theta_\infty = 0.048$), the scour process is slow.

(4) The present model predicts the scour reasonably well with an over-prediction of about 10% on the scour depth.

Fig. 8 shows the scour profiles for the live bed case $\theta_\infty = 0.065$. The scour process is like the clear water case, however it also has some differences. The following points can be observed.

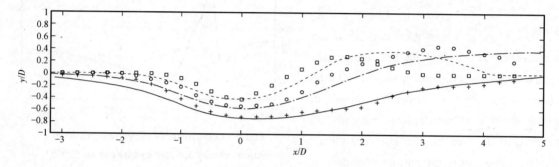

Fig. 8 Development of bed profile, live bed case ($\theta_\infty = 0.065$)

(1) The numerical model again over-predicts the sand accretion behind the pipe. But it is slighter than the clear water case. It also disappears later.

(2) The numerical scour process is a little faster than the experiment. The difference between them is smaller than the difference in the clear water case. It confirms that the slower the process is, the bigger the initial gap has an effect on the scour process.

(3) The present model predicts the scour of this case better than the clear water case. There is little difference between numerical and experimental results near the equilibrium state.

(4) The scour of live bed case reaches the equilibrium state earlier. This is mainly due to the larger Shields parameter.

Based on the results of the two cases and the comparison between them, it is easy to draw the conclusion that the conclusion that the present numerical model is able to predict the scour process accurately. First of all, the numerical model predicts the scour depth a little larger. The larger the Shields parameter is, the smaller the over-prediction is. Secondly, the numerically predicted profile is more different from the experiment at the early stage, which is due to the initial specified gap. Finally, the predicted scour profile is much the same as the experiment at the equilibrium state.

CONCLUSIONS

A 2-D numerical model is developed with Fluent to predict the scour process under uniform flows. The model uses the $k-\varepsilon$ model to compute the flow with turbulence. A convection diffusion equation is introduced into the model to compute the suspended sediment transport. A user defined function with a sand slide model is used to calculate the scour and morphological change. The meshes near the bed are updated with a dynamic meshing method. Based on the numerical results obtained in this study, the following conclusions can be drawn:

(1) The suspended sediment transport model employed in this study works very well for the net entrainment test case.

(2) The present numerical model overestimates the sand accretion behind the pipe at the early stage. This is mainly due to that the $k-\varepsilon$ model smoothens the fluctuations produced by vortex shedding, which weakens the interaction between the vortices and the bed. But this does not seem to affect the prediction of the scour profile at the late stage.

(3) The larger the Shields parameter is, the faster the equilibrium state is reached. It is expected that a larger Shields parameter produces a larger shear stress on the bed, therefore causes a larger sediment transport rate.

(4) The present numerical model gives very good predictions on the equilibrium profile. Along with the good pre-process and post-process modules of Fluent, this model can be used for engineering and further researches.

The model presented in this study is able to simulate the scour around a pipeline rather accurately. Along with the great pre-process and post-process capacity of Fluent, it can make it easier for engineers to know how the scour below a pipeline develops. With fluid-solid coupling capability provided by ANSYS, it would be easier to study the interaction between pipeline vibration and scour.

REFERENCES

Brors B. 1999. Numerical Modeling of Flow and Scour at Pipelines. [J]. Hydraul. Eng. ,125(5):511-523.

Chao J L, Hennessy P V. 1972. Local Scour Under Ocean Outfall Pipelines[J]. Water Pollut. Control Fed. ,44(7): 1443-1447.

Cheng L, Yeow K, Zhang Z, et al. 2009. Three-dimensional Scour Below Offshore Pipelines in Seady Currents[J]. Coast. Eng. ,56:577-590.

Chiew Y M. 1991. Prediction of Maximum Scour Depth at Submarine Pipelines[J]. Hydraul. Eng. ,117(4):452-466.

Gao F, Yang B, Wu Y, Yan S. 2006. Steady Current Induced Seabed Scour Around a Vibrating Pipeline[J]. Applied Ocean Research,28:291-298.

Hansen E A, Fredsoe J Mao Y. 1986. Two-directional Scour Belowpipelines[C]. Proc. 5th Int. Symp. on Offshore Mech. and Arctic Engrg. ,vol. 3. American Society of Mechanical Engineering,Tokyo,Japan:670-678.

Kjeldsen S P, GjOrsvtk O, Brtngaker K G, Jacobsen J. 1973. Local Scour Near Offshore Pipelines[C]. Second Internat. Port and Ocean Engineering under Arctic Conditions,Conf. Iceland:308-331.

Leeuwestein W, Bijker E W, Peerbolte E B, Wind H G. 1985. The Natural Selfburial of Submarine Pipelines[C]. Proc 4th int. Conf. on Behaviour of Offshore Structure. Elsevier Science:717-728.

Leeuwestein W, Wind H G. 1984. The Computation of Bed Shear in a Numerical Model[C]. Proc 19th International Conference on Coastal Engineering,Houston,TX,2:1685-1702.

Li F, Cheng L. 1999. A Numerical Model for Local Scour Under Offshore Pipelines[J]. Hydraul. Eng. ,125(4):

400 – 406.

Liang D, Cheng L, Li F. 2005. Numerical Modeling of Flow and Scour Below a Pipeline in Currents Part II [J]. Scour Simulation. Coast. Eng. ,52:43 – 62.

Mao Y. 1986. The Interaction Between a Pipeline and an Erodible Bed[D]. PhD thesis, Technical University of Denmark, Lyngby, Denmark.

Sumer B M, Fredsoe J. 1990. Scour Below Pipelines in Waves[J]. ASCE J. Water. Port Coast. Ocean Eng. ,116, (3):307 – 323.

Sumer B M, Mao Y, Fredsoe J. 1988. Interaction Between Vibrating Pipe and Erodible Bed[J]. Journal of Waterway, Port, Coastal, and Ocean Engineering,114:81 – 92.

Van Beek, F A, Wind H G. 1990. Numerical Modelling of Erosion and Sedimentation around Pipelines [J]. Coast. Eng. ,14:107 – 128.

Van Rijn, L C. 1986. Mathematical Modeling of Suspended Sediment in Non Uniform Flows[J]. Hydraul. Eng. , 112(6):433 – 455.

Van Rijn, L. C. 1987. Mathematical Modeling of Morphological Processes in the case of Suspended Sediment Transport[J]. ASCE J Hydr Div,113(3):981 – 994.

Wu Y, Chiew Y M. 2010. Three-Dimensional Scour at Submarine Pipelines in Unidirectional Steady Current[C]. Proceedings of the Fifth International Conference on Scour and Erosion. San Francisco:45 – 55.

Zhao M, Cheng L. 2010. Numerical Investigation of Local Scour Below a Vibrating Pipeline Under Steady Currents [J]. Coast. Eng. ,57:397 – 406.

Study on the Burst Limit States of Submarine Pipeline under Internal Corrosion

Zhigang Li[1], Xianwei Hu[2], Yu Zhang[2], Xuan Gao[2]

[1] Offshore Oil Engineering Co., Ltd., Tianjin, China
[2] Offshore Oil and Gas Research Center, China University of Petroleum, Beijing, China

Abstract Burst limit states are fundamental to the study of submarine pipeline under long internal corrosion. In this paper, a new burst limit function is proposed. Firstly, analytical solution of burst limit function of intact pipeline is worked out. Considering longitudinal internal corrosion, burst limit function is modified. Additionally, the modified limit function is further perfected by multiple regression analysis. Finally, the above function is compared to former models, verifying its accuracy in dealing with submarine pipeline under long internal corrosion.

Key Words Submarine pipeline; Long internal corrosion; Burst limit state; Multiple regression analysis

NOMENCLATURE

σ_y yield strength of pipe material
σ_u tensile strength of pipe material
σ_e equivalent stress based on von-Mises criterion
σ_f flow stress
σ_θ hoop stress
σ_r axial stress
σ_z longitudinal stress
ε_e equivalent strain based on von-Mises criterion
ε_θ hoop strain
ε_r axial strain
ε_z longitudinal strain
D outside diameter of pipeline
t thickness of pipeline
L length of internal corrosion defect
d depth of internal corrosion defect
p_f burst pressure of corroded pipeline
p_i burst pressure of intact pipeline
P internal pressure of corroded pipeline
M bulging stress magnification factor

INTRODUCTION

Internal corrosion is a main cause of submarine pipeline failure, especially for those with relatively long service time. In order to analyze its residual strength and reliability, burst limit states should be firstly worked out. The following are some important limit state functions from codes and papers.

(1) ASME B31G(2009).

ASME B31G is firstly proposed in 1984. However, it is confirmed to be relatively conservative by Kiefner(1990), after which ASME B31G is further modified. It is still widely used in the analysis of residual strength of corroded pipeline.

Original B31G:

$$p_f = \begin{cases} 1.11 \dfrac{2\sigma_y t}{D}\left(\dfrac{1-(2/3)(d/t)}{1-(2/3)(d/t)M^{-1}}\right) & \dfrac{L^2}{Dt} \leq 20 \\ 1.11 \dfrac{2\sigma_y t}{D}\left(1-\dfrac{d}{t}\right) & \dfrac{L^2}{Dt} > 20 \end{cases}$$

Where

$$M = \left(1 + 0.893 \dfrac{L^2}{Dt}\right)^{1/2} \tag{1}$$

Modified B31G:

$$p_f = \dfrac{2(\sigma_y + 68.95)t}{D}\left(\dfrac{1-0.85(d/t)}{1-0.85(d/t)M^{-1}}\right)$$

Where

$$M = \begin{cases} \left(1 + 0.6275\dfrac{L^2}{Dt} - 0.003375\left(\dfrac{L^2}{Dt}\right)^2\right)^{1/2} & \dfrac{L^2}{Dt} \leq 50 \\ 0.032\dfrac{L^2}{Dt} + 3.3 & \dfrac{L^2}{Dt} > 50 \end{cases} \tag{2}$$

(2) DNV-RP-F101(2004).

In DNV-RP-F101(2004), the bursting pressure is for capacity estimation considering the rectangular defect.

$$p_f = 1.05 \dfrac{2t\sigma_u}{(D-t)}\dfrac{(1-d/t)}{(1-(d/t)M^{-1})}$$

Where

$$M = (1 + 0.31(L/\sqrt{Dt})^2)^{1/2} \tag{3}$$

(3) T. A. Netto et al. (2005).

Based on finite element method and experiment, T. A. Netto constructed a burst limit function, which takes a full consideration of defect length, depth and width.

$$\frac{p_f}{p_i} = 1 - 0.9435\left(\frac{d}{t}\right)^{1.6}\left(\frac{1}{D}\right)^{0.4} \tag{4}$$

For $\qquad w/D \geqslant 0.0785, 0.1 \leqslant d/t \leqslant 0.8, l/D \leqslant 1.5.$

(4) B. N. Leis etc (1997).

B. N. Leis and D. R. Stephens constructed a burst limit function with a combination of defect size and profile through parametric analyses.

$$p_f = \frac{2t\sigma_u}{D}\left(1 - \frac{d}{t}M\right) \tag{5}$$

Where

$$M = 1 - \exp\left\{-0.157\frac{L}{[D(t-d)/2]^{1/2}}\right\}$$

1 SOLUTION OF BURST LIMIT STATES

1.1 Burst Pressure of Intact Pipeline

Assume the material of pipeline is power-law hardening and stress-strain model can be illustrated in Eq. (6).

$$\sigma = K\varepsilon^n \tag{6}$$

Assume the pipeline is thin thickness, the equivalent stress and strain can be worked out based on von-Mises criterion.

$$\sigma_\varepsilon = (\sigma_\theta^2 - \sigma_\theta\sigma_z + \sigma_z^2)^{1/2}$$
$$\varepsilon_\varepsilon = \left(\frac{2}{3}(\varepsilon_\theta^2 - \varepsilon_r^2 + \varepsilon_z^2)\right)^{1/2} \tag{7}$$

When engineering stress achieves its limit value σ_u, pipeline appears necking. Accordingly, coefficient K can be obtained:

$$K = \sigma_u\left(\frac{\varepsilon}{n}\right)^n \tag{8}$$

In addition, coefficient n is related to yield-to-tensile ratio.

$$\left[\frac{e(0.002 + \sigma_y/E)}{n}\right]^n = \frac{\sigma_y}{\sigma_u} \tag{9}$$

Based on Eq. (6) to Eq. (9), the burst pressure of intact pipeline can be worked out in Eq. (10):

$$p_i = \left(\frac{1}{\sqrt{3}}\right)^{n+1}\frac{4t}{D-t}\sigma_u \tag{10}$$

1.2 Burst Pressure of Pipeline with Long Internal Corrosion
1.2.1 Expression of Burst Pressure

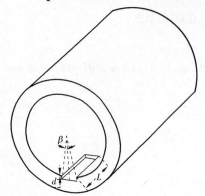

Fig. 1 Defect caused by internal corrosion

The defect caused by long internal corrosion is illustrated in Fig. 1.

In terms of long internal corrosion, the length of defect is no less than $\sqrt{20Dt}$. That is to say, the length of defect is no longer sensitive to the burst pressure of pipeline. Accordingly, radius – thickness ratio, defect depth ratio and width ratio are considered to be the main influence factor of burst pressure of pipeline. The modified burst pressure of pipeline with long internal corrosion is illustrated in Eq. (11).

$$p_f = p_i M \qquad (11)$$

Where

$$M = f\left(\frac{d}{t}\right) g\left(\frac{\beta}{\pi}\right)$$

1.2.2 Numerical Solution of Burst Pressure

Burst pressure of pipeline with long internal corrosion is calculated through finite element method. In terms of a certain issue, say D is 0.508mm, t is 0.00635mm, L is 0.89916mm, d is 0.00216mm, β is 0.01875rad, σ_y is 414MPa, σ_u is 565MPa, p is 90MPa and material is API – X60. Assume that limit state is achieved when Mises-stress of the whole defect is over flow stress. Here, σ_f is defined to be $0.9\sigma_u$. The following is the simulation of this issue through ABAQUS.

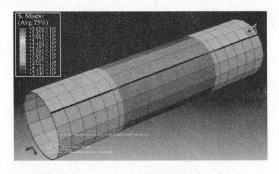

Fig. 2 Stress distribution from outer-view

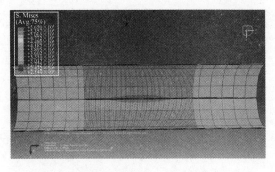

Fig. 3 Stress distribution from inner-view

Fig. 4 to Fig. 6 are three situations of stress distribution along radius direction. The inner pressures of pipeline are separately 8.213MPa, 9.352MPa and 10.419MPa.

Comparing among the above three figures, the last one accords to the definition of limit state. That is to say, burst pressure of this issue is 10.419MPa. Similarly, burst pressure of other issues can be worked out.

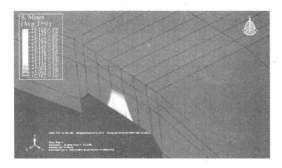
Fig. 4 Stress distribution when p is 8.213MPa

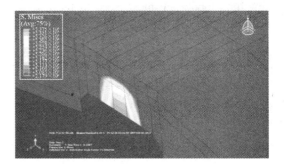
Fig. 5 Stress distribution when p is 9.352MPa

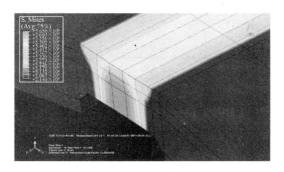
Fig. 6 Stress distribution when p is 10.419MPa

1.2.3 Regression of Burst Pressure

Based on multiple regression through Origin, the impact factor of defect depth and width can be worked out.

$$f\left(\frac{d}{t}\right) = 0.253 + 21.351 e^{-12.955(\frac{d}{t})}$$

$$g\left(\frac{\beta}{\pi}\right) = 0.03\left(\frac{\beta}{\pi}\right)^2 - 0.11\left(\frac{\beta}{\pi}\right) + 1.3$$

Accordingly, Eq. (11) can be written as follows.

$$p_f = \left(\frac{1}{\sqrt{3}}\right)^{n+1} \frac{4t}{D-t} \sigma_u \times (0.253 + 21.351 e^{-12.955(\frac{d}{t})}) \times \left(0.03\left(\frac{\beta}{\pi}\right)^2 - 0.11\left(\frac{\beta}{\pi}\right) + 1.3\right)$$

(12)

2 VERIFICATION OF BURST PRESSURE

In order to make an appropriate comparison among the above-mentioned models, a series of burst test values are chosen as criteria (Freire et al. 2006). Information of pipeline and defect is listed as follows: pipeline material is API-X60, whose yield strength is 4.78MPa and tensile strength is 5.42MPa.

Based on the defects in Table 1, burst pressure related to the above limit models are worked out. Moreover, compared to the test values, errors were calculated, shown in Fig. 7. Results from new limit model are most close to the test values. Accordingly, we can safely conclude that new limit model is more appropriate to pipeline under long internal corrosion defects.

Table 1 Defects

No.	D/m	L/m	d/t	β/π
1	0.508	0.500	4.664×10^{-1}	5.971×10^{-2}
2	0.508	0.500	4.507×10^{-1}	5.971×10^{-2}
3	0.508	0.500	4.568×10^{-1}	5.971×10^{-2}
4	0.324	0.256	6.806×10^{-1}	9.363×10^{-2}
5	0.324	0.306	6.905×10^{-1}	9.363×10^{-2}
6	0.324	0.350	6.869×10^{-1}	9.363×10^{-2}
7	0.324	0.395	6.869×10^{-1}	9.363×10^{-2}
8	0.324	0.500	6.813×10^{-1}	9.363×10^{-2}
9	0.324	0.528	6.848×10^{-1}	9.363×10^{-2}
10	0.324	0.433	6.731×10^{-1}	9.363×10^{-2}
11	0.324	0.467	6.710×10^{-1}	9.363×10^{-2}
12	0.324	0.484	6.813×10^{-1}	9.363×10^{-2}

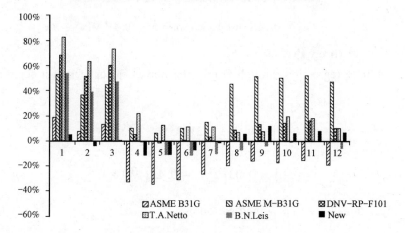

Fig. 7 Error comparison among limit models

CONCLUSIONS

A new limit model is proposed through analytical deduction, finite element analysis and multiple regressions. Through comparing with other limit models and test values, the new limit model is verified to be more appropriate to burst pressure of submarine pipeline under long internal defects.

REFERENCES

ASME B31G. 2009. Manual for Determining the Remaining Strength of Corroded Pipelines. The American Society of Mechanical Engineers.

DNV - RP - F101. 2004. Corroded pipeline. Det Norske Veritas.

Freire J L F, et al. 2006. Part 3: Burst Tests of Pipeline with Extensive Longitudinal Metal Loss[J]. Experimental

Techniques:60 – 65.

Kiefner J F, Vieth P H. 1990. Evaluating Pipe 1: New Method Corrects Criterion for Evaluating Corroded Pipe[J]. Oil & Gas Journal,88(32):56 – 59.

Leis B N, Stephens D R. 1997. An Alternative Approach to Assess the Integrity of Corroded Line Pipe, Part II-Alternative Criterion[C]. Proceedings of the Seventh International Offshore and Polar Engineering Conference, Honolulu, USA:624 – 641.

Netto T A, et al. 2005. The Effect of Corrosion Defects on the Burst Pressure of Pipelines[J]. Journal of constructional steel research,1185 – 1204.

VIV Prediction of a Long Tension-Dominated Riser using a Wake Oscillator Model

Jijun Gu[1], Yu Zhang[2], Yi Wang[2], Menglan Duan[2], Carlos Levi[1]

[1] Laboratório de Tecnologia Oceânica-LabOceano,
Universidade Federal do Rio de Janeiro (UFRJ), Rio de Janeiro, Brazil
[2] Offshore Oil/Gas Research Center, China University of Petroleum, Beijing, China

Abstract The vortex-induced vibration (VIV) of flexible long structures with combined in-line and cross-flow motion has been studied using a wake oscillator in this paper. The mean top tension was evaluated analytically and verified with experimental results, good agreement was observed to present its validity. Then the nonlinear coupled dynamics of the in-line and cross-flow VIV of a long tension-dominated riser were analyzed with the consideration of variation of mean top tension. The in-line and cross-flow resonant frequencies, lift and drag coefficients, dominant mode numbers, amplitudes and instantaneous deflections are reported and compared with experimental results, excellent agreements are observed. The comparison of mode numbers between the calculation with and without consideration of variation of mean top tension showed that the proposed analytical solution of mean top tension can produce a better prediction of multi-mode VIV.

Key Words Wake oscillator model; Riser; VIV prediction; Top tension; Multi-mode

INTRODUCTION

Deep water, string-like, marine risers subject to strong ocean currents, suffer from vortex-induced vibrations (VIV), where vortex shedding interacts with the structural properties of the riser, resulting in large amplitude vibrations in both in-line (IL) and cross-flow (CF) directions. When the vortex shedding frequency approaches the natural frequency of a marine riser, the cylinder takes control of the shedding process causing the vortices to be shed at a frequency close to its natural frequency. This phenomenon is called vortex shedding lock-in or synchronization. Under lock in conditions, large resonant oscillations will reduce the fatigue life significantly.

Despite continued research work on the mathematical model of description of the VIV, the lock-in phenomenon has remained elusive. Computational Fluid Dynamic codes have been developed and are capable of capturing some insight of the VIV, but the large requirement of the computing capacity gives rise to the limitation of modeling the deepwater riser, which in some cases the length has exceeded more than 1,500 meters.

Alternatively, the semi-empirical model for VIV response analysis has been used prevalently in the engineering. These models normally use the hydrodynamic force coefficients as a database, such as drag coefficient, lift coefficient, added mass coefficient and hydrodynamic damping coefficient. These coefficients are obtained from rigid cylinder model tests with forced motions. All these models are based on the assumption that in one mode there is a dominant resonant frequency. But for the deepwater riser, the shear flow and travelling wave will cause the multi-mode and the dense spectrum of the response frequencies. However, a safety factor has been used widely in the design of the risers.

A second type of semi-empirical model is the wake oscillator model which was brought up by Birkoff and Zarantanello(1957). Instead of direct application of the measured fluid forces to the equation of structural motion, wake oscillator models couple the equation of structural motion with a nonlinear oscillator equation that describes the CF fluid force. Three different coupling terms, including acceleration, velocity, and displacement coupling, were evaluated by Facchinetti et al. (2004a). Violette et al. (2007) and Xu et al. (2008) performed the VIV prediction along the spanwise extent of a slender structure under uniform flow, non-uniform flow and linearly sheared flow, the good agreement was achieved. The wake oscillator has been extended to predict IL dynamic response beside the CF response in recent literatures(Ge et al. 2009,2011). Guo et al. (2008) and Li et al. (2010) performed dynamic response analysis of VIV of a long flexible riser with the consideration of influence of top tension and internal flow, and the comparison with experimental results presented a successfully application of such wake oscillator. A comparison between laboratory measurements and blind predictions of eleven different numerical models was carried out by Chaplin et al. (2005a) to analyze the capability of different numerical tools to predict VIV. The response prediction of Orcina Wake Oscillator code was between 85% and 105% of the corresponding measurements, which has the same capability of predicting CF displacements and curvatures as empirical models(such as VIVA, VIVANA, VICoMo, SHEAR7 and ABAVIV), but better than codes based on CFD(such as Norsk Hydro, USP, DeepFlow and VIVIC). Wake oscillator also has been shown to be able to model VIV of vortex-induced waves along cables(Facchinetti et al. 2004b; Violette et al. 2007).

Although much work has been done for investigation of VIV prediction of long slender structures using wake oscillator model, but some of the important factors have not been evaluated well, such as the variation of top tension. The purpose of the present study is to present an oscillator model that couple the structural model both in the IL and CF motion and implement an influence analysis of top tension quantitatively. The comparison between the model and experiment from literature will be carried out.

1 MODEL DESCRIPTION

The dynamic response of one degree of freedom oscillation of a cylinder was successfully simulated by a wake oscillator model(Gu et al. 2012). As a extend work, we implement the dynamic analysis of two degree of freedom oscillation by the same wake oscillator model. The $x - y$ trajectory shape was observed as a figure-of-eight (Jauvtis and Williamson, 2004; Sanchis et al. 2008), and the frequency of IL vibration is typically twice CF frequency. It may be reasonable to expect that the mode number of IL also should be twice CF mode number. In the IL displacement, we use the same wake oscillator model. The model sketch of the coupled structure and wake oscillators combined in-line and cross-flow motion(see Fig. 1).

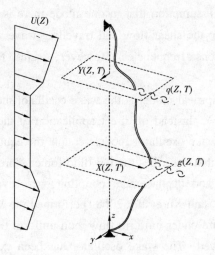

Fig. 1 Model of coupled structure and wake oscillators combining in-line and cross-flow motion

The coupled fluid-structure dynamical system are represented as:

$$\frac{\partial^2 y}{\partial t^2} + \left(\frac{r_s}{m\Omega_f} + \frac{\gamma}{\mu}\right)\frac{\partial y}{\partial t} - c^2 \frac{\partial^2 y}{\partial z^2} + b^2 \frac{\partial^4 y}{\partial z^4} = Mq \qquad (1)$$

$$\frac{\partial^2 q}{\partial t^2} + \varepsilon(q^2 - 1)\frac{\partial q}{\partial t} + q = A_s \frac{\partial^2 y}{\partial t^2} \qquad (2)$$

$$\frac{\partial^2 x}{\partial t^2} + \left(\frac{r_s}{m\Omega_f} + 2\frac{\gamma}{\mu}\right)\frac{\partial x}{\partial t} - c^2 \frac{\partial^2 x}{\partial z^2} + b^2 \frac{\partial^4 x}{\partial z^4} = N\left(C_{Dm} + \frac{C_{Do}}{2}g\right) \qquad (3)$$

$$\frac{\partial^2 g}{\partial t^2} + 2\varepsilon(g^2 - 1)\frac{\partial g}{\partial t} + 4g = B_s \frac{\partial^2 x}{\partial t^2} \qquad (4)$$

$$\frac{C_{Dm}}{C_{Do}} = 1 + 1.043(2Y_{rms}/D)^{0.65} \qquad (5)$$

The dimensionless tension c, bending stiffness b and mass number M, N are given by:

$$c^2 = \frac{T_{top}}{m\Omega_f^2 L^2}, b^2 = \frac{EI}{m\Omega_f^2 L^4}, M = \frac{C_{Lo}}{2}\frac{1}{8\pi^2 St^2 \mu}, N = \frac{1}{8\pi^2 St^2 \mu} \qquad (6)$$

where q, g are reduced fluctuating lift and drag coefficients, $q = 2C_L/C_{Lo}$, $g = 2C'_D/C_{Do}$; C_L denotes lift coefficient. We separate the drag coefficient C_D into two terms: one is the mean drag coefficient C_{Dm} and the other is the fluctuating drag coefficient C'_D. The C_{Dm} is the time averaged mean of the drag coefficient C_D, and C'_D is obtained by removing the time averaged mean C_{Dm}, i.e. $C_D = C_{Dm} + C'_D$. The coefficients C_{Lo}, C_{Do} denote amplitude of fluctuating lift and drag coefficients for a fixed rigid cylinder subjected to vortex shedding. The values of $C_{Do} = 1.5$ and $C_{Lo} = 0.8$ were selected in all cases (Sumer and Fredsøe, 1997; Blevins, 1990). Ω_f and Ω_i are CF and IL reference angular frequencies, and $\Omega_i = 2\Omega_f$.

The values of the van der Pol parameter ε and scaling parameter A_s can be derived from experimental results from Facchinetti et al(2004a). Under the acceleration coupling model, the value of ε is set as 0.3 according to a best-fitting on the lock-in bands for synchronization of vortex shedding with transverse cylinder vibration. The value of the combined parameter $A_s/\varepsilon = 40$ is proposed from a least-squares interpolation between lift magnification and the imposed structure motion amplitude, thus setting $A_s = 12$. $B_s = 3$ is an empirical coefficient in present simulation.

The time-averaged drag coefficient on a cylinder vibrating at or near the vortex shedding frequency is also a function of CF vibration amplitude(Blevins,1990), and some of them have been evaluated by Huang and Sworn(2011). The expression widely used for the increase in the drag coefficient with vibration is shown in Eq.(5) which is based on the results from field experiment with long flexible cylinders carried out by Vandiver(1983). This relation could couple the IL drag force with the CF response. Y_{rms} is the standard deviation of the anti-node displacement in diameters.

The riser model was pin-ended, hence displacements and curvatures were zero at each end with the following set of boundary conditions:

$$y(0,t) = 0, y(1,t) = 0, x(0,t) = 0, x(1,t) = 0 \, \forall t$$

$$\frac{\partial^2 y(0,t)}{\partial z^2} = 0, \frac{\partial^2 y(1,t)}{\partial z^2} = 0, \frac{\partial^2 x(0,t)}{\partial z^2} = 0, \frac{\partial^2 x(1,t)}{\partial z^2} = 0 \, \forall t \tag{7}$$

The relation between initial top tension and the mean top tension was derived by Gu et al. (2012) as follows:

$$T_{mean} = T_{top} + \frac{EA_c(\rho u^2 D C_{Dm} L)^2}{16(EI\pi^2 + T_{mean}L^2)} \tag{8}$$

where T_{top} and T_{mean} being the initial and mean top tensions applied at the top of the riser model, A_c is the wall cross section area. The algebra Eq.(8) could be calculated to get the relationship of T_{mean} versus Reduced velocity u_r which is defined by $u_r = u/f_n D$, where f_n is the fundamental natural frequency of the model.

The coupled Eq.(1) to Eq.(5) are numerically integrated in time and space using a standard centered finite difference method of the second order in both domains. As initial conditions, a random noise with amplitude of order $O(10^{-3})$ is applied to the fluid variable q and g. Zero displacement and zero velocity initial conditions for both IL and CF are applied to the structure. The first time derivative of the fluid variables is also set zero, as initial condition(Violette et al. 2007). For the spatial discretization, 100 points are used for the simulation and a dimensionless time step of 0.01 is used. The integration is carried for dimensionless time t of 100,000 with 10 cycles. The initial values of $C_{Dm} = 0.001$ have been selected in all simulations. The Strouhal number $St = 0.17$ is chosen based on the research done by Larsen(2005).

2 COMPARISON WITH EXPERIMENTAL DATA

A series of numerical calculations were performed as direct simulation of the experiments in uniform flow conducted by Huarte(2006), these simulations could assess the capability of the wake oscillator model described above. The experiments were carried out at Delft Hydraulics in the Delta Flume, which was used as a towing tank. The overall layout of the experiments is shown in Fig. 2. The tank is 230 m in length, 5m in width and, in these tests the water depth was 6.5m. It is equipped with a heavy carriage whose maximum speed in both directions is 1.0m/s. The flow profile was stepped with water in the lower part of the riser model. The lower 5.94 m of its length was subjected to a uniform current; the rest remained in still water. Since the riser model was not fully surrounded by water, the mean drag force only applied on the submerged part in the numerical simulation. The summary of main parameters used in the experiment is given in Table 1.

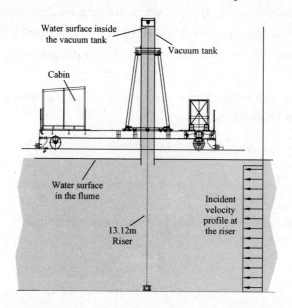

Fig. 2 Model of coupled structure and wake oscillators combined in-line and cross-flow motion. Figure from Chaplin et al. (2005b)

Table 1 Summary of main parameters of the experiment (From Huarte, 2006)

Total length $L(m)$	13.12
Outer diameter $D(m)$	0.028
Submerged Length $L_s(m)$	5.94
Bending stiffness $EI(N \cdot m)^2$	29.88
Mass ratio μ	3
Aspect ratio Λ	470
Top tension $T_{top}(N)$	350 ~ 2,000
Flow speed $u(m/s)$	< 1
Reynolds number Re	2,800 ~ 28,000
Damping ratio ξ	0.0026 ~ 0.014

8 cases have been conducted in their experiment, ranging from low mode response to high mode response in order to show the main features of the computed loads. Table 2 shows the results for all 8 cases: resonant frequencies, dominant mode number in IL and CF, as well as the mean drag coefficient (time-averaged) and root-mean-square (RMS) of lift coefficient.

Table 2 Summary of experimental results (From Huarte, 2006)

Case	u (m/s)	Re	T_{top} (N)	Dominant modes			
				n_y	f_y	n_x	f_x
1	0.31	7,622	1,175	2	1.66	4	3.40
2	0.55	13,649	1,922	3	3.33	6	6.63
3	0.60	14,851	1,538	4	3.52	7	7.04
4	0.85	20,928	1,922	4	4.75	7	9.45
5	0.49	12,191	810	4	3.00	8	6.00
6	0.63	15,559	810	5	3.66	9	7.33
7	0.75	18,422	810	6	4.71	11	9.43
8	0.85	20,900	810	7	5.32	12	10.67

The variation of the mean top tension have been compared between the experimental data from Huarte (2006) and the numerical prediction from Eq. (8) as shown in Fig. 3. The dashed lines are the corresponding fit curved simulated by Eq. (8), good agreements are observed. The effect of the top tension will be more discussed in next section.

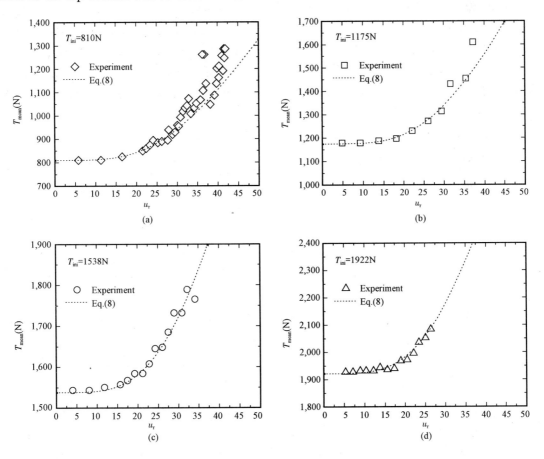

Fig. 3 Comparison of the mean top tension obtained by Eq. 8 and from experiment

Fig. 4 depicts a typical result at midspan of IL and CF motions in each simulation. The power spectrum of IL motion was calculated by removing the mean displacement. It is clear that the vibration of CF has larger amplitude but lower frequency than IL vibration, which agrees with the experimental observation.

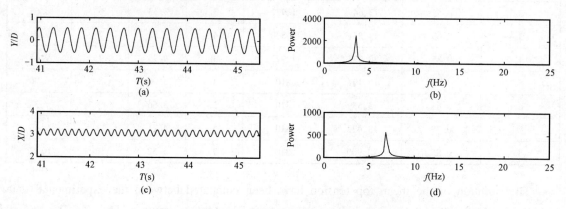

Fig. 4 Time trace of x and y motions at midspan
(a) for CF motion;
(b) and (d) are the frequency spectrum of CF and IL vibration case 5;

The comparison of resonant frequencies and mode numbers are given in Fig. 5 and Fig. 6. Most of the resonant frequencies from numerical simulation are perfectly in agreement with measurement, except in the case 2 and case 4, where the IL resonant frequencies are 85% and 110% of the measured frequencies, respectively. It is noticed that only 4 points have 1 mode number discrepancy. This prediction of the mode numbers is better when compared with that published by Xu et al(2008) and Ge et al(2009,2011).

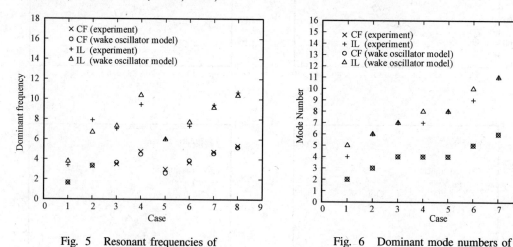

Fig. 5 Resonant frequencies of IL and CF motions for all 8 cases

Fig. 6 Dominant mode numbers of IL and CF motions

Fig. 7 shows the CF and IL maxima amplitudes without mean. Most of the simulations are in good agreement with experiment; exceptions are case 2, case 4 in CF direction and case Fig. 8 in CF and IL directions. As can be seen from Fig. 7, normally, CF amplitude lays between 100% to 117%, and IL amplitude between 100% to 125% of the corresponding measurement. Fig. 8

shows IL maxima amplitudes with mean. Most of the IL amplitude lays between 76% to 100%, exception case 1.

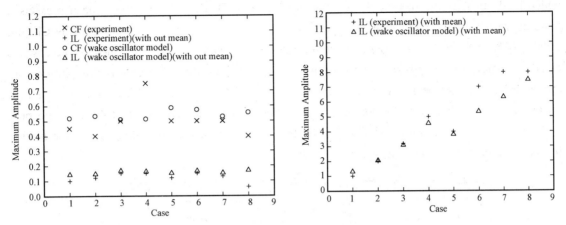

Fig. 7 CF maxima amplitudes and IL maxima amplitudes without mean

Fig. 8 IL maxima amplitudes with mean

The comparisons of deflection and mode shape between experiments and numerical calculations are given in Fig. 9, for case 3, and case 10, for case 5. The typical instantaneous IL deflections from measurement are shown in first column with the mean deflected shape and without it in second column; RMS of the IL in third column; instantaneous CF deflections in forth column; RMS

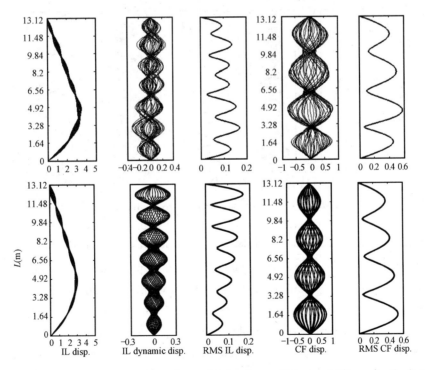

Fig. 9 The comparison between experiment (upper) and wake oscillator model (lower). the instantaneous IL deflections are shown in first column with the mean deflected shape and without it in second column; RMS of the IL third column; instantaneous CF deflections forth column; RMS of the CF fifth column, case 3

of the CF in fifth column. The lower row shows the corresponding results from the numerical simulation. The CF deflected shape as shown in case 3 and case 5 depicts a standard standing wave which is identical to the measurement. The instantaneous IL deflection without mean presents a travelling wave more intense than the measurement. The travelling wave was captured by Chaplin et al. (2005a) with the same experiment set-up but different case. As he mentioned, in the absence of any knowledge of distribution of added mass, and neglecting the effect of variations in tension, the mode shapes have been assumed to be sinusoids. The fact that all contributing modes defined in this way are neither in phase nor in anti-phase with each other, and the fact there are no pure nodes in the profiles (as shown in Fig. 10 second column, second row), indicated that the motion in both directions is a travelling wave.

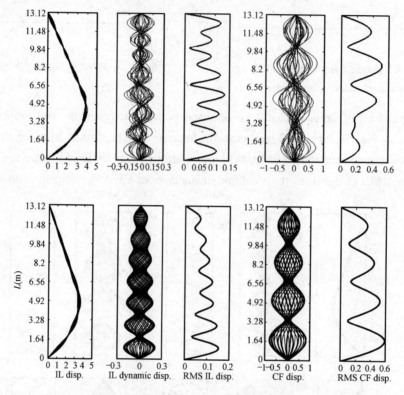

Fig. 10 The comparison between experiment (upper) and wake oscillator model (lower). the instantaneous IL deflections are shown in first column with the mean deflected shape and without it in second column; RMS of the IL third column; instantaneous CF deflections forth column; RMS of the CF fifth column, case 5

It was noticed that in all cases the maxima IL mean displacements were placed at the lower part of cylinder model, this is confirmed both by the experiment and numerical simulation, the probable reasons were: (1) the drag force exerted on the lower part of the riser model; (2) the less axial tension force in the lower part of flexible cylinder, placed vertically.

The mean top tension would increase when the flow velocity increased as we yielded the relation in Eq. (8), Fig. 3 verified the good availability. Fig. 11 shows the CF and IL model numbers versus reduced velocity in terms of different initial top tension 810N, 1175N, 1538N, 1922N. The solid square and cycle symbols represented the experimental results. $T_{mean} = T_{top}$ means the mean top tension is equal to the initial top tension when the towing speed is changed, i. e. the variation of mean top tension is not considered in all simulations, T_{mean} is variate means all the simulations will consider the variation of mean top tension based on the Eq. (8).

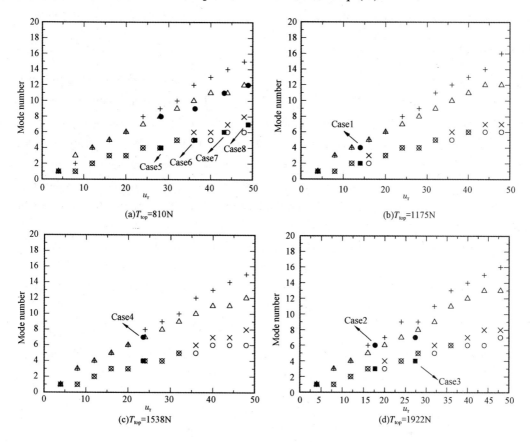

Fig. 11 Mode number versus reduced velocity
○, △—CF and IL mode number when T_{mean} is variate;
×, +—CF and IL mode number when $T_{mean} = T_{top}$;
■, ●—experimental result from Huarete(2006)

From Fig. 11 it is noticed that when the reduced velocity u_r is less than 25, the CF and IL mode numbers are almost the same in terms of different consideration of mean top tension; when the reduced velocity u_r is larger than 25, the CF and IL mode numbers of $T_{mean} = T_{top}$ are more or less higher than that of T_{mean} is variate. The discrepancy of mode numbers increase with the reduced velocity increased, and the experimental results are in better agreement with the simulation of T_{mean} is variate than that of $T_{mean} = T_{top}$, such as case 5 to case 9 in the plot of $T_{top} = 810N$, case 2, case 3 in the plot of $T_{top} = 1922N$. It makes sense as we know increasing axial tension will increase the natural frequency of cylinder model, then the frequency of excitation (vortex shedding frequency) have to be increased consequently in order to obtain the same mode number, that means a higher towing speed is needed. In another word, when the towing speed is the same, the frequency of excitation only can be "controlled" in lower natural frequency of cylinder model which should be applied with a lower top tension.

Meanwhile, as shown in all of the plots, the experimental results agreed better when T_{mean} is variate than $T_{mean} = T_{top}$, so the variation of mean top tension should not be negligible in the case of long flexible cylinder simulation, especially at the range of high reduced velocity. That also might be one of the factors that induced the mode numbers overestimation from Xu et al(2008).

CONCLUSIONS

In this paper, a wake oscillator model coupled both in IL and CF motion has been considered to predict the nature of VIV of long structures. The proposed analytical solution of variation of mean top tension can produce a better prediction of multi-mode VIV. The comparisons of the dominant resonant frequencies, mode numbers and displacements between experimental data and the numerical simulation have shown that the present model could predict some aspects of long flexible cylinder VIV. The variation of mean top tension is obviously another important factor to affect the prediction of mode numbers. When the mean top tension increases, the mode numbers of response decrease.

Due to the simplicity of the wake oscillator model, all the results presented in this paper required a short period of computational time, which is more practicable than the CFD method. And the capability of wake oscillator model could help engineer in the real application on offshore.

ACKNOWLEDGEMENTS

The author would like to thank the Brazilian National Research Council(CNPq) and the National Basic Research Program of China(973 Program) Grant No. 2011CB013702 for the financial support of this research.

REFERENCES

Blevins R D. 1990. Flow-Induced Vibration[M]. Krieger Publishing Company, Malabar, Florida.

Chaplin J R, Bearman P W, Cheng Y, Fontaine E, Graham J M R, Herfjord K, Huera-Huarte F J, Isherwood M, Lambrakos K, Larsen C M, Meneghini J R, Moe G, Pattenden R, Triantafyllo M S, Willden R H J. 2005a. Blind Predictions of Laboratory Measurements of Vortex-induced Vibrations of a Tension Riser[J]. Journal of Fluids and Structures, 21(1):25 – 40.

Chaplin J R, Bearman P W, Huera-Huarte F J, Pattenden R J. 2005b. Laboratory Measurements of Vortex-induced Vibrations of a Vertical Tension Riser in a Stepped Current [J]. Journal of Fluids and Structures, 21(1):3 – 24.

Facchinetti M L, de Langre E, Biolley F. 2004a. Coupling of Structure and Wake Oscillators in Vortex-Induced Vibrations[J]. Journal of Fluids and Structures, 19(2):123 – 140.

Facchinetti M L, de Langre E, Biolley F. 2004b. Vortex-induced Travelling Waves Along a Cable[J]. European Journal of Mechanics B-Fluids, 23(1):199 – 208.

Ge F, Long X, Wang L, Hong Y. 2009. Flow-induced Vibrations of Long Circular Cylinders Modeled by Coupled Nonlinear Oscillators[J]. Science in China Series G-Physics Mechanics & Astronomy, 52(7):1086 – 1093.

Ge F, Lu W, Wang L, Hong Y S. 2011. Shear Flow Induced Vibrations of Long Slender Cylinders with a Wake Oscillator Model[J]. Acta Mechanica Sinica, 27(3):330 – 338.

Gu J j, An C, Carlos L, Su J. 2012. VIV Prediction of Long Flexible Cylinders Modeled by a Coupled Nonlinear Oscillator: Integral Transform Solution[J]. Journal of Hydrodynamics, Ser. B, In press.

Guo H y, Li X m, Liu X c. 2008. Numerical Prediction of Vortex-induced Vibrations on Top Tensioned Riser in

Consideration of Internal Flow[J]. China Ocean Engineering,22(4):675–682.

Huang S,Sworn A. 2011. Some Observations of Two Interfering VIV Circular Cylinders of Unequal Diameters in Tandem[J]. Journal of Hydrodynamics Ser. B,23(5):535–543.

Huarte F J H. 2006. Multi-mode Vortex-induced Vibrations of a Flexible Circular Cylinder[D]. Ph. D. thesis,Imperial College London,Prince Consort Road,London SW7 2BY.

Jauvtis N,Williamson C H K. 2004. The Effect of two Degrees of Freedom on Vortex-induced Vibration at Low Mass and Damping[J]. Journal of Fluid Mechanics,509:23–62.

Larsen C M,K K. 2005. Empirical Model for the Analysis of Vortex Induced Vibrations of Free Spanning Pipelines[C]. In:EURODYN Conference. Paris,France.

Li X m,Guo H y,Meng F S. 2010. Nonlinear Coupled In-Line and Cross-Flow Vortex-induced Vibration Analysis of Top Tensioned Riser[J]. China Ocean Engineering,4:749–758.

Sanchis A,Saelevik G,Grue J. 2008. Two-degree-of-freedom Vortex-induced Vibrations of a Spring-mounted Rigid Cylinder with Low Mass Ratio[J]. Journal of Fluids and Structures,24(6):907-919.

Sumer B M,Fredsøe J. 1997. Hydrodynamics Around Cylindrical structures[J]. Advanced Series on Ocean Engineering,12.

Vandiver J. 1983. Drag Coeffcient of Long Flexible Cylinders[C]. In:Offshore Technology Conference. No. 4490. Houston,TX,USA.

Violette R,de Langre E,Szydlowski J. 2007. Computation of Vortex-induced Vibrations of Long Structures Using a Wake Oscillator Model:Comparison with DNS and Experiments[J]. Computers & Structures,85(11–14): 1134–1141.

Xu W H,Zeng X H,Wu Y X. 2008. High Aspect Ratio(L/D)Riser VIV Prediction Using Wake Oscillator Model [J]. Ocean Engineering,35(17–18):1769–1774.

UNDERWATER SYSTEMS

An Overview of Deepwater Subsea Manifold Vertical Connectors

Hong Guo[1], Penghui Bai[2], Biwei Ren[2], Shichao Lu[2]

[1] CNOOC Research Institute, Beijing, China

[2] Offshore Oil/Gas Research Center, China University of Petroleum, Beijing, China

Abstract Different types of subsea manifold connectors from foreign companies are surveyed and classified in two ways: First, collet connector vs. clamp connector; Second, mechanical connector vs. hydraulic connector. And the characteristics and applicability of each type of connector are well analyzed. The article might be helpful to the development of China's deepwater subsea manifold vertical connectors.

In this sample paper, we describe and provide the formatting guidelines for submissions to the SUTTC Conference Proceedings. Simply download this template from the web, and insert your information where applicable. These guidelines and template streamline the production process, promote uniformity of appearance, improve overall esthetics. By conforming to the specs of this template, your paper will reflect the look of this document.

Key Words Deepwater; Subsea vertical manifold connector; Characteristics; Applicability

INTRODUCTION

Subsea production system is a vital part of exploiting deepwater crude oil and gas, which mainly includes subsea well, Christmas tree, manifold, PLET and PLEM. The production system collects crude oil and gas from every well and transports it by riser to the platform. Subsea connection system is used to connect all the subsea production system equipments, such as Christmas tree vs. manifold, pipelines vs. Christmas tree, and PLET vs. PLEM. Connection system is the guarantee of production safety. A connection system consists of jumpers, subsea connectors and PLET/PLEM. Connectors are very important parts. Currently many foreign companies are capable of designing and manufacturing subsea connectors. China, however, is a late starter in terms of deepwater exploration and production, hence the need to import most of the subsea equipments, including connectors. Given the country's desire of accelerating deepwater resource development, it is highly necessary to improve research on subsea connectors. This article will focus on subsea vertical manifold connectors.

1 A SNAPSHOT OF DOMESTIC AND FOREIGN TECHNOLOGIES

1.1 A Snapshot of Domestic Technology

So far, Chinese oil and gas companies could operate 300 meters under the sea. But China is still in its initial stage of deepwater connection technology development. For instance, it depends on divers to connect different equipments, and it could not independently design and produce connectors. Harbin Engineering University is reported to have made a connector prototype, but its research program is still at a preliminary stage.

1.2 A Snapshot of Foreign Technology

Foreign deepwater connector technology is much more mature, and a wide range of products are available at the international marketplace, such as MAX series of FMC company, CVC series of CAMERON company and HydroTechTM series of Oil States company. Subsea manifold vertical connectors can be categorized as collet connector vs. clamp connector according to their different working modalities. Collect connectors can be further classified into mechanical vs. hydraulic according to their different driving power. Table 1 shows a collection of different kinds of connectors. Different manifold vertical connectors have different characteristics and applicability, which will be analyzed in detail in this paper.

Table 1 Different kinds of subsea manifold connectors made by different companies

	Vertical mechanical collect connector	Vertical hydraulic collect connector	Vertical clamp connector
FMC	✓	✓	✓
Cameron	✓		
GE(Vetco Gray)	✓	✓	✓
Aker Solutions			✓
Oil States		✓	

2 A THOROUGH ANALYSIS OF DIFFERENT KINDS OF SUBSEA VERTICAL MANIFOLD CONNECTORS

2.1 Vertical Collet Connectors vs. Vertical Clamp Connectors

Vertical collet connectors are now widely used. Fig. 1 shows a typical vertical collet connector. It works as follows: connecting the end portion of jumper(called upper hub) and the interface of Christmas tree or manifold(called low hub) with 10 to 20 locking parts, using hydraulic cylinder to push down the locking ring along the surface of the locking part, and finally making the circumferential locking parts clamp and connecting two pipes together.

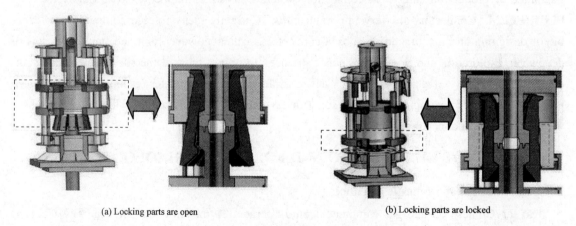

(a) Locking parts are open (b) Locking parts are locked

Fig. 1 Vertical collet connector

Vertical clamp connectors are less often used. Fig. 2 shows a typical clamp connector with two and three pieces of clamps which are locked by hydraulic bolt which ROV could use tools to twist or open. Fig. 3 shows the working principle of clamp connector. Its working principle as follows: the interface of the clamp connector and the hub is designed to a certain angle cone, when the bolts is twisted and clamp would make the two hubs get close to each other and add force to the seal ring. So the two pipes are connected.

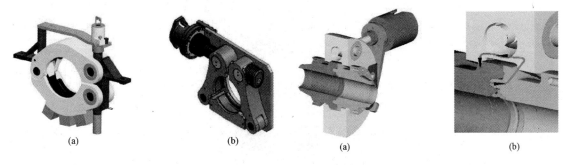

    (a)   (b)   (a)   (b)

    Fig. 2 Clamp connector   Fig. 3 The working principle

The above two kinds of connectors both have been used in engineering for more than 20 years and proved to be reliable. In terms of structure and connecting method, clamp connectors are simpler and more cost-effective than collect connectors, but require more complicated and time-consuming installation. Moreover, clamp connectors make adding insulating layer easier. For instance, Troika oil field in Mexico chooses clamp connectors for the sake of thermal insulation.

These two kinds of connectors are the major connectors in deepwater subsea connection system. Compared with the connectors of flange bolts in the shallow water, these two have some common advantages. At the same time they have its strengths and weaknesses. Table 2 summarizes the strengths and weaknesses of the two collectors.

Table 2 Collect vs. clamp connectors

		Collect	Clamp
Strength		No threaded connections	Low demand for the connection tool
		Available for horizontal or vertical connection	Simple and compact structure, light weight
		Small hole(4in to 12in) is preferred for vertical connection	Low cost totally
		Mechanical connectors have higher reliability during work	No hydraulic devices are left under the sea after installation of the connectors
			Big hole is good (more than 12in) for horizontal connection
			Making adding insulating layer easier
Weakness		Complicated structure and high cost	The decline of threaded connection performance will seriously affect the performance of the connectors during work
		Requiring complicated installment tools	Few application under 1000 meters
Similarity		Allow certain biased alignment	
		Friendly ROV operation panel	

2.2 Mechanical Collet Connectors VS. Hydraulic Collet Connectors

Collect connectors can be further divided into two types: mechanical and hydraulic, according to the different position of the hydraulic cylinders driving the collect. If the hydraulic cylinders are embedded in the installation tool, the connector is mechanical (see Fig. 4). By contrast, if the hydraulic cylinders are part of the connector itself, the connector is a hydraulic one.

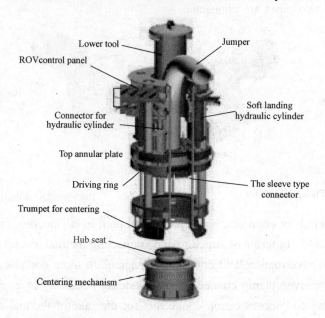

Fig. 4 Mechanical collet connector of CAMERON

Mechanical connectors have lower cost and yet higher reliability, since no hydraulic devices are left under the sea after installation of the connectors. Hydraulic connectors feature faster installation and no need for complicated installation tools, but they are less reliable and more expensive. Table 3 summarizes their differences.

Table 3 Mechanical v s. hydraulic connectos

		Mechanical	Hydraulic
Strength		Simple and compact structure	Do not need running tool
		No hydraulic devices are left under the sea after installation of the connectors	High connection speed
		Installation tools can be used again	Making adding insulating layer easier
		When using an installation tool connects more than three connectors, mechanical connectors can show a higher efficiency	
Weakness		The need for complicated installation tools which generally are more complicated	Volume and cost far outweigh the mechanical connectors
		Installment tool needs to be retrieved after installment. Lower connection efficiency than hydraulic	The decline of hydraulic devices performance will seriously affect the performance of the connectors during work
		Thermal insulation for connectors, a separate subsea insulation work is needed	

3 KEY TECHNOLOGY

After investigating the structure and installation process of the vertical collet connectors, the key technologies during design have been summarized below:

(1) Accurate alignment during the installation process. The connector axis and the hub axis cannot keep parallel completely duo to the ocean current during the installation process, so When the connector is located on the hub, the connector body and the hub may form a tilt angle. How to make the connector and the hub aligned accurately and ensure the connector is effective and reliable in certain angle range is a key technology.

(2) The structure of the locking part. A lot of practice proved that the corner and the mesh surface of the locking part are prone to failure, especially in high-temperature high-pressure condition the connector suffer a variety of loads like stretch, reverse, bend, thermal stress and so on. Therefore, the structure of the locking part cannot failure during work.

(3) The hydraulic system. The operation of locking and unlocking between the locking parts and the hub is completed by hydraulic actuators which provide hydraulic pressure. The control of the force scale and the flow is a key factor during design and operation.

(4) Design of the locking ring. The locking parts' lock and unlock is fulfilled through the locking ring's moving up and down. The locking parts need to take complex loads whose transfer make the locking ring must have a lot of structure strength. And the locking ring would have constant friction with the locking parts, which requires great material stiffness. Moreover, the locking ring would have contact with subsea water, which needs its material must have very good corrosion resistance performance. And synthetic material performance requirements make the cost of a single connector high. how to make the locking ring meet the requirements in the structure and the function and its material cost reduction is a key technology.

(5) Seal system. Seal of the connector is usually using metal to metal seal. Seal test is needed after installation to ensure the reliability of the connector. The seal structure design not only ensures convenient installation of the sealing part but also completes seal test and the sealing part have a long life. besides, how to replace the sealing part quickly when the seal is out of control needs consideration.

(6) Material. The environment that the connector is located in has a high requirement with the material. How to search for a economical material which could meet the demand is a key technology during design.

CONCLUSION

Deepwater subsea manifold connectors are critical to safety of subsea oil and gas production. There are two categories of vertical connectors: collect vs. clamp, and collect connectors can be further divided into mechanical v s. hydraulic. A number of foreign companies are capable of designing and manufacturing a wide range of subsea connectors, with established technology and reliable performance. This paper has analyzed in detail the different kinds of subsea connectors, with a focus on their distinctive structure, characteristics and applicability. It might serve as some reference to China's development of deepwater subsea manifold connectors.

REFERENCES

Andrew Jefferies. 2000. Deep Sea Development Services Inc. Gemini Subsea System Design[C]// Offshore Technology Conference Proceedings. OTC Paper No. 11866.

D S McKeehan, T G Freet. 1993. DeeDwater Flowline Tie-ins and Jumpers: What works best? [C]// Offshore Technology Conference Proceedings. OTC Paper No. 7269.

FMC Technologies. Torus Tie-in System[EB/OL]. (2011-11-13)[2011-11-13].

Sanjay K Reddy, B Mark Paull, Bjorn E Hals. 1996. Diverless Hard-Pipe Connection Systems for Subsea Pipelines and Flowlines[C]//Offshore Technology Conference Proceedings. OTC Paper No. 7269.

Analysis of Specification on Offshore Drilling Unit

Jianjun Wang [1]; Caixia Li [2]; Dongshi Wang[1]; Menglan Duan[2]; Jianbo Li [1], Weiwei Gao[2]

[1] China Oilfield Services Limited, Hebei, China
[2] Offshore Oil/Gas Research Center, China University of Petroleum, Beijing, China

Abstract As the basic analysis of the specification on offshore drilling unit is lacked in China, the international and domestic standards of offshore drilling unit which including machinery, survey, safety and environmental protection and other regulations are researched in this paper. The importance and applicability of the standards for the drilling unit are also discussed, and the analyzing, summarizing and classifying of the rule laid a solid foundation for the application of specification on offshore drilling unit.

Key Words Offshore drilling units; Specification; Machinery; Survey

INTRODUCTION

Oil is the vital lifeline of the national economy, as the increasing fierce competition in the global oil resources; the offshore oil has become the core in the world competition. Offshore drilling unit, as the indispensable part, has become more perfect in technically and standardized management. Offshore drilling unit can be divided into eight categories, there are drill barge, inland barge, drill ship, jack-up, unit rig, semi-submersible, submersible and tender. Now, the drilling units are increasing to 1370 (see Fig. 1). The jack-up and semi-submersible drilling rigs are the main part as commercial unit.

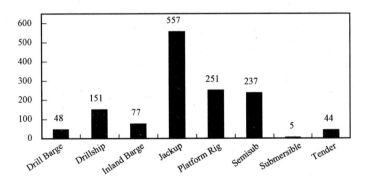

Fig. 1 Global offshore drilling unit type and quantity

Note: The data is come from RIGZONE sides; "Now" refers to September 10, 2012.

Constraints from regulations and standards are increasing along with the drilling unit towards to overseas gradually, and the existing laws and regulations cannot meet our actual requirements. With the analysis of specifications for international offshore drilling industry, the standard system gives real-time operating basis to meet the operation requirements on the drilling unit.

This research has significantly introduced the category of the existing laws, standards, specifications and regulations. With the detail analysis of the specification, its main content: mechanical equipment, survey, environmental conservation and personal safety are discussed.

1 SPECIFICATION CATEGORY

Offshore drilling specification is a series of provisions formulated about its design, manufacture, construction, testing and other technical matters in the process of production and construction. It can be divided into international conventions, classification rules, industry regulations and other standard.

1.1 International Conventions

International conventions refer to the international multilateral treaty about the political, economic, technological, and other aspects. It sets rights and obligations for the main body of the convention. As marine drilling specifications, SOLAS, MARPOL, and other conventions are formulated and promulgated by the International Maritime Organization(IMO).

The IMO is a subsidiary body of the United Nations, it focused on two goals: one is to promote the safety of international shipping and maritime, the other is to prevent marine pollution from ships. It has formulated a unified specification to promote shipping and technical cooperation, to enhance maritime security, to improve the efficiency of vessel traffic, to prevent and control the marine pollution, and to deal with legal issues. The international conventions or amendments cover all aspects of the ship, such as the crew, the safety of maritime navigation, sea freight, damages, and pollution prevention.

The organization consists of an assembly, a council and five main committees: the maritime safety committee(MSC); the marine environment protection committee(MEPC); the legal committee(LC); the technical co-operation committee(TCC) and the facilitation committee(FC) and a number of sub-committees support the work of the main technical committees. International conventions contain the IMO resolutions and circulars; the IMO resolutions consist of MSC and MEPC resolutions, and the circulars consist of MSC, MEPC and united circulars. The main responsibilities of the organization are described in Table 1.

Table 1 The mission of the IMO organization

Organization	Regular Sessions	Consist	Mission
Assembly	Once every two years	All Member States	The highest Governing Body, approving the work programmers, voting the budget, determining the financial arrangements and elects the Council
Council	Once every two years	All Member States	The Executive Organ
Maritime Safety Committee	At least once every year	All Member States	The highest technical body, consider any matter that directly affecting maritime safety
Legal Committee	At least once every year	All Member States	Deal with any legal matters and perform any duties
Marine Environment Protection Committee	At least once every year	All Member States	consider any matter that concerned with prevention and control of pollution from ship.

Continued

Organization	Regular Sessions	Consist	Mission
Technical Co-operation Committee	At least once every year	All Member States	Consider any matter that concerned with the implementation of technical co-operation projects
Facilitation Committee	At least once every year	All Member States	Deals with the work in eliminating unnecessary formalities and matters that concerned with the facilitation of international maritime traffic

IMO has issued 58 international conventions, protocols and amendments, including 47 has come into force; 37 has accepted by China including 6 only apply to the Hong Kong Special Administrative Region. Table 2 describes the international conventions, protocols and amendments issued by IMO and come into force in China (excluding the 6 only apply to the Hong Kong Special Administrative Region).

Table 2 The international conventions, protocols and amendments issued by IMO and come into force in China

Abbr.	Contracted number	Effective date	Effective date in China
IMO(IMCO)1948	168	1958 – 3 – 17	1973 – 3 – 1
SOLAS	159	1980 – 5 – 25	1980 – 5 – 25
SOLAS PROT 1978	114	1981 – 5 – 1	1983 – 3 – 17
SOLAS PROT 1988	93	2000 – 2 – 3	2000 – 2 – 3
ISPS CODE	–	2004 – 7 – 1	2004 – 7 – 1
COLREG1972	152	1977 – 7 – 15	1980 – 1 – 7
MARPOL73/78	149	1983 – 10 – 2	1983 – 10 – 2
MARPOL PROT1997	56	2005 – 5 – 19	2006 – 8 – 23
FAL1965	114	1967 – 3 – 5	1995 – 3 – 17
LL1966	159	1968 – 7 – 21	1974 – 1 – 5
LL PROT1988	90	2000 – 2 – 3	2000 – 2 – 3
TONNAGE1969	150	1982 – 7 – 18	1982 – 7 – 18
INTERVENTION1969	86	1975 – 5 – 6	1990 – 5 – 24
INTERVENTION1973	53	1983 – 3 – 30	1990 – 5 – 25
CLC PROT1976	53	1981 – 4 – 8	1986 – 12 – 28
CLC PROT1992	118	1996 – 5 – 30	2000 – 1 – 5
CSC 1972	78	1977 – 9 – 6	1981 – 9 – 23
PAL 1974	32	1987 – 4 – 28	1994 – 8 – 30
PAL PROT 1976	25	1989 – 4 – 30	1994 – 8 – 30
INMARSATC	92	1979 – 7 – 16	1979 – 7 – 16
INMARSATOA	87	1979 – 7 – 16	1979 – 7 – 16
STCW 1978	153	1984 – 4 – 28	1984 – 4 – 28
STCW CODE	153	1997 – 2 – 1	1997 – 2 – 1
SAR 1979	95	1985 – 6 – 22	1985 – 7 – 24
SUA 1988	153	1992 – 3 – 1	1992 – 3 – 1
SUA PROT 1988	141	1992 – 3 – 1	1992 – 3 – 1

Continued

Abbr.	Contracted number	Effective date	Effective date in China
SALVAGE 1989	57	1996 – 7 – 14	1996 – 7 – 14
COSPAS SARSAT	22	1988 – 8 – 30	1992 – 11 – 18
OPRC1990	97	1995 – 5 – 13	1998 – 6 – 30
BUNKERS2001	40	2008 – 11 – 21	2009 – 3 – 9
LDC1972	85	1975 – 8 – 30	1985 – 12 – 14

1.2 Classification Rules

Classification society is the organization to provide classification service, classification standards, technical specifications, and survey and technical guidance; and promote the personal and property safety and environmental protection. There are more than 50 classification societies in the world.

International Association of Classification Societies (IACS) is composed by 13 major classification societies, established to improve the international maritime safety standards. Each classification society has its own rules for the survey requirement, and drilling unit has to comply with the requirements for keeping the classification. The drilling unit may join a high international recognition classification society except for CCS, such as ABS or DNV, to increase competition. China is difficult to meet the requirements in equipment, technology and other aspects as DNV's rules are stringent, and ABS's rules are more fit reality and practicability, more drilling units are joined into ABS classification in recent years.

IACS is the only non – governmental organization which could formulate and apply the standard in IMO. It consists of 13 full members (see Table 3).

Table 3 IACS full members

No.	Abbr.	Name
1	ABS	American Bureau of Shipping
2	BV	Bureau Veritas
3	CCS	China Classification Society
4	CRS	Croatian Register of Shipping
5	DNV	Det Norske Veritas
6	GL	Germanischer Lloyd
7	IRS	India Register of Shipping
8	KR	Korean Register of Shipping
9	LR	Lloyd's Register
10	NK	Nippon Kaiji Kyokai
11	PRS	Polski Rejester Statkow
12	RINA	Registro Italiano Navale
13	RS	Russian Maritime Register of Shipping

1.3 Industry Regulations

The offshore drilling unit can be divided into the domestic service and overseas service drilling units according to operating conditions. Industry regulations is the offshore oil and gas industry standards, which are developed by China National Offshore Oil Corporation, and issued by the National Development and Reform Commission to describe the specific implementation of operating process for domestic operating drilling units. It can divide into mandatory standards(SY) and recommended standards(SY/T). The industry standards promulgated by the American Petroleum Institute(API) are applied for overseas drilling units.

As the oil and gas industry regulations promulgated by the API become more value, more and more offshore oil and gas industry standards are adopting API standards as guiding standards to narrow the gap with foreign countries. Nowadays, API has issued more than 500 standards and recommended practices, and referenced by more than 100 national standards including ISO, the International Organization of Legal Metrology. It has two main features:

(1) Practicality. API specifications provide the standard of design and process in addition to the device performance, covering oil refining, oil production, fire protection, environmental and other aspects. API standards have no rig standards, and no separate component standards, but the main components of the rig are ruled in several categories. API standards are tending to be the use standard, provisioned on design, material and testing from the user, and accepted by the user and manufacture parties.

(2) Commonality. API standards are from practice, give the standards involved size, survey and acceptance, and the important formula, and provide different use and maintenance requirements according to the product specific condition. There are accepted voluntary and complied with for its applicability in the oil industry, although it is not forced and do not undertake any law liability for this standard.

API standards are widely used for its commonality and practicality, dividing into four categories as standards, specifications, recommended practices and bulletins(see Table 4).

Table 4 API specification category

Name	Abbr.	Content
Standard	Std	Documents about the design and manufacture of petroleum equipment in the oil or relevant industry
Specification	Spec	Documents about the communication between Buyers and manufacturers, the equipment in different manufacturers and different production time, the exchange between materials, and etc.
Recommended Practice	RP	Recommend documents about the project, design, construction, installation, test and operation in oil or relevant industry
Bulletin	Bull	Release the documents when the proceeding is not suitable for operation specification or recommended practice in the oil and relevant industry

1.4 Other Standard

Offshore drilling unit specification also involves corporation standards according to their own conditions in addition to international conventions, classification rules and industry regulations. The enterprises are listed in Table 5 for the offshore drilling unit standard in China.

Table 5 The enterprise standard

Name	Abbr.
China National Petroleum Corporation standard	Q/CNPC
China National Offshore Oil Corporation standard	Q/HS
China Petroleum and Chemical Corporation standard	Q/SH
PetroChina Company Limited standard	Q/SY

2 SPECIFICATION ANALYSIS

2.1 Mechanical Equipment

Mechanical equipments for offshore drilling unit contain rotating machinery, pumps and piping systems, and electrical installations. It has a general description and requirements in the classification and construction rules, and giving the specific requirements in the API specification. For example, it not only provide the performance, design, materials, testing and inspection, welding requirements, but also provide the specific operating conditions and requirements of the pressure, temperature and downhole fluid for the drill through equipment in API 16A.

Mechanical equipment is a major part during the offshore drilling unit operation, six categories are divided as hull structure, drilling equipment, unit equipment, crew equipment, machinery component and platform common systems (see Table 6).

Table 6 The category and specification for offshore drilling mechanical equipments

Name	Content	Specification
Hull structure	Pontoons, columns, legs, trussed frames, main deck, super-structure and deck house, hull outfitting, and etc.	ABS Rules for Building and Classing Mobile Offshore Drilling Units
		DNV Rules for Classification of Ships
Drilling equipment	Derrick with components, drill floor equipment, well control equipment, drill string and downhole equipment, and etc.	API Spec 4F Specification for Drilling and Well Servicing Structures
		API Spec 16A Specification for Drill Through Equipment
Platform equipment	Maneuvering machinery and equipment, navigation and searching equipment, communication equipment, anchoring, towing and equipment, and etc.	Q/HS 5003 specification of communication outfit for offshore platform
		API RP 2SK Design and Analysis of Stationkeeping Systems for Floating Structures

Continued

Name	Content	Specification
Crew equipment	Lifesaving, protection and medical equipment, ventilation, air-conditioning and heating system, sanitary system with discharges, and etc.	IMO International Life-Saving Appliance Code
		CCAR Rules on Operation Certification of Small Aircraft Commercial Transport Operators
Machinery components	Transmissions, foils, boilers, steam and gas generators, fuel system, cooling system, and etc.	ASME Boiler and Pressure Vessel Code
		GB755-87 General Requirements for Rotating Electrical Machines
Platform common systems	Ballast and bilge system, gutter pipes outside accommodation, electrical power supply, electrical ventilators, and etc.	GB50160 Fire Prevention Code of Petrochemical Enter Prise Design
		CCS Safety Rules for Mobile Offshore Units

2.2 Survey

The survey is the inspection, auditing, testing and qualification from the inspection organizations in various conditions to ensure to navigate and operate safely. it can be divided into construction survey, periodical survey, initial survey (see Table 7) according to the different operation status, the survey for the drilling unit should according with the classification and operation survey requirements, and implemented accordance with the requirements of the classification survey and related regulations strictly.

Table 7　The survey category and application for ship

No.	Category	Application
1	Classification	New or reclassification ship
2	Construction	Construct or reconstruct ship
3	Periodical	Service ship
4	Initial	The ship from foreign flag to Chinese flag
5	Pre-operation	The drill ship, mobile unit with foreign flag engaged in drilling, development work
6	Towing	Mobile unit, floating dock and other large facilities sailing before drag towing
7	Temporary	As the accident influent the seaworthiness performance Change the limited usage and navigation area the ship of the certificate issued by a ship survey organization is failure

The class ship must be various surveyed regularly or irregularly for keeping the class; Table 8 gives the survey requirements and period.

Table 8　The survey category and period for keeping class

Category	Application	Survey interval
Annual survey	Classification ship	Within 3 months before or after each anniversary of the completion, commissioning or special survey
Intermediate survey	Classification ship	The 2nd or 3rd annual survey after completion, commissioning or special survey

Continued

Category	Application	Survey interval
Docking survey	Classification ship	Twice in any five-year period, with an interval not exceeding three years between successive surveys
Special survey	Classification ship	Within 5 years from the date of completion, commissioning or special survey
Boiler survey	Classification ship	Internally at least twice within 5 years at an interval not exceeding 3 years, externally at the time of the annual survey
Continuous survey	Classification ship	Within 5 years from the date of completion, commissioning or special survey

2.3 Environmental Conservation

The environmental protection on offshore drilling units is preventing oil pollution, hazardous substances pollution, and air pollution during the operating process of drilling units, and generally divided into four categories:

(1) Prevention of pollution by oil; oil fouling caused by petroleum in any form including crude oil, fuel oil, oil refuse and refined products. The specification gives contingency plans, and provides the oil pollution emission standards in common and special condition.

(2) Prevention of pollution by harmful substances; the used hazard substance from marine resources or human health during normal operation on the drilling unit, mainly refers to the toxic liquid. The specification gives the residue emission standards of harmful substances, and gives contingency plan which causing marine pollution.

(3) Prevent pollution by the ship including sewage and garbage; specification gives the sewage emission standards and connector standard, and provisions for garbage of the special requirements in the special regional.

(4) Prevent air pollution from ships; specification gives the gas emission standard caused by nitrogen oxides, sulfur oxides, volatile organic compounds as well as the burning on ship.

2.4 Personal Safety

Offshore drilling unit is a closed space for its structure and the external environment, a fire or explosion may be occurrence along with the serious environmental pollution and personnel damage once oil spilling. The personal safety specification can be divided into two categories, lifesaving and fire protection; such as the international convention for the safety of life at sea (SOLAS convention), the LSA code (LSA) and international fire safety system rules (FSS) issued by the IMO. Table 9 describes the detail appliance for the personnel safety.

All drilling units must be observed the personal safety standard strictly as the international mandatory regulations in every countries.

Table 9 The detail appliance for the personnel safety

No.	Life – saving appliances	Content
1	Personal Life – Saving Appliances	Lifebuoys, Lifejackets, Immersion suits, Anti-exposure suits, Thermal protective aids, and etc.
2	Visual Signals	Rocket parachute flares, Hand flares, Buoyant smoke signals, and etc.
3	Survival Craft	Inflatable liferafts, Rigid liferafts, Partially enclosed lifeboats, Totally enclosed lifeboats, Free-fall lifeboats, Fire-protected lifeboats, and etc.
4	Rescue Boats	Rigid rescue boats, Inflated rescue boats
5	Launching And Embarkation Appliances	Launching and embarkation appliances, Marine evacuation systems
6	Other Life-Saving Appliances	Line-throwing appliances, General alarm and public address system

CONCLUSIONS

The article describes the IMO convention, the classification specifications, API standards and industry standards for the offshore drilling unit in China. Mechanical equipment specifications, survey specifications, standards for environmental protection and personal safety standards for the drilling unit were introduced. Application of offshore drilling unit was summarized briefly; it is significance for the basic research of offshore drilling units in China.

The research about regulations and standards for domestic offshore drilling unit is a infrastructure problem, this study could provide the necessary theoretical basis and technical guidance for toward international and exploring the international market, and promote the process of the modernization and internationalization in offshore drilling field.

REFERENCES

Chen Di, Jiang Yuli, Li Li. 2010. Connotation and Application API Standard on Petroleum Machinery Products [J]. International Petroleum Economics, 5:60 – 63.

David Pinder. 2001. Offshore Oil and Gas: Global Resource Knowledge and Technological Change [J]. Ocean & Coastal Management, 44:579 – 600.

International Life-saving Appliance Code [S]. International Maritime Organization Standard.

Liu Gongchen, E Hailiang. 2008. The Implementation of the IMO Conventions on Pollution Prevention from Ships in China [J]. China Maritime Safety, 6:4 – 6.

Wang Dingya, Ding Liping. 2010. The Technology Present Situation and Development Tendency of Marine Drilling Unit [J]. China Petroleum Machinery, 38(4):69 – 72.

Wei Jingtian, Ma Yanling. 2009. IMO's 60 Years for Regulating and Serving the International Shipping——A Review of IMO's History of 60 Years [J]. China Maritime Safety, 10:58 – 62.

Zhao Hongshan, Liu Xinhua, Bai Liye. 2010. Development Tendency of Offshore Petroleum Drilling Equipment in Deep Water [J]. Oil Field Equipment, 39(5):69 – 74.

Application of Explicit Quasi-Static Analysis in the Design of Subsea Connector

Yi Hong[1], Kai Tian[2], Menglan Duan[2]

[1] Research Institute, Beijing, China
[2] Offshore Oil/Gas Research Center, China University of Petroleum, Beijing, China

Abstract Subsea connector is widely used in the deep water field development. In the design of this machine FEA(Finite element analysis) is cost saving and effective. But because of the complicated contact conditions between parts of the connector, it is hard to reach convergence. Through the comparison between contact algorithms of explicit and implicit analysis it is found that explicit which is a dynamic solving method is applicable for the FEA of subsea connector. Considering the relatively slow locking speed explicit quasi-static analysis is the best choice.

Key Words Explicit analysis; Quasi-static analysis; Subsea connector

INTRODUCTION

Subsea connection system (Fig. 1) is an important part of the subsea production system (Wang, Duan and Feng, 2011). It is widely used to connect jumper with manifold or charismas tree. Now all the mature products in the market are from FMC, CAMERON, Aker Solution and Oil State. If China wants to develop its own subsea technology, ability of manufacturing subsea connectors is of necessity.

Fig. 1 Subsea Connection System (FMC technologies, 2000)

As a complex machine structure which is operated unmanned in deep water, design of such structure deserves special attention. Prototype is time consuming and waste lots of time. So Finite Element analysis is a better choice. The contact within this structure is important for its function but simulation of contact is always difficult. This paper will illustrate the difficulty and make a comparison with the two methods used to solve the contact problem. Some suggestions are made too.

1 PROBLEM

Using Finite Element Method to simulate the structure and analyze its capability is a trend in the modern industry. It can detect design problems before a model test done which saves lots of money and time. But for the subsea connection system there is a difficulty in the design with FEM. As the Fig. 2 every locking segment has three contact surfaces which are for the actuator for the upper hub and for the down hub. In addition the gasket also has two contact surfaces for upper hub and for the down hub. If there are 18 locking segments totally in the subsea connector, 56 contact pairs are needed to be created in the model.

These contact pairs promote the simulation difficulty in two aspects(Wang,2003).

(1) The areas in contact are always changing when the connector is locked up.

(2) Nonlinearity in contact condition, including: ① The contact pair can't penetrate into each other; ② Normal component of the contact force can only be pressure; ③ Friction exists on the tangential contact. These constraints are inequality constraint with strong nonlinearity.

Fig. 2 Contact pairs in subsea connector (Cameron,2000)

2 SOLUTION METHOD SELECTION

Usually there are two methods solving the contact problem in the finite element analysis: implicit and explicit.

For the implicit analysis method trying to establish the accurate conditions for so many surfaces at the same will induce severe incontinuous iteration. To eliminate the difficulty it is recommended to force the components into contact state with initial boundary conditions before the loads are applied. For example small displacements are usually used to help build the contact conditions. In this way large Interference contact and steep change in the contact force can be avoided. Reasonable contact conditions will be established stably.

But for the complicated mechanical structure of subsea collet connector interference contact is unavoidable in the analysis. When the connector is being locked up the contact force on every surface is changing. What made it worse is that dozens of locking segment are going to experience this procedure at the same time. To establish all the contacts smoothly at the beginning means a lot of time and workloads.

For the explicit analysis central difference method is commonly used:

$$u^{t+\Delta t} = u^t + \dot{u}^t \Delta t + \frac{1}{2}\ddot{u}^t \Delta t^2 \tag{1}$$

$$\dot{u}^{t+\Delta t} = \dot{u}^t + \frac{1}{2}(\ddot{u}^t + \ddot{u}^{t+\Delta t})\Delta t \tag{2}$$

Where $u, \dot{u}, \ddot{u}$ are the displacement velocity and acceleration vector; Δt is the time increment.

Finite element solution function is:

$$M\ddot{u}^{t+\Delta t} = Q_L^{t+\Delta t} + Q_c^{t+\Delta t} - F_{t+\Delta t}^{t+\Delta t} \tag{3}$$

Where M is lump mass matrix; Q_c is the equivalent contact force vector on node; Q_T is the load vector on node; F is the internal force vector.

For the U. L. format the calculation procedure is as follows (Wang, 2003):
(1) set $t = 0$ and Δt is determined;
(2) M is formed and M^{-1} is obtained;
(3) $\ddot{u}^t = M^{-1}(Q_L^t - F_t^t)$;
(4) $u^{t+\Delta t} = u^t + \dot{u}^t \Delta t + \frac{1}{2}\ddot{u}^t \Delta t^2$;
(5) calculate $Q_L^{t+\Delta t}$;
(6) Cauchy stress vector $\widehat{\tau}^{t+\Delta t}$ and internal force $F_{t+\Delta t}^{t+\Delta t}$;
(7) contact force $F_{t+\Delta t}^{At+\Delta t}$;
(8) Search for contact node pair $(P, Q)_k (k = 1, 2, \cdots, n_c)$;
(9) Judge the contact condition of every contact node pair;
(10) $Q_c^{t+\Delta t} = \sum_{k=1}^{n_c} (Q_c^{t+\Delta t})_k$;
(11) $\ddot{u}^{t+\Delta t} = M^{-1}(Q_L^{t+\Delta t} + Q_c^{t+\Delta t} - F_{t+\Delta t}^{t+\Delta t})$;
(12) $\dot{u}^{t+\Delta t} = \dot{u}^t + \frac{1}{2}(\ddot{u}^t + \ddot{u}^{t+\Delta t})\Delta t$;
(13) set $t = t + \Delta t$;
(14) If $t < T$ return to step (4), or the calculation is over.

In step (6), $\widehat{\tau}^{t+\Delta t}$ is Cauchy stress vector. It is replaced with the second kind Piola – Kirchhoff vector $\widehat{S}_0^{t+\Delta t}$ for the T. L. format. At the same time $F_{t+\Delta t}^{t+\Delta t}$ in step (6) is replaced with $F_0^{t+\Delta t}$.

From the above procedure it can be seen obviously that the iteration procedure is avoided in every increment. Meanwhile the inversion of asymmetric stiffness matrix owing to the friction slide contact is also avoided. The reason is that the stress vector and contact force vector is computed based on the quantity of state such as displacement at the end of the previous step. So it can be placed on the right end of the recurrence equation as part of equivalent load. As a result, the slide friction contact on the locking surface between locking segment and up/down hub or between locking segment and actuator can be simulated better in explicit analysis method.

3 CALCULATION EXAMPLE WITH A TYPICAL SUBSEA CONNECTOR

To prove the analysis better, ABAQUS/EXPLICIT is used to simulate a typical subsea connector (Zhuang, 2009). The model (Fig. 3) is imported from SOLIDWORKS a famous CAD design software. Simulation of the locking procedure is done by giving the actuator a displacement downward.

The calculation is finished in 23 hours. For such a complex machine structure the cost of time is within tolerance. The bearing capacity analysis of the connector can also be done on this model using same method. To increase the analysis speed further two methods is presented:

(1) Mass scaling. The stable extreme increment is expressed as:

$$\Delta t = \frac{L^e}{c_d} \quad (4)$$

Where L^e is the characteristic element length; c_d is the expand wave speed of material. When the Poisson's ration is zero

$$c_d = \sqrt{\frac{E}{\rho}} \qquad (5)$$

Where ρ is the density of material.

When the density ρ increases n^2 times the wave speed decreases n times and stable time increment will increase n times. But it should be noted that the change of mass must be limited to certain level so that the results will not be too far away from the reality. Usually the level is set to be 15%.

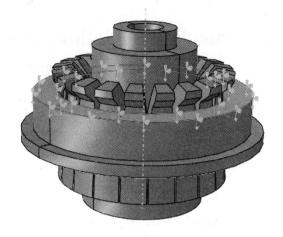

Fig. 3 Typical subsea connector

(2) Restart analysis is also helpful when the analysis involves several steps with different displacement definitions. The afterwards analysis can be done on the basis of previous results.

CONCLUSION

In the FEA of subsea connector, simulation of connect pair is a major difficulty. The common used implicit method is not applicable here. Through comparison between the contact algorithms of implicit and explicit method, it is proved that explicit method is applicable for the subsea connector simulation. To improve the solving efficiency two suggestions: mass scaling and restart analysis are made.

ACKNOWLEDGEMENTS

This work was financially supported by the National Science and Technology Major Project (2011ZX05026 - 003 - 02 Prototype Development of Subsea Connector for manifolds) and National High Technology Research and Development Program (863 Program, 2012AA09A205 Key Technical Research of Subsea Tree and Complete Sets of Equipment Development).

REFERENCES

Cameron. 2000. Cameron Stack Equipment. Cameron cooperation.
FMC technologies. 2010. MAX Vertical Tie-in System, FMC technologies.
Wang Xucheng. 2003. Finite Element Method[M]. Beijing: Tsinghua University Press.
Wang Y Y, Duan M L, Feng W, et al. 2011. Investigation on Installation Methods of Deepwater Manifolds and Their Applications to LW3 - 1 Gas Field in South China Sea[J]. The ocean engineering, 29(3). 23 - 30.
Zhuang Zhuo. 2009. Finite Element Analysis and Application Based on ABAQUS[M] Beijing: Tsinghua University Press.

Calculation Method of Gas Hydrate Formation Inhibitor Concentration for Qiong Dongnan Deepwatwer Drilling

Ke Ke, Qiang Zhang, Minsheng Wang

Sinopec Research Institute of Petroleum Engineering, Beijing, China

Abstract Sinopec has planned to develop the oil and gas district of Qiong Dongnan deepwater area. Gas hydrate would be formed easily under high pressure and low temperature condition in some well section when gas kick happened and cause many problems during drilling. Temperature and pressure distribution and phase equilibrium condition determination method of Qiong Dongnan deepwater area have been established with consideration of Qiong Dongnan sea water temperature distribution and different well section, drilling condition and hydraulic parameters. The hydraulic formation region and inhibitor concentration prediction model are established based on the temperature-pressure distribution and hydrate phase equilibrium condition. According to the calculation example of a deepwater well in Huaguang district of Qiong Dongnan area, the law about influence parameters on hydrate formation region has been analyzed and recommended inhibitor concentration program is provided.

Key Words Deepwater drilling; Gas hydrate; Phase equilibrium condition; Inhibitor concentration

INTRODUCTION

Natural gas hydrate is a kind of crystal lattice ice-shaped substance formed by water and natural gas under high pressure and low temperature. Natural gas hydrate formation brings serious operation and safety problems to deepwater drilling. The risk of natural gas hydrate formation is more serious with water depth increasing in deepwater drilling. The average water depth of Qiong Dongnan deepwater area is deeper than 1,000m. During drilling operation in this area, when natural gas from oil and gas reservoir enters wellbore, the gas hydrate is easy to be formed due to the high pressure and low temperature in deepwater wellbore and conduits, which will bring new difficulties and challenges to drilling and production parameter design, well control, and riser design. Therefore, it is necessary to predict the formation region of gas hydrate in order to effectively control its forming. Several domestic and abroad researchers have carried out much work on the forming condition of gas hydrate, that is temperature and pressure condition, and sound prediction model was formed. Based on these models, Wang and Sun has founded an approach to predict the formation area of hydrate, but it is unavailable to exactly determine the inhibitor concentration hydrate that can inhibit the hydrate formation. By establishing the theoretical prediction model of gas hydrate formation region and by using the computer calculation, the paper presents the sea water temperature profile in Qiong Dongnan district, variation of pressure field and temperature field in the process of deepwater drilling, and the inhibitor concentration prediction method. At last, According to the calculation example of a deepwater well in Huaguang district of Qiong Dongnan area, the law about influence parameters on hydrate formation region has been analyzed and recommended inhibitor concentration program is provided.

1 CALCULATION MODEL FOR QIONG DONGNAN SEAWATER TEMPERATURE PROFILE ESTABLISHMENT

The vertical temperature distribution could be divided into three layers: (1) Mix-layer. In this layer, the water depth is less than 100m, the water temperature distributed homogenously and vertical gratitude is small because the seawater mixture resulted by the convection, wind and the wave. (2) Temperature leaping layer. This layer is between the mixture layer and the constant temperature layer, in this layer, the seawater temperature will reduce dramatically with the water depth decreasing, the vertical water temperature gratitude is high. (3) Constant temperature layer. In this layer the water temperature change is little and between 2 to 6 centigrade degree.

There is a certain relationship between seawater temperature and water depth. According to the temperature leaping layer results, the max depth of the lower boundary of this layer is 200m, the temperature of the seawater deeper than this layer can't be impacted by the surface water temperature, so a fitting formula can be used for the seawater temperature in which the depth is deeper than 200m.

The seawater temperature fitting formula is as follow based on the Levitus database:

$$T_{sea} = a_2 + (a_1 - a_2)/[1 + e^{(h+a_0)/a_3}] \quad h > 200m \quad (1)$$

Where $a_1 = 39.398, a_2 = 2.307, a_0 = 130.137, a_3 = 402.732, T_{sea}$ is seawater temperature, ℃; h is water depth, m.

The seawater temperature less than 200m is more complicated attributed to the geological situation of the seabed, wave and tide, climate and the different sun radiation strength in different area. The average seawater temperature less than 20m of Bohai is 12℃ while it in Huanghai is 16℃, in Donghai is around 22℃ and in Nanhai is around 26℃. The mix layer depth of Qiong Dongnan area in winter and fall can reach 50m and 100m while it in spring and summer is little. Assuming there is a steady temperature gratitude in leaping layer, the fitting formulas are as follows:

Spring:
$$T_{sea} = \frac{T_S(200 - h) + 13.68h}{200} \quad 0 \leqslant h < 200m$$

Summer:
$$T_{sea} = T_S \quad 0 \leqslant h < 20m$$

$$T_{sea} = \frac{T_S(200 - h) + 13.7(h - 20)}{180} \quad 20 \leqslant h < 200m$$

Fall:
$$T_{sea} = T_S \quad 0 \leqslant h < 50m$$

$$T_{sea} = \frac{T_S(200 - h) + 13.7(h - 50)}{150} \quad 50 \leqslant h < 200m$$

Winter:
$$T_{sea} = T_S \quad 0 \leqslant h < 100m$$

$$T_{\text{sea}} = \frac{T_S(200-h) + 13.7(h-100)}{100} \quad 100 \leq h < 200\text{m} \tag{2}$$

Where, T_{sea} is seawater temperature, ℃; T_S is seawater suface temperature, ℃; h is depth, m. the fitting results are shown in Fig. 1 as follow.

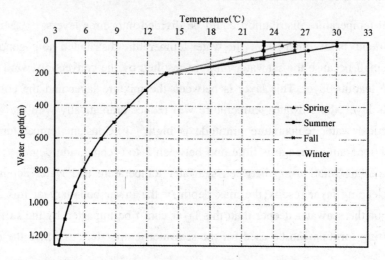

Fig. 1 Seawater temperature profiles of four seasons in Qiong dongnan deepwater area

2 HYDRATE INHIBITOR CONCENTRATION CALCULATION MODEL IN WELLBORE DURING DEEPWATER DRILLING

The Hydrate inhibitor concentration calculation model in wellbore during deepwater drilling can be gained based on wellbore temperature and pressure distribution and the hydrate formation thermodynamic equation. So the hydrate inhibitor concentration determination model includes three parts: hydrate phase equilibrium condition determination, wellbore temperature and pressure field calculation.

2.1 Hydrate Phase Equilibrium Condition Determination With The Hydrate Inhibitor

In the crystal structure of gas hydrate, there exists the chemical potential balance relation among water phase, gas phase, and crystal lattice. According to the thermodynamic balance theory, we can obtain the thermodynamic equation of gas hydrate.

$$\frac{\Delta\mu_0}{RT_0} - \int_{T_0}^{T_H} \frac{\Delta H_0 + \Delta C_K(T_H - T_0)}{RT_H^2} dT_H + \int_{p_0}^{p_H} \frac{\Delta V}{RT_H} dp_H$$

$$= \ln\left(\frac{f_w}{f_{wr}}\right) - \sum_{i=1}^{l} M_i \ln\left(1 - \sum_{j=1}^{L} \theta_{ij}\right) \quad i = 1,2,\cdots,l; j = 1,2,\cdots,L \tag{3}$$

$$\ln\left(\frac{f_w}{f_{wr}}\right) = \ln x_w \tag{4}$$

If hydrate inhibitor is added:

$$\ln\left(\frac{f_w}{f_{wr}}\right) = \ln(x_w \gamma_w) \tag{5}$$

Where, $\Delta\mu_0$ is aqueous chemical potential difference in pure water and empty hydrate crystal lattice at standard condition, J/mol; R is universal gas constant, J/(mol·K); T_0 is standard condition temperature, K; T_H is phase temperature of hydrate formation, K; ΔH_0 is enthalpy difference between pure water and empty hydrate crystal lattice, J/kg; ΔC_K is heat absorption capacity difference between pure water and empty hydrate crystal lattice, J/(kg·K); p_H is phase pressure of hydrate formation, Pa; p_0 is standard condition pressure, Pa; ΔV is specific volume difference between pure water and empty hydrate crystal lattice, m³/kg; f_w is aqueous fugacity in abundant water, Pa; f_{wr} is aqueous fugacity at reference state (T_H, p_H), Pa; l is number of hydrate species; M_i is cavity number of No. i hydrate in unit water molecule; L is component number of hydrate; θ_{ij} is volume percentage of molecule j in i cavity; x_w is water mole fraction in abundant water, nondimension; y_w is water activity coefficient in abundant water, nondimension.

Eq. (3) to Eq. (5) are hydrate phase equilibrium condition determination equations when hydrate inhibitor is added. Sodium chloride and ethanol are two kinds of hydrate inhibitor often used, the impact of different Sodium chloride and ethanol concentration has been analyzed which are shown in Fig. 2 and Fig. 3. These figures indicate that the hydrate phase equilibrium pressure increase at the same temperature with hydrate inhibitor compared with no hydrate inhibitor situation, and higher the inhibitor concentration, higher the hydrate phase equilibrium pressure. In addition, with the same concentration, ethanol has better effect for hydrate formation inhibition.

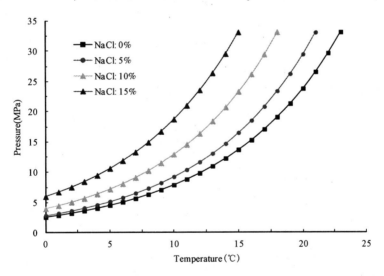

Fig. 2 The impact of sodium chloride inhibitor concentration on methane hydrate phase equilibrium condition

2.2 Wellbore Pressure Field Equation

High pressure is a necessary factor in gas hydrate formation. Also, the pressure field should be calculated accurately when predicting gas hydrate formation region in wellbore. Only when the gas enters wellbore, gas hydrate may form. And once gas invasion occurs, the whole wellbore will be in a multiphase environment with gas, liquid, and solid. In order to calculate the pressure field accurately, we must consider the influence of multiphase fluids and establish the multiphase flow governing equation. Then, the pressure distribution in the wellbore will be obtained by solving the equation.

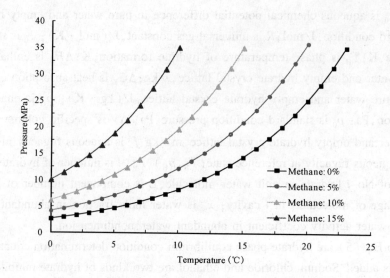

Fig. 3　The impact of ethanol inhibitor concentration on methane hydrate phase equilibrium condition

Continuity equation:
Gas phase

$$\frac{d}{dt}(AE_g\rho_g) + \frac{d}{ds}(AE_g\rho_g v_g) = q_{pg} \quad (6)$$

Liquid phase

$$\frac{d}{dt}(AE_m\rho_m) + \frac{d}{ds}(AE_m\rho_m v_m) = 0 \quad (7)$$

Drill cuttings

$$\frac{d}{dt}(AE_c\rho_c) + \frac{d}{ds}(AE_c\rho_c v_c) = 0 \quad (8)$$

Momentum conservation equation:

$$\frac{d}{ds}(AE_m\rho_m v_m^2 + AE_c\rho_c v_c^2) + Ag\cos\alpha(E_m\rho_m + E_c\rho_c) + \frac{d(Ap)}{ds} + A\left|\frac{dp}{ds}\right|_{fr} = 0 \quad (9)$$

Where, t is time, s; A is cross-sectional area of annulus, m²; ρ_g is density of produced gas, kg/m³; E_g is volume fraction of produced gas, nondimension; s is the distance between any point in wellbore and bottom hole, m; v_g is uphole velocity of produced gas, m/s; q_g is gas invasion rate, kg/s; ρ_m is density of fluid phase, kg/m³; E_m is volume fraction of liquid phase, nondimension; v_m is uphole velocity of liquid phase, m/s; ρ_c is density of cuttings, kg/m³; E_c is volume fraction of cuttings, nondimension; v_c is uphole velocity of cuttings, m/s; q_c is cuttings production rate, kg/s; g is acceleration of gravity, m/s²; α is well deviation angle, rad; p is annular pressure, Pa; p_f is on way friction loss, Pa.

2.3 Temperature Field Equations

Low temperature is another necessary factor in gas hydrate formation. Wellbore temperature can be affected by many factors, including environment temperature, static formation temperature, wellbore structure, circulation parameters, etc.

Considering wellbore and circulation system as a thermodynamic system, temperature field equations in wellbore are derived according to principle of energy conservation.

Temperature field equation in annulus under mudline:

$$\frac{\partial}{\partial t}(\rho_g E_g C_{pg} T_a) + (\rho_1 E_1 C_1 T_a)]\bar{A} - \left[\frac{\partial(w_g C_{pg} T_a)}{\partial z} + \frac{\partial(w_1 C_1 T_a)}{\partial z}\right]$$

$$= 2\left[\frac{1}{A'}(T_{ei} - T_a) - \frac{1}{B'}(T_a - T_t)\right] \quad (10)$$

Temperature field equation inside drill string under mudline:

$$\frac{\partial}{\partial t}(\rho_1 C_1 T_t)\bar{A}_t + \frac{\partial(w_1 C_1 T_t)}{\partial z} = \frac{2}{B'}(T_a - T_t) \quad (11)$$

where, A_t are area of annual and inner drilling string respectively, m^2; Temperature field equation over the mudline is similar with the equation under the mudline, the only difference is the expression of A': $A' = \frac{1}{2\pi r_{ro} U_a}$。

2.4 Hydrate Inhibitor Concentration Calculation Method

The steps are as follow:

(1) Hydrate inhibitor concentration calculation method when pump is off:

① According to the temperature field calculation method, calculate the wellbore temperature field 1 hour after the pump being off, draw the temperature-pressure curve L_{PT}。

② Choose a kind of hydrate inhibitor and assume the initial concentration C_0, according to the hydrate phase equilibrium condition determination method when hydrate inhibitor is added and the wellbore pressure field calculation result, the phase equilibrium curve L_{C_0} can be gained。

③ If the depth D_0 of the common area made by L_{C_0} and L_{PT} more than 1m (as shown in Fig. 4), assume the new inhibitor concentration $C_1 = 2C_0$; if these two curves don't have common area ($D_0 < 0$), assume the new inhibitor concentration $C_1 = \frac{C_0}{2}$. According to step (1), the phase equilibrium curve L_{C_1} with the new concentration can be gained, and then the new depth D_1 of common area made by the new phase equilibrium curve L_{C_1} and the temperature-pressure curve L_{PT}.

④ According to the the Eq. (12), calculate the new hydrate inhibitor concentration repeatedly until the depth D_{i+1} of the common area made by the curve L_{C_i} and L_{PT} satisfies the Eq. (13), the concentration C_{i+1} is what we want to gain.

$$\begin{cases} C_{i+1} = \dfrac{C_i + C_{i-1}}{2} & D_i < 0 < 1m < D_{i+1} \\ C_{i+1} = 2C_i & D_i > 1m \end{cases} \quad (12)$$

$$0 \leq D_{i+1} \leq 1m \quad (13)$$

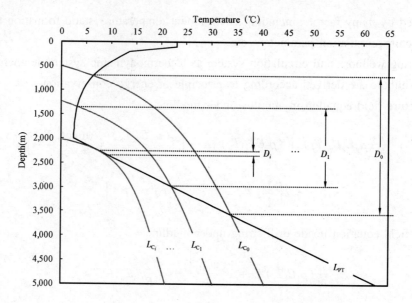

Fig. 4 Steps of inhibitor concentration calculation

(2) Hydrate inhibitor concentration calculation method when circulating.

The Hydrate inhibitor concentration calculation method when circulating is similar with the method when pump is off, the only difference is that the temperature and pressure field calculation should be based on the circulating parameters.

3 CASE STUDY OF HUA GUANG X IN QIONG DONGNAN DISTRICT

3.1 Basic Well Data

Basic data of well Hua Guang X is shown in Table 1.

Table 1 Basic data of well Hua Guang X

Well name	Huang Guang X	Water depth(m)	1,260
Type	Vertical well, Wildcat well	TD(m)	4,650
Pore pressure (Pa)	1.24	Formation temperature gradient (℃/100m)	3.1
Surface seawater temperature (℃)	25	Mudline temperature (℃)	3.3
Natural gas compostion	Methane:93.46%, Ethane:5.40%, Propane 1.14%		

Hydraulic parameters are shown as Table 2.

Table 2 Hydraulic parameters of well Hua Guang X

Hole diameter(mm)	Mud density(g/cm^3)	Displacement(L/s)
660.4	1.03	75
444.5	1.12	70
311.15	1.21	63
215.9	1.27	37

Casing Program of the well Hua Guang X is shown as Fig. 5.

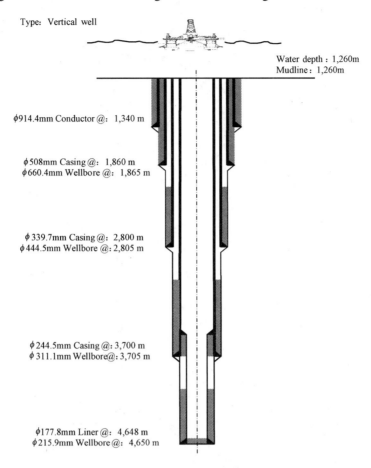

Fig. 5　Hua Guang X well casing program

3.2　Hydrate Formation Prediction in 660.4mm Wellbore

Fig. 6 shows the prediction results of hydrate formation in wellbore during drilling process without inhibitor. According to the simulation, in 660.4mm wellbore, if natural gas enters wellbore, the hydrate formation section is 355 ~ 1,865m when pump is off while it is 355 ~ 1,865m when drilling is processed.

Methanol is often used as a typical kind of hydrate inhibitor, Fig. 7 shows the result of hydrate formation in 660.4mm wellbore when methanol inhibitor is added. It can been inferred from Fig. 7 that 8.5% molar concentration methanol is needed at least to prevent hydrate formation when drilling while 18.3% when pump is off.

3.3　Hydrate Formation Prediction in 444.5mm Wellbore

Fig. 8 shows the prediction results of hydrate formation in 444.5mm wellbore during drilling process without inhibitor. According to the simulation, in 444.5mm wellbore, if natural gas enters wellbore, the hydrate formation section is 350 ~ 1,861m when pump is off while it is 408 ~ 1,202m during drilling.

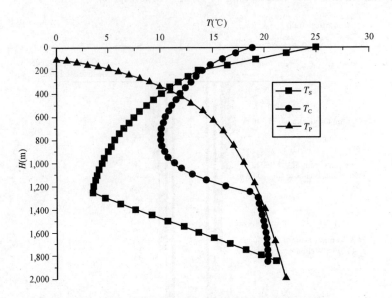

Fig. 6 Hydrate formation prediction in 660.4mm wellbore

T_S is wellbore temperature when pump is off; T_C means wellbore temperature when circulating; T_P is hydrate phase equilibrium curve

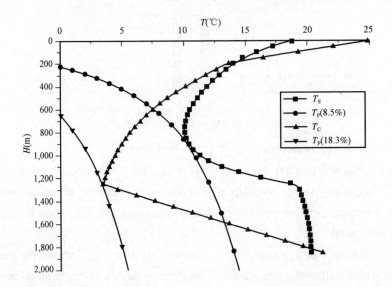

Fig. 7 Hydrate formation prediction in 660.4mm wellbore when methanol is added

T_C means wellbore temperature when pump is off; T_S means wellbore temperature when circulating;

$T_P(8.5\%)$ is hydrate phase equilibrium curve when molar concentration is 8.5%;

$T_P(18.3\%)$ is hydrate phase equilibrium curve when molar concentration is 18.3%

Fig. 9 shows the result of hydrate formation in 444.5mm wellbore when methanol inhibitor is added. It can been inferred from Fig. 9 that 8% molar concentration methanol is needed at least to prevent hydrate formation when drilling while 18.1% when pump is off.

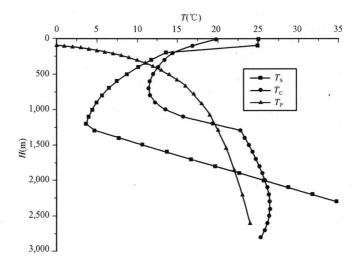

Fig. 8 Hydrate formation prediction in 444.5mm wellbore

T_S is wellbore temperature when pump is off; T_C means wellbore temperature when circulating; T_P is hydrate phase equilibrium curve

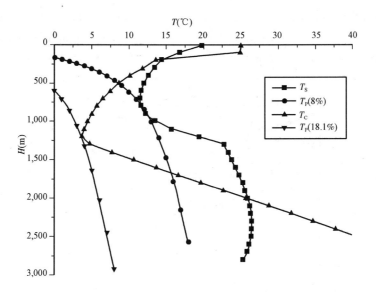

Fig. 9 Hydrate formation prediction in 444.5mm wellbore when methanol is added

T_C is wellbore temperature when pump is off; T_S means wellbore temperature when circulating; $T_P(8\%)$ is hydrate phase equilibrium curve when molar concentration is 8%; $T_P(18.1\%)$ is hydrate phase equilibrium curve when molar concentration is 18.1%

3.4 Hydrate Formation Prediction in 311.1mm Wellbore

Fig. 10 shows the prediction results of hydrate formation in 311.1mm wellbore during drilling process without inhibitor. According to the simulation, in 311.1mm wellbore, if natural gas enters wellbore, the hydrate formation section is 315 ~ 1,850m when pump is off while it is 406 ~ 1,162m during drilling.

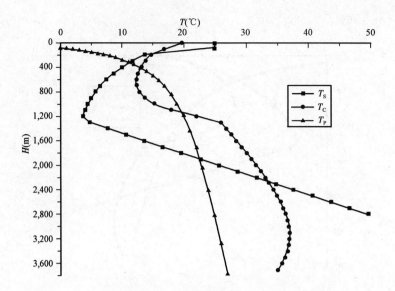

Fig. 10 Hydrate formation prediction in 311.1mm wellbore

T_S is wellbore temperature when pump is off; T_C means wellbore temperature when circulating;

T_P is hydrate phase equilibrium curve

Fig. 11 shows the result of hydrate formation in 311.1mm wellbore when methanol inhibitor is added. It can been inferred from Fig. 9 that 6.5% molar concentration methanol is needed at least to prevent hydrate formation when drilling while 17.9% when pump is off.

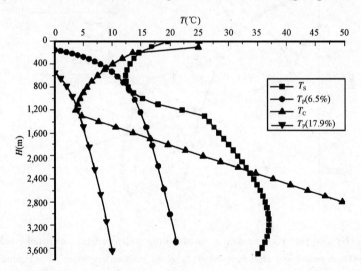

Fig. 11 Hydrate formation prediction in 311.1mm wellbore when methanol is added

T_C is wellbore temperature when pump is off; T_S means wellbore temperature when circulating;

$T_P(6.5\%)$ is hydrate phase equilibrium curve when molar concentration is 6.5%;

$T_P(17.9\%)$ is hydrate phase equilibrium curve when molar concentration is 17.9%

3.5 Hydrate Formation Prediction in 215.9mm Wellbore

Fig. 12 shows the prediction results of hydrate formation in 215.9mm wellbore during drilling process without inhibitor. According to the simulation, in 215.9mm wellbore, if natural gas enters wellbore, the hydrate formation section is 310 ~ 1,855m when pump is off while it is 405 ~ 1,103m during drilling.

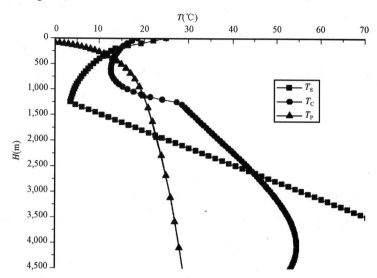

Fig. 12 Hydrate formation prediction in 215.9mm wellbore

T_S is wellbore temperature when pump is off; T_C means wellbore temperature when circulating;
T_P is hydrate phase equilibrium curve

Fig. 13 shows the result of hydrate formation in 215.9mm wellbore when methanol inhibitor is added. It can been inferred from Fig. 13 that 6% molar concentration methanol is needed at least to prevent hydrate formation when drilling while 17.5% when pump is off.

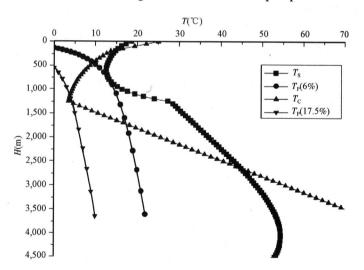

Fig. 13 Hydrate formation prediction in 215.9mm wellbore when methanol is added

T_c is wellbore temperature when pump is off; T_S is wellbore temperature when circulating;
$T_P(6\%)$ is hydrate phase equilibrium curve when molar concentration is 6%;
$T_P(17.5\%)$ is hydrate phase equilibrium curve when molar concentration is 17.5%

The results of methanol concentration that is needed to prevent hydrate formation in wellbore are shown in Table 3.

Table 3 The results of methanol concentration that is needed to prevent hydrate formation in wellbore

Wellbore (mm)	Pump is off (Long Time After Pump Being Off)		Deilling Stuation	
	Hydrate Formation Section(m)	Methanol Inhibitor Concentration(%)	Hydrate Formation Section(m)	Methanol Inhibitor Concentration(%)
660.4	355~1,865	18.3	436~1,865	8.5
444.5	350~1,861	18.1	408~1,202	8
311.15	315~1,858	17.9	406~1,162	6.5
215.9	310~1,855	17.5	405~1,130	6

CONCLUSION

The method for hydrate formation section prediction and inhibitor concentration determination is proposed based on the wellbore temperature and pressure field calculation method, hydrate phase equilibrium condition determination approach and the statistic result of seawater temperature of Qiong Dongnan district. The inhibitor concentration to prevent hydrate formation in wellbore of the well Hua Guang X is analyzed. The prediction results could be benefit for the well plan to prevent the hydrate forming in the wellbore.

REFERENCES

Barker J W, Gomez R K. 1987. Formation of hydrates during deepwater drilling operations. SPE 16130.

Du Qingjun, Chen Yueming, Li Shuxia, et al. 2007. Mathematical Model for Natural Gas Hydrate Production by Heat Injection[J]. Petroleum Expoloration and Development, 34(4): 470 – 473, 487.

Ebeltoft H, Yousif M. 1997. Hydrate control during deep water drilling: overview and new drilling fluids [S]. SPE 38567.

Englezos P, Bishoi P R. 1988. Prediction of Gas Hydrate Formation Conditions in Aqueous Eletrolyte Solution[J]. AICHE. J., 34 (3): 1718 – 1721.

Hao Yongmao, Bo Qiwei, Chen Yueming, et al. 2006. Laboratory Investigation of Pressure Development of Natural Gas Hydrates[J]. Petroleum Exploration and Development, 33(2): 217 – 220.

Knott T. 2001. Holding Hydrates at Bay[J]. Offshore Engineer, 45(2): 29 – 34.

Li Yuxing, Feng Shuchu. 1999. Method of Identifying Gas Hydrate Formation in Pipeline[J]. Natural Gas Industry, 19(2): 99 – 102.

Liu Yanjun, Liu Xiwu, Liu Dameng, et al. 2007. Applications of Geophysical Techniques to Gas Hydrate Prediction [J]. Petroleum Expoloration and Development, 34(5): 566 – 573.

Long Zhihui, Wang Zhiming, Fan Jun. 2006. A Dynamic Modeling of Underbalanced Drilling Multiphase-flow and Numerical Calculation[J]. Petroleum Expoloration and Development, 33(6): 749 – 753.

Sun Zhigao, Guo Kaihua. 2002. Measurement and Prediction of Gas Hydrate Phase Equilibrium Containing Ethylene Glycol and Salts[J]. Journal of Shanghai Jiaotong University, 36(10): 1509 – 1512.

Thierry Botrel, Patrick Isambourg, Total Fina Elf. 2001. Off Setting Kill and Choke lines Friction Losses, a New Method for Deep Water Well Control. SPE 67813.

Wang Zhiyuan, Sun Baojiang. 2008. Prediction of Gas Hydrate Formation Region in the Wellbore of Deepwater Drilling [J]. Petroleum Expoloration and Development, 35(6): 731 – 735.

Xue Wanjun. 1991. Trough and trench of South China Sea[J]. South China Sea Study, (3):118 – 125.

Zuo Y X, Guo T M. 1991. Extension of the Patel-Teja Equation of State to the Prediction of the Solubility of Natural Gas in Formation Water[J]. Chem. Eng. Sci., 46(5): 3251 – 3258.

Casing Program Design and Optimization for Deepwater Drilling and Application in JDZ Block of West Africa

Baoping Lu[1], Minsheng Wang[1], Ke Ke[1], Zhichuan Guan[2]

[1] Sinopec Research Institute of Petroleum Engineering, Beijing, China
[2] China University of Petroleum, Qingdao, China

Abstract JDZ Block in west of Africa is a strategic oversea deepwater exploration and development project for Sinopec who worked as an operator at the first time in this ultra-deepwater drilling project. Casing program design and optimization is the key point in drilling engineering plan. Deepwater drilling has the characteristics of special jet in technique for conductor; open hole drilling for surface casing section; low formation fracture pressure gradient in shallow formation, and higher uncertainty in formation pressure information. Therefore there are significant differences for casing design and optimization compared with shallow water and onshore drilling. The subsea wellhead mechanical stability and conductor setting depth, surface casing setting depth, casing program and setting depth under formation pressure uncertainties were detailed for deepwater casing design and optimization based on investigation of deepwater well structure. The paper presents cases study in JDZ block in west of Africa, the results show that approach suggested in this paper could be fitter for deepwater drilling than conventional methods, it can lower risk possibility and improve the operation efficiency.

Key Words Deepwater drilling; JDZ block; West of Africa; Casing design and optimization; Risk analysis

INTRODUCTION

Deepwater drilling has the characteristics of special jet in technique for conductor, open hole drilling for surface casing section, low formation fracture pressure gradient in shallow formation, and higher uncertainty in formation pressure information. Therefore there are significant differences for casing design and optimization compared with shallow water and onshore drilling. Some foreign scholars such as Dumans, Nobuo Mortia, Sergio A. B. da Forntura and Q. J. Liang have suggested separately mechanical analysis method for subsea wellhead, conductor jetting depth determination method and risk assessment method for deepwater drilling casing program design. In china, based on those methods, Prof. Zhichuan Guan, Baoping Lu and Jin Yang have separately established conductor jetting depth and hydraulic parameters determination method based on conductor and subsea wellhead load capacity theory, surface casing point determination method for open hole drilling section, casing point determination and quantitative risk assessment and optimization method for casing program. According to these latest research results, the paper presents a casing program design method which is consist of conductor jetting depth determination, surface casing point determination, casing point design and risk analysis. The case study with this method in JDZ block in deepwater area of West Africa has been presented in this paper. The results showed that the new method could be more suitable for deepwater drilling than conventional method, it can enhance the drilling efficiency.

1 CASING PROGRAM COMMON USED IN DEEPWATER DRILLING

Casing programs common used in the main deepwater area over the world (GOM, West Africa, Brazil and East Canada) are shown in follow Table 1.

Table 1 Casing programs common used in deepwater drilling

No.	Type	Layer1	Layer2	Layer3	Layer4	Layer5	Layer6
1	Wellbore(mm)		444.5 or 406.4	311.1			
	Casing(mm)	914.4 or 762.0	238.1	339.7			
2	Wellbore(mm)		444.5	311.1			
	Casing(mm)	914.4	355.6	244.5			
3	Wellbore(mm)		660.4	444.5	311.1		
	Casing(mm)	914.4 or 762.0	508	339.7 or 346.1	238.1 or 311.1		
4	Wellbore(mm)		660.4	444.5	311.1		
	Casing(mm)	914.4	508	355.6	244.4		
5	Wellbore(mm)		660.4	444.5	311.1	215.9	
	Casing(mm)	914.4 or 762.0	508	339.7 or 346.1	238.1 or 244.4	177.8 or 193.6	
6	Wellbore(mm)		660.4	444.5 ream to 508	444.5	311.1	215.9
	Casing(mm)	914.4 or 762.0	508	406.4	339.7 or 346.1	238.1 or 244.4	177.8 or 193.6

In short, casing program of deepwater drilling has follow characteristics:

(1) In order to ensure the profit of investment and the coordination to completion or testing equipments to wellbore, the minimum casing size (or tube casing) could be no less than 177.8mm (the wellbore size is 193.7mm).

(2) There is high uncertainty in formation pressure information in deepwater drilling. For most exploration wells, especially deepwater wells, the lack of offset wells' data and complicated pressure system will lead to insufficient understanding of the pressure system. This has made the pressure profile acquired by engineers has some uncertainties, and will result in different casing program designs. So the current casing program design method which is based on the deterministic formation pressure profile could not satisfy deepwater drilling engineering requirements, the casing program designed by this method could result in longer operation time and cost increase.

(3) The surface casing size is 508mm, it limit the following casing size. The surface casing section is drilled in open hole way and the drilling fluid in this section is seawater, so the surface casing setting depth is limited a lot by formation pressure.

(4) It is more difficult to solve shallow hazard in deepwater drilling than shallow water or onshore drilling, we could not solve this problem by increasing casing layers only. The shallow problems in open-hole drilling well section could not be dealt with long-time drilling fluid density ad-

justment, and in this section the BOP haven't been installed yet, so the well control measurement could not be set enough. Now, φ193.7mm or φ244.5mm pilot hole is usually drilled to explore whether the shallow hazard exists. The φ660.4mm, φ609.6mm and φ558.8mm backup casings could only be used to deal with mild shallow problems, and running casings in open hole drilling well section needs a lot of time because the aiming operation, so it can increase the drilling cost a lot. Therefore, more backup casing layers plan is not recommended in surface casing section.

(5) Reaming-while-drilling is used commonly in deepwater drilling. This kind of operation is usually adopted in the drilling plan or is the emergency measurement when complex problems occurred during drilling.

(6) Top-bottom design method is commonly applied in casing program design. The formation fracture pressure is lower than onshore wells, so using this method could leave enough casing layer space for following drilling or back up casings to deal with unexpected problems.

2 DEEPWATER DRILLING CASING PROGRAM DESIGN METHOD

2.1 Conductor Jetting Depth Determination

In order to prevent conductor sinking or assure the jetting depth is not over deep, the total load on the conductor should equal or close to the in-time bearing capacity at a certain recovery time after disturbing, as shown in following:

$$\varepsilon_d < Q_t - Q_w < \varepsilon_u$$

Where, ε_d is reasonable lower limit safety value, kN; ε_u is reasonable upper limit safety value, kN; Q_t is conductor in-time bearing capacity at time t, kN; Q_w is total load on the conductor, kN.

The solution flow is: (1) Calculate in-time bearing capacity profile according to conductor frictional resistance and end resistance profile at a certain depth under the mudline obtained from ocean floor soil sampling; (2) Assume the conductor setting depth x_j, and then calculate the total load Q_{wj} with depth x_j, determine the bearing capacity Q_{tj} with depth x_j according to in-time bearing capacity curve, if $\varepsilon_d < Q_{tj} - Q_{wj} < \varepsilon_u$, then the x_j is the reasonable setting depth, otherwise, assume a new setting depth x_j and repeat this process until the design requirement is met. The working flow is shown in Fig. 1.

2.2 Surface Casing Setting Depth Determination

Surface casing section of deepwater well usually uses open-hole drilling and takes seawater as drilling fluid. According to a large number of literature materials about deepwater wells design, current surface casing setting depth is mainly determined based on: (1) formation pore pressure. The surface casing point could be the depth where the pore pressure starts to become abnormal high. (2) Lithology and structure. The surface casing point could be lithological setting point. (3) For exploration well surface casing point could be less than 800m under the mud line. Fred R. Holasek has established determination method for φ508mm surface casing setting point based on calculating the difference between formation pore pressure and drilling fluid density, but this kind of method is too empirical. Zhichuan Guan has suggested that the surface casing point could be deeper by weighting the drilling fluid density, and the setting depth is limited by four factors: (1)

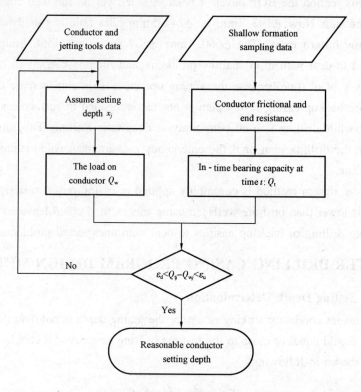

Fig. 1 Working flow for conductor setting or jetting depth determination

the capacity of mud pool and the mud treatability of the rig;(2) formation fracture pressure;(3) mud pump power;(4) the empirical maximum setting depth: 800m.

2.3 Casing Program Design With Pressure Uncertainties

The working flow of the casing program design under pressure uncertainties is shown in Fig. 2.

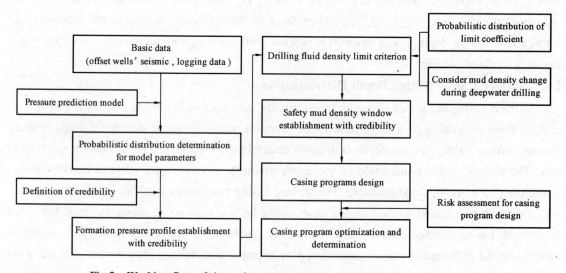

Fig 2 Working flow of the casing program design under pressure uncertainties

(1) Establish the pressure profile with credibility information. First, according to offset wells' seismic data, logging data and drilling report materials, choose reasonable formation pressure prediction model, get the probability distribution and probabilistic parameters by using statistics theory, and then, based on the well's seismic data, the formation pressure profile with credibility could be established.

(2) Establish the safety mud window with credibility. In conventional casing program design method, all drilling fluid limit coefficients (swab pressure coefficient S_b, surge pressure coefficient S_g, additional mud density increment $\Delta\rho$, kick tolerance S_k and so on) are deterministic values. With the new method, each design coefficient could be described as a probabilistic distribution which can be set by drilling or logging data statistics and fitting. In deepwater drilling, the mud density will change under the low temperature condition, and this factor could be taken into consideration for deepwater casing program design, so an additional limit coefficient S_w is added, then the drilling fluid density upper limit changes to:

$$\rho_m \leq p_f - S_g - S_f - S_w$$

Where, S_b means swab pressure coefficient, g/cm^3; S_g means surge pressure coefficient, g/cm^3; S_f means formation fracturing pressure additional increment, g/cm^3; S_w means safety deepwater drilling fluid density increment, g/cm^3; p_f means formation fracture pressure gradient, g/cm^3; ρ_m means drilling fluid density, g/cm^3. So the safety mud window with credibility could be established though the probabilistic distribution of limit coefficients and limit criterion.

(3) Casing layers and setting depth determination. The result by the new design method consists of several casing programs because the safety mud window with credibility is not made up of tow single lines but two bands with probabilistic information. According to risk assessment model, we can analyze each casing program's risk and optimize one casing program into drilling program.

3 JDZ BLOCK CASE STUDY IN DEEPWATER AREA OF WEST AFRICA

3.1 JDZ OBO -1 General Information

The method introduced above has been applied in well Obo - 1. The well is an exploration well located in JDZ block 1 of the deepwater area in the west of the Africa (as shown in Fig. 3). The average water depth is 1,650 ~ 1,800 m (as shown in Fig. 4). This well has a few offset wells such as Akpo - 1, Akpo - 2 and Akpo - 3 whose logging data could be helpful for the formation pressure prediction of the well obo - 1. The formation pressure profile with credibility has been established through the interval velocity data of the well Obo - 1 and the different kinds of logging data of wells in the Akpo block.

3.2 Conductor Jetting Depth Determination

(1) Basic data for calculation

Relative basic data for calculation is shown as Table 2 and Table 3.

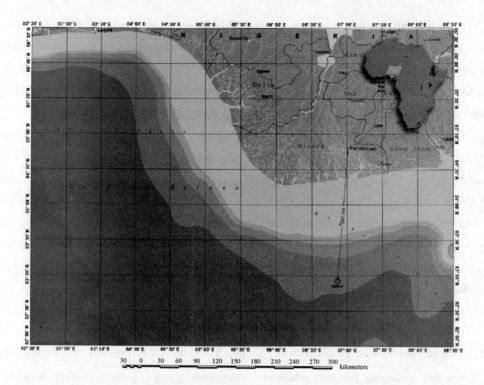

Fig. 3　Well Obo－1 location

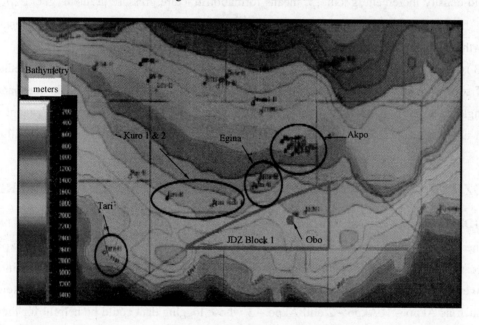

Fig. 4　Water depth distribution of JDZ Block

Table 2　Well basic data list

Parameters	Value	Parameters	Value
Water depth(m)	1,750.0	BOPs weight(kN)	2,000.0
Conductor OD(mm)	914.4	Load on wellhead(kN)	3,000.0
Conductor thickness(mm)	25.4	Steel density(kg/m^3)	7,850.0

Table 3 Soil parameters list

Depth from mudline (m)	10	20	30	40	50	60	70	80	90	100
Undrained shear strength (kPa)	20.0	35.0	45.0	60.0	80.0	95.0	95.0	95.0	95.0	95.0
Submerged unit weight (kN/m^3)	7.0	7.5	8.0	8.5	9.0	9.5	10.0	10.0	10.0	10.0

(2) Setting depth determination

Obo–1 well's conductor frictional and end resistance results are shown as Fig. 5.

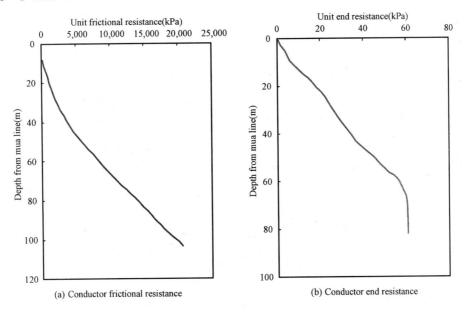

Fig. 5 Obo–1 Conductor frictional and end resistance profile

Assuming formation bearing capacity coefficient is 0.06 and the time after jetting is 12h, we can get the maximum bearing capacity Q_u curve and in-time bearing capacity curve varying with the depth x (as Fig. 6). By using iteration, the total load on conductor $Q_{wj} = 1,255.8$ kN when its length is 80m, and when the depth from mudline is 80m, the bearing capacity is 1,385.5kN, the load and the bearing capacity have met the design requirement, so the conductor setting depth is 80m.

In addition, the conductor setting depth has been determined when formation bearing capacity coefficient is 0.06, the time after jetting is 24h, and when formation bearing capacity coefficient is 0.1, the time after jetting is 12h or 24h. The reasonable depth is from 65m to 85m. For the safety consideration, the setting depth is 80m.

3.3 Casing Program Design

(1) Formation pressure profile with credibility establishment.

The formation pressure profile with credibility could be established through offset wells' interval velocity and logging data, and all the lines were smoothed by FTT method (as shown in Fig. 7).

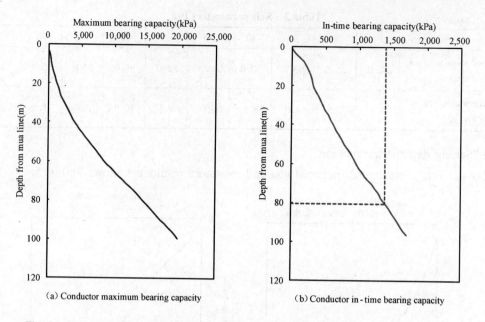

(a) Conductor maximum bearing capacity (b) Conductor in-time bearing capacity

Fig. 6 Obo – 1 well's conductor maximum bearing capacity and in-time bearing capacity
(formation bearing capacity coefficient is 0.06, the time after jetting is 12h)

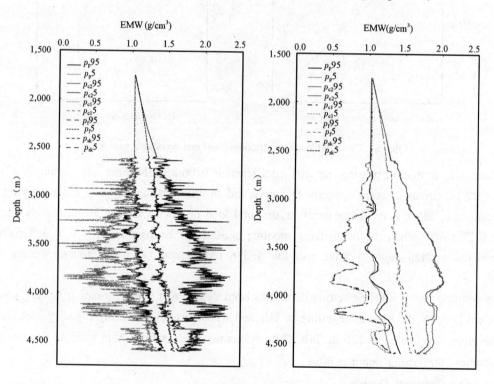

Fig. 7 Obo – 1 well's formation pressure prediction and smoothing result (2,600~4,750m)

The formation pressure profile with credibility includes pore pressure p_p, minimum collapse pressure p_{c1}, fracture pressure p_f and maximum collapse pressure p_{c2}. As shown in the Fig. 7, $p_p 5$ means the pore pressure with the cumulative probability 5%, $p_p 95$ means the pore pressure with the cumulative probability 95%, so the credibility of the pore pressure profile is 90%, it means the probability of the pore pressure between p_p5 and p_p95 is 90%. The meaning of $p_{c1}5, p_{c1}95, p_f5, p_f95, p_{c2}5$ and $p_{c2}95$ are similar with p_p5.

(2) Safety mud density window establishment.

Using the safety mud density limitations, the safety mud density profile with credibility could be established through Monte Carlo simulation (as shown in Fig. 8). Similar with the meaning of the formation pressure, in Fig. 8, L_k95 means the lower limit curve of mud density to prevent kicking with cumulative probability 95%, $L_{c1}95$ means the lower limit curve of mud density to prevent collapse with cumulative probability 95%, $L_{c2}95$ means the upper limit curve of mud density to prevent collapse with cumulative probability 95%, L_L95 means the upper limit curve of mud density to prevent lost circulation with cumulative probability 95%, $L_{sk}95$ means the upper limit curve of mud density to prevent sticking with cumulative probability 95%, similarly, L_k5 means the lower limit curve of mud density to prevent kicking with cumulative probability 5%, and other symbols have the similar meanings. Then the lower and upper limit mud density profile with credibility and the safety mud density window with credibility could be established.

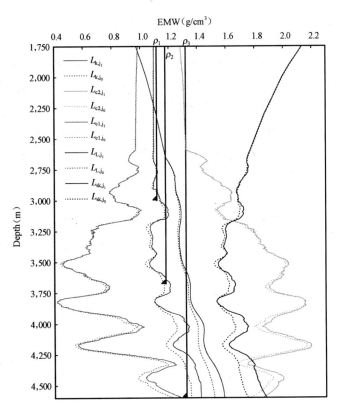

Fig. 8 Safety mud density window with credibility and the original casing point design of Obo-1

(3) Casing program design.

The casing program by *top to bottom* and *bottom to top* method could be constructed (as shown in Table 4 and Table 5).

Table 4 Design result by using *top to bottom* method

Casing program 1	Setting point(m)	Credibility(%)	Casing program 2	Setting point(m)	Credibility(%)
Surface casing	2,485~2,520		Surface casing	2,500~2,520	
Technical casing 1	2,950~3,074	90	Technical casing 1	3,021~3,074	90
Technical casing 2	3,874~3,909	90	Technical casing 2	3,891~3,909	90
Technical casing 3	4,110~4,600	90	Tubing(or open hole)	4,600	90
Tubing(or open hole)	4,600	90			90

Table 5 Design result by using *bottom to top* method

Casing layer	Setting point(m)	Credibility(%)
Surface casing	2,394~2,427	90
Technical casing 1	2,500~2,580	90
Technical casing 2	3,021~3,400	90
Technical casing 3	3,891~3,977	90
Tubing(or open hole)	4,600	90

As shown in Table 4, when the setting point of surface casing, technical casing 1, technical casing 2 in casing program 1 deeper than 2,500m, 3,021m and 3,891m separately, the setting point of technical casing 3 in casing program 1 could reach TD, and then the casing program 2 is generated. According to drilling data of the well Obo-1, we can get the actual well structure which is shown in Table 6. It can be seen that the actual well structure is included in the design results.

Table 6 Obo-1 well actual casing program

Hole size(mm)	Casing type	Casing size(mm)	Setting point (RKB)(m)
660.4	Conductor	914.4	1,829.4
660.4	Surface Casing	508	2,517.7
444.5	Technical casing	346	3,012.9
311.1 ream to 368.3	Technical casing	298.4	3,834.7
311.1	Technical casing	244.5	4,136.0
215.9	Open hole	Open hole	4,687.8

(From Obo-1 drilling data, ChevronTexaco Company)

Using the risk assessment model, the risk assessment for the original casing program (see Table 7) by Chevron of this well is conducted (as shown in Fig. 9~Fig. 11).

Table 7 Original casing program design of well obo−1 (except conductor)

Casing layer	Setting point (m)	Design mud density (g/cm³)
Surface casing	2,410	1.03 (sea water)
Technical casing 1	3,000	$\rho_1 = 1.13$
Technical casing 2	3,730	$\rho_2 = 1.20$
Tubing (or open hole)	4,600	$\rho_3 = 1.35$

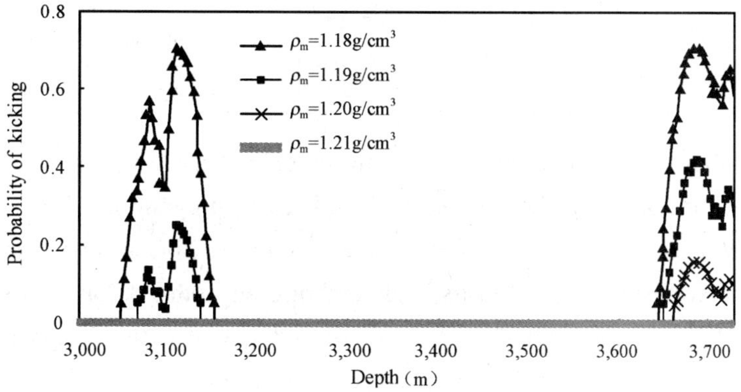

Fig. 9 Risk probability of kicking from 3,000m to 3,730m

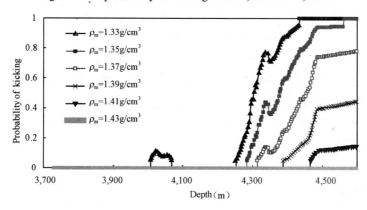

Fig. 10 Risk probability of kicking from 3,730m to 4,600m

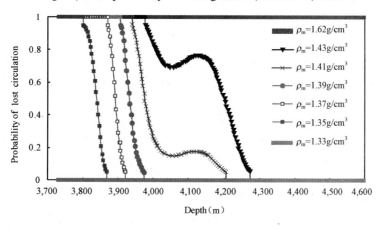

Fig. 11 Risk probability of lost circulation from 3,730m to 4,600m

Table 8 Risk assessment result of original casing program design of well Obo – 1
(conductor is not included)

Original designed hole section (m)	Original designed mud density(g/cm³)	Hole section with risk (m)	Risk kinds	Max risk probability (and depth) [%(m)]
1,830 ~ 2,410	1.03	N/A	N/A	0
2,410 ~ 3,000	1.13	N/A	N/A	0
3,000 ~ 3,730	1.20	3,663 ~ 3,730	Kicking	16.2 (3,693)
3,730 ~ 4,600	1.35	3,730 ~ 3,867	Lost circulation	100 (3,730 ~ 3,800)
		4,281 ~ 4,600	Kicking	100 (4,562 ~ 4,600)

The risk probability of original designed casing program (Table 7, Fig. 8) is generated. From the result (as shown in Table 8), the original designed casing program has 100% risk sections. As shown in the Fig. 9, in the original designed 3rd spud section, the hole section with kicking risk can be shortened by increasing the mud density without bring any other kinds of risk, and when the mud density is 1.21 g/cm³, the kicking risk is eliminated. It means the designed 3rd spud section could be 100% safe when the mud density increases to 1.21 g/cm³.

In the original designed 4th spud section, similar with the 3rd spud section, the hole section with kicking risk could be shortened or even to be eliminated by increasing the mud density (as shown in Fig. 10), but with the mud increase, the risk probability of lost circulation in this section become higher and the hole section with risk of lost circulation become longer (as shown in Fig. 11).

Discussion above shows that the original designed casing program could not satisfy the engineering requirements. For drilling the well safely, an additional casing layer should be added to the original designed casing program. Considering the lithology requirements of this well, the recommended casing program is displayed in Table 9, the highest probability of risk is only 17.5% at 4,120m.

Table 9 Recommended casing program design result of well Obo – 1

Casing layer	Setting point (m)	Designed mud density(g/cm³)	Hole section with risk(m)	Risk kinds	Max risk probability (and depth) [%(m)]
S. C.	2,495	1.03	N/A	N/A	0 (N/A)
T. C. 1	2,950	1.13	N/A	N/A	0 (N/A)
T. C. 2	3,874	1.21	N/A	N/A	0 (N/A)
T. C. 3	4,110	1.35	N/A	N/A	0 (N/A)
Tubing (or open hole)	4,600	1.41	4,110 ~ 4,205	Lost circulation	17.5 (4,120)
			4,464 ~ 4,600	Kicking	14.7 (4,600)

S. C. —Surface Casing; T. C. —Technical Casing.

According to the drilling report of Well Obo – 1, serious kicking was encountered at 4489m and then the mud density was increased to 1.62 g/cm³ (13.5 ppg) and resulted in serious lost circulation at 4,257m and 4,489m. These situation accorded with the predicted results (Fig. 11). The actual casing program is shown in Table 9, it indicates that the recommended casing program design is rational and safer then original design result.

From the daily report, for dealing with the kicking, lost circulation and doing reamer operation, additional 180h was expended in the drilling of Obo - 1. If the recommended program is applied, with the high daily rent of the deepwater drilling ship (more than 600 thousand dollars per day), the cost of the whole well drilling could be obviously lower than original casing program.

SINOPEC as a qualified operator has participated in drilling for well B - 1 in deep water area of West Africa. The methods discussed above have been applied successfully in casing program design of this well. According to the design suggestion, SINOPEC changed the original casing program planned by another company who was also operator for this well. The practice proved that the new plan is better than the old one, it mitigate drilling problems dramatically. The actual drilling time is 6 days less than plan and more than 7 million US dollars were saved in drilling.

CONCLUSION

(1) Conductor jetting is widely used in deepwater drilling. The approach based on foundation soil mechanics to determine the setting depth of conductor through calculating the frictional and end resistance is becoming a common way to design conductor jetting depth.

(2) For deepwater exploration wells, the aim of formation pressure prediction is not an accurate value but a possible range because the lack of useful data. A new approach introduced in this paper can construct the formation profile with credibility. Be different with the traditional formation pressure prediction results, every kind of formation pressure profile with credibility is not a single curve but a range including probability information. This can be helpful for engineers to understand formation pressure trend better.

(3) Using the casing program design method with the consideration of pressure uncertainties, the design result of the casing point is a range of depth not a depth value any more.

(4) The risk assessment model can not only be used to assess the casing setting point selection, but also to identify the vulnerable hole sections. It provides great convenience for drilling designers to prevent drilling incidents and make emergency plan, and also make drilling operations easier and safer than the traditional method.

(5) Casing program design is a dynamic process. Pre-drilling emergency plans should be made in the event of possible problems. During the drilling process, the original plan should be adjusted, amended or optimized in-time according to the real time drilling data and type incidents. With respect to the well in this paper, for example, when the severe drilling incidents occurred, the original plan should have been adjusted. Adding a new set of casing is recommended if it can satisfy the requirements of well completion. The increased cost is much less than that spent on dealing with the drilling incidents that otherwise occurred during the 44 days. In deep water drilling, considering the expensive cost of the offshore drill ship, this method will show more advantages.

REFERENCES

Rocha L A S, Junqueira P, Roque J L. 2003. Overcoming Deep and Ultra Deepwater Drilling Challenges[R]. OTC 15233.

Juiniti R, Salies J, Polillo A. 2003. Campos Basin: Lessons Learned and Critical Issues to be Overcome in Drilling and Completion Operations[R]. OTC 15221.

Shaughnessy J, Daugherty W, Graff R, et al. 2007. More Ultra Deepwater Drilling Problems[R]. SPE 105792.

Yang Jin, Cao Shijing. 2008. Deepwater Drilling Technology Development Review[J]. Oil drilling and production technology, 30(2): 10 - 13.

Charles D, Whitson C D, McFadyen M K. 2001. Lessons Learned in the Planning and Drilling of Deep, Subsalt Wells in the Deepwater Gulf of Mexico[R]. SPE/IADC 71363.

Stephen A Rohleder, W Wayne Sanders, Gray L Faul. 2003. Challenges of Drilling an Ultra-deep Well in Deepwater-spa prospect[R]. SPE/IADC 79810.

Jenkins R W, Schmidt D A, Stokes D, Ong D. 2003. Drilling the First Ultra Deepwater Wells Offshore Malaysia [R]. SPE/IADC 79807.

Willson S M, Edwards S, Heppard P D, Li X, Coltrin G, Chester D K, Harrison H L, Cocales B W. 2003. Wellbore Stability Challenges in the Deep Water, Gulf of Mexico: Case History Examples from the Pompano Field[R]. SPE 84266.

Dumans C F F. 1995. Quantification of the Effect of Uncertainties on the Reliability of Wellbore Stability Model Prediction[D]. Tulsa: Univ. of Tulsa.

Nobuo Mortia. 1995. Uncertainty Analysis of Borehole Stability Problems[R]. SPE30502.

Sergio A B, Dafontoura, Bruno B Holzberg, Edson C Teixira, Marcelo Frydman. 2002. Probabilistic Analysis of Wellbore Stability During Drilling[R]. SPE78179.

Liang Q J. 2002. Application of Quantitative Risk Analysis to Pore Pressure and Fracture Gradient Rrediction [R]. SPE77354.

Guan Zhichuan, Ke Ke, Lu Baoping. 2009. An Approach to Casing Program Design with Formation Pressure Uncertainties. Journal of China University of Petroleum, 33(4): 71 - 75.

Guan Zhichuan. 2009. Analysis on Lateral Load 2 Bearing Capacity of Conductor and Surface Casing for Deepwater Drilling. ACTA PETROL EI SINICA, 30(2): 285 - 290.

Yang Jing. 2002. Study on the Minimum Diving Depth of Offshore Drilling Riser [J]. Oil drilling and production technology, 24(2): 1 - 3.

Cunha J C. 2004. Innovative Design for Deepwater Exploratory Wells[R]. IADC/SPE 87154.

Philippe Jeanjean. 2002. Innovative Design Method for Deepwater Surface Casings[R]. SPE 77357.

Baker J W. 1997. Wellbore Design with Reduced Clearance Detween Casing Strings[R]. SPE/IADC 37615.

Samuel G Robello, Adolfo Gonzales, Scot Ellis, Issa Kalil. 2002. Multistring Casing Design for Deepwater and Ultradeep HP/HT Wells: a New Approach[R]. IADC/SPE 74490.

Fred R Holasek. 2006. Determination of 20 in Conductor Setting Depth in Deepwater Wells in the Krishna-Godavari Basin Offshore India[R]. IADC/SPE103667.

Chu Daoyu. 2010. The Practice and Knowledge of Drilling Deepwater Well in West Africa [J]. Offshore Oil, 30 (4): 111 - 116.

Fred R Holasek. 2002. Evaluation of Deepwater Drilling Prospects Using New Concepts to Identify, Quantify and Mitigate (IQM) Risks for Well Design[R]. IADC/SPE 74489.

L F de A e Souza, Xia J. 2005. The Challenging Design Aspects of the High-pressure Tieback SCRs for the K2 Development in the Gulf of Mexico[R]. OTC 17564.

Experimental Study on Tethered Underwater Robot

Jiaming Wu[1], Dewen Yang[1], Li Wu[2], Zhiquan Ma[2], Zhilin Chen[2], Jinghua Ye[1], Ying Xu[1]

[1] Department of Naval Architecture and Ocean Engineering,
South China University of Technology, Guangzhou 510640, China
[2] Guangzhou Panyu Lingshan Shipyard Ltd., Guangzhou 511473, China

Abstract A new kind of tethered underwater robot is proposed. The robot is equipped with two water jet thrusters which are served as main control mechanisms for the robot, thus the trajectory and attitude of the robot in multiple degrees can be maneuvered according to the commands of users. Laboratory experimental investigation on the behavior of this kind of underwater robot in a test water tank is reported the hydrodynamic and control performances of the robot in the test water tank are described. In our paper the electromechanical control system to the robot is also introduced. The results of the study indicate that a flexible attitude and trajectory control of a tethered underwater robot can be achieved with the water jet thrusters proposed in this research.

Key Words Underwater; Tethered underwater robot; Model test; Hydraulic jet propulsion; Multi-degree control

INTRODUCTION

A tethered underwater robot is an undersea survey apparatus which is extensively used for underwater observation and research. The robot usually consists of a main body mounted with several control mechanisms for the robot attitude and trajectory control and an umbilical cable is connected to the robot for transmission of the control signals and power. With the increasing development of underwater survey activities, the demand of tethered underwater robots for conducting undersea exploring operations is growing. The working nature of detecting devices in an underwater robot requires that the robot should have the ability to neatly adjust its trajectory and attitude in six degrees of freedom during a control operation. Therefore how can we manipulate the underwater robot simply and rapidly according to the control requirement to the robot with definite control mechanisms is a fundamental key to develop an economical and practical underwater robot. For a tethered underwater robots used nowadays, different degree controls to the robot are usually accomplished by different control propellers respectively (Wu et al, 2005; Fang et al, 2007; Shim et al, 2010; Avila and Adamowski, 2011). The main drawback of this kind of control manner lies that multiple control propellers and complicated control manipulations are needed in order to maintain effective control to the robot attitude, which will undoubtedly increase the complexity of manipulation operation and difficulty for the user to control the robot, and also confine its underwater application because of the complex control mechanism for manipulating the robot. Therefore, how to simplify the underwater control operation while at the same time to manipulate the tethered underwater robot stably and neatly becomes a more and more concerned topic by researchers and users.

In this paper a new type of tethered underwater robot is proposed. The investigated underwater robot is composed of a main body, two water jet thrusters and a cylindrical shape buoyancy chamber fixed above the main body. It is expected that with such a design, the objective of a rapid trajectory manipulation and a relative stable attitude control with a relatively simple control way to the robot during a surveying operation can be obtained. The underwater robot discussed in this research is a prototype intended to apply in field underwater survey operation, this paper focuses laboratory experiments in a test water tank to examine the hydrodynamic and control behaviors of this kind of underwater robot.

1 OUTLINE OF THE UNDERWATER ROBOT IN THE EXPERIMENT

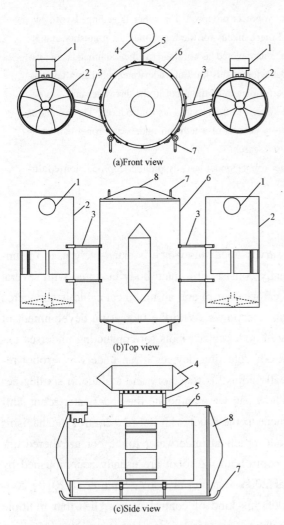

Fig. 1 A schematic drawing of the underwater robot
1—end plate driven motor; 2—water jet thruster;
3—thruster supporting frame; 4—buoyancy chamber;
5—buoyancy chamber frame; 6—robot main body;
7—robot bracket; 8—nose section

The underwater robot investigated in our experimental study is shown in Fig. 1. The robot consists of a robot main body being cylindrical in shape with two hemispherical nose sections at both ends of the cylindrical portion, two water jet thrusters acting as attitude and trajectory control mechanisms for the robot being attached at two sides of the main body horizontally, to create a righting moment or a favorite roll damping for the robot when it is in an underwater survey operation, a cylindrical shape buoyancy chamber is fixed above the main body. An umbilical cable is connected to the robot for transmission of the control signals and power.

The main body has a cylindrical shape contour whose fundamental function is to house different kinds of ocean survey instruments such as the physical monitors of inspection sonar, optical camera or the chemical sensors of salinity, fluorescence, dissolved oxygen, pH values etc according to different underwater survey purposes. Input and output ducts on the body allow water to flow through the onboard sensors so that sea water parameters can be collected if seawater parameter sensors are mounted on the vehicle. The advantage of the cylindrical shape main body is that a small drag force in a turning motion can be obtained due to its round shape contour; a large payload capacity is available with such a space chamber. The principal dimensions of the three major parts of underwater robot are given in Table 1, and the overall parameters of the underwater robot are presented in Table 2.

Table 1 Principal dimensions of the three major parts

Main body:	
Diameter(mm)	290
Length(mm)	540
Water jet thrusters:	
Diameter(mm)	180
Length(mm)	410
Buoyancy chamber:	
Diameter(mm)	80
Length(mm)	300

Table 2 Overall parameters of the underwater robot

Length(mm)	540
Width(inc. two thrusters)(mm)	855
Height(inc. buoyancy chamber)(mm)	430
Weight(kg)	
Air	30
Water	5

The water jet thruster's principal function is to provide necessary propulsion forces for the underwater robot running in six degrees of freedom according to different control commands. The designed water jet thrusters in this research allow that the robot can be driven in multi-degrees of freedom by the only two water jet thrusters which can be achieved by changing direction of propulsion force via different combinations of water jet directions of the water jet thrusters.

The basic structure of the water jet thrusters is shown in Fig. 2. The water jet thruster consists of a cylindrical shell, a propeller, a propeller motor, a thimble, a thimble driven motor, a round control end plate, and a end plate driven motor etc. There is a partial annular opening in the middle section of the cylindrical shell, whose function is to allow the water jet orifice on the thimble turning along the opening round the axis of the cylindrical shell to produce a necessary propulsion force in different required directions, the arc length of the opening holds the 80% of the shell perimeter, the rest part of the middle section remains un-cutting to maintain the integrality of the cylindrical shell structure and also to provide a closed portion for robot's control purpose. The schematic drawings of the thimble and the cylindrical shell are presented in Figs. 3 and Fig. 4. The parameters of motors in the water jet thruster are given in Table 3.

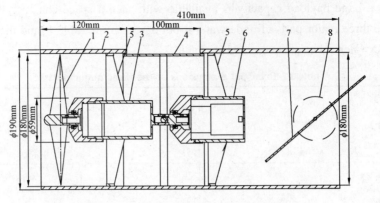

Fig. 2 Basic structure of the water jet thruster

1— propeller;2—cylindrical shell;3—propeller motor;4—thimble;5—motor frame;
6—thimble driven motor;7—end plate;8—end plate driven motor

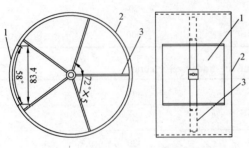

Fig. 3 The thimble in the water jet thruster
1—water jet orifice;2— thimble;3—stiffened frame

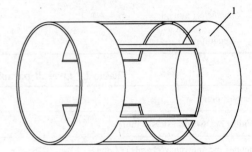

Fig. 4 The cylindrical shell with openings

Table 3 The parameters of motors in the water jet thruster

Propeller motor	
Type	DC motor
Power(W)	30
Rotating speed(r/min)	800
Thimble driven motor	
Type	Stepping motor
Step angle(°)	1.8
Static torque(g/cm)	3.5
Location torque(g/cm)	0.2
End plate driven motor	
Type	Stepping motor
Step angle(°)	1.8
Static torque(g/cm)	2.0
Location torque(g/cm)	0.15

In order to manipulate the robot traveling in multiple degrees, we can actuate the thimble driven motor, and/or the end plate driven motor to govern the turning positions of the thimble and round control end plate so that needed control propulsion forces for fulfilling the control requirement can be obtained, the forward velocity of the robot can also be governed by adjusting the rotating speed of the propeller motor.

For example, if a longitudinal propulsion force is required for the purpose of manipulating the robot, we can actuate the end plate driven motor to open up the end plate in an opening position, turn on the thimble driven motor to adjust the water jet orifice on the thimble to a position where the water jet orifice is blockaded by the un-cutting portion at the middle section of the cylindrical shell as shown in Fig. 5. Since there is only longitudinal water jet along the axis of the cylindrical shell, and no water jet from the water jet orifice is produced in this condition, only a longitudinal propulsion force is engendered. If the angle of the end plate is adjusted to a position between a fully opened and a closed ones as shown in Fig. 2, then the direction of water jet from the cylindrical shell can be changed to a required one in horizontal plane by means of governing the end plate angle, thus the required direction of the propulsion force in horizontal plane can be obtained accordingly. If we turn the position of water jet orifice to a downward direction (the six o'clock position) meanwhile turn the end plate to a closed position as shown in Fig. 6, then there is only a downward direction of water jet and an upward direction of propulsion force can be produced.

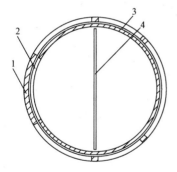
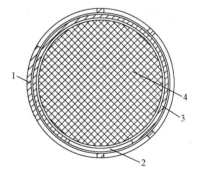

Fig. 5 The combination to produce a longitudinal force with the end plate opened
1—cylindrical shell; 2—water jet orifice; 3—thimble; 4—round control end plate in open position

Fig. 6 The combination to produce an upward force with the end plate closed
1—cylindrical shell; 2—water jet orifice; 3—thimble; 4—round control end plate in closed position

Other directions of control propulsion forces can also be gained in a similar way by controlling the direction of water jet orifice on the thimble and/or turning position of the end plate which can be achieved through a joint control operation on the three control motors, i.e. the propeller motor, thimble driven motor and end plate driven motor.

The fundamental advantage of the proposed tethered underwater robot over other types of tethered underwater robots is that the robot has a more stable attitude because a buoyancy chamber

Fig. 7   The photo of the underwater robot developed in this research

is fixed above the main body which provides a bigger righting moment and reduces the difficulty for the robot to maintain its stable motion during a control operation, and that the robot can be maneuvered in a more simple way since a multiple degree propulsion can be conducted by only two water jet thrusters with different turning position combinations of water jet orifice and round control end plate in the thrusters. Fig. 7 presents the photo of the underwater robot developed and tested in this research.

2 DESCRIPTION OF THE EXPERIMENT

The experiments are conducted in a test water tank to examine the hydrodynamic and control behaviors of the underwater robot proposed in this paper under given control manipulations. The tank dimension is 120 m (Length) ×8 m (Width) ×4.4 m (Height), the tank is filled with fresh water to a depth of 4 m. It is equipped with a two track power-driven towing carriage in the tank. There is a control room on the carriage where the computers for control operations and treatment of experimental data are set there. In this experiment, the carriage dose not run, and only stop in the middle section of the length. In the tests the propulsion forces for the underwater robot trajectory control are provided mainly by the two water jet thrusters fitted at two sides of the robot. The running signals to the two thrusters are delivered by the control commands which are transmitted down via an umbilical cable from the central computer in the control room.

The active control operations to the underwater robot are mainly implemented by manipulating the actions of the water jet thrusters. The operations are implemented in terms of the command signals from the central computer in the control room. The command signals are processed by a chip microprocessor before they reach the control modules to actuate the thrusters. The diameter, disk area ratio, ratio blade numbers of the propeller in the thrusters are 175 mm, 0.5, 1.0 and 4 respectively.

Command signals of above control apparatus communicate through signal wires between the underwater robot and the computers in the control room. All the signal wires are tired together by adhesive plastic tapes to form a composite umbilical cable.

3 EXPERIMENTAL OBSERVATIONS

The experiments are carried out with above described test facilities to examine the hydrodynamic and control behaviors of the proposed underwater robot. To demonstrate the maneuvering capacity on the trajectory and attitude control to the underwater robot in different degrees of freedom by the proposed water jet thrusters, experimental results of the underwater robot in several control operations under different control manipulations by the propulsion forces issued from the two water jet thrusters are presented.

In order to manipulate the tethered underwater robot to travel underwater in different patterns of motions, the directions of water jet orifices on the thimbles and the turning positions of the end plates in the robot should be controlled through relevant driving motors to corresponding positions so that the required control forces to propel the robot can be obtained. For example, if the motions of the robot in forward straight, backward straight, ascending, descending, lateral movement and turning in a circle are needed to be driven, the direction of water jet orifices on the thimbles and the turning positions of the end plates in the robot adjusted by controlling the propeller motor, thimble driven motor and end plate driven motor during these motions should be controlled to the conditions as shown in Table 4.

Table 4 Parameters of the control mechanism of the robot during some patterns of motions

	Left			Right		
	Jet orifice direction (o'clock position)	End plate	Propeller motor	Jet orifice direction (o'clock position)	End plate	Propeller motor
Forward straight	3	Opened	Running	9	Opened	Running
Backward straight	3	Opened	Reversal Running	9	Opened	Reversal Running
Ascending	6	Closed	Running	6	Closed	Running
Descending	12	Closed	Running	12	Closed	Running
Right lateral movement	9	Closed	Running	9	Closed	Running
Turning	9	Opened	Stop	9	Opened	Running

Remark: (1) Definition of right and left in this table is based on Fig. 1; (2) The un-cutting portion at the middle section of the left cylindrical shell as shown in Fig. 1 is at 3 o'clock position; while that of the right is at 9 o'clock position.

Parts of the screenshots of the video when the robot was under the controlled tests in the test water tank are presented in Fig. 8 and Fig. 9. Fig. 8 shows the screenshots when the robot was in forward straight control test, and Fig. 9 provides the ones when the robot was in turning control operation. From our experimental observations, it is found that with the proposed tethered underwater robot, the objective of a rapid trajectory manipulation and a relative stable attitude with a relatively simple control way for the robot can be achieved.

CONCLUSIONS AND DISCUSSIONS

A new type of tethered underwater robot is proposed. The underwater robot is mainly composed of a cylindrical shape contour main body, a buoyancy chamber and two water jet thrusters. It is found from the test water tank experiments that the proposed tethered underwater robot possesses the following characteristic:

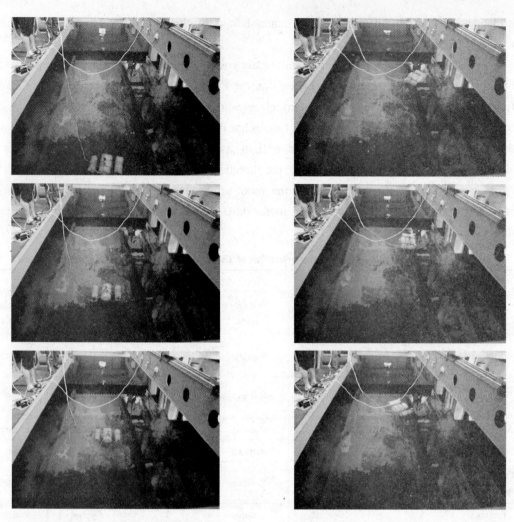

Fig. 8 The robot in forward straight control test   Fig. 9 The robot in turning control operation

(1) Multiple degree controls to the robot can be achieved with a relatively simple control way. There are only two propellers in our robot, the control operations to the robot in multiple degrees of freedom can be made by a joint control operation on the water jet orifices, turning positions of the end plates and the propeller motors. Compared with the conventional tethered underwater robot, the difficulty of design for the control system to the proposed robot is reduced greatly. Meanwhile, the degrees of freedom for the robot's control can be added.

(2) Much better attitude stability of the robot during its control operation can be maintained. The buoyancy chamber above the main body of the robot provides a greater restoring moment which produces a greater damping for the robot's rolling and pitching motions. Thus, a stronger self stability of the robot during the robot survey operations can be guaranteed. This nature gives the users of the robot an easy way to keep the robot being stable while it is manipulated.

(3) Versatile applications are presented. Since all the control mechanisms are installed outside of the main body of the robot, much more underwater detecting instruments can be fitted in the chamber of the main body, and the interferences between the control mechanisms and the detecting instruments can be reduced.

ACKNOWLEDGMENTS

The research was sponsored by the National Natural Science Foundation of China under Grant No. 10772068, the Industry-University-Research Foundation of State Education Ministry and Guangdong Province of China under Grant No. 2010B090400501 and the Science and Technology Project of Panyu District, Guangzhou, China under Grant No. 2011 – Special Funds 02 –4. 13.

REFERENCES

Avila J P J, Adamowski J C. 2011. Experimental Evaluation of the Hydrodynamic Coefficients of a ROV through Morison's Equation[J]. Ocean Engineering, Vol. 38:2162 –2170.

Fang M C, Hou C S, Luo J H. 2007. On the Motions of the Underwater Remotely Operated Vehicle with the Umbilical Cable Effect[J]. Ocean Engineering, 34:1275 –1289.

Shim S, Jun H H, Lee P M, Baek H, Lee J. 2010. Workspace Control System of Underwater Tele-Operated Manipulators on an ROV[J]. Ocean Engineering, 37:1036 –1047.

Wu J M, Ye J W, Yang C, Chen Y M, Tian H P, Xiong X H. 2005. Experimental Study on a Controllable Underwater Towed System[J]. Ocean Engineering, 32:1803 –1817.

Maintenance and Repair Methods of Deepwater Equipments

Jing Wang[1], Ling Ma[1], Xiangneng Ma[1]

[1]China Ship Scientific Research Center, Wuxi, China

Abstract Subsea production systems are widely used in deepwater oil and gas fields. There are 80 percent of deepwater production systems are subsea production systems in the Gulf of Mexico. Due to the high water depth, the emergency repair is difficult to operate when there is something wrong with the equipments on the seabed. Deepwater pipeline, subsea production trees and subsea BOPs are the equipments with a high failure rate. This paper summarized several maintenance and repair methods of these equipments.

Key words Emergency repair; Deepwater pipeline; Subsea production system; BOP

INTRODUCTION

The South China Sea is rich in oil and gas resources and natural gas hydrate resources, the petroleum geological reserves of it is about 230 to 300 million tons, which is about one-third of the total amounts of oil and gas resources of china. It's worth mentioning that 70 percent of the resources in south china sea are in the deepwater where the water depth is over 500 m. So the development of the deepwater oil and gas fields is an inevitable trend, and higher requirerments of the techniques and equipments for emergency repair are raised.

1 MAINTENANCE AND REPAIR OF DEEPWATER PIPELINE

Being known as "the lifeline of oil and gas field", the subsea pipeline is the main facility for offshore oil and gas development. Pipeline damage makes a big influence on oil and gas production. The unpredictability of the pipeline damage and the high water depth make it difficult to operate emergency repair.

Generally, the damage can be divided into the big damage and the small damage. When the length of the damaged pipeline is less than its radius and it makes no effect on the structural integrity of the pipeline it is called small damage. When the length of the damaged pipeline is bigger than its radius it is called big damage. If the length of the damaged pipeline is less than 240ft it is called the short bending deflection, otherwise called the long bending deflection (Killeen, 2006; Stress Subsea, 2005; Wei, 2002). It should take different measures to repair different damage.

Before choosing the right repair technique the following steps should be carried out:

Isolate the pipeline damage;

Survey the damaged area;

Define the location where the pipeline damaged;

Define the severity of the damage;

Uncover and map the damaged section;

Choose the appropriate repair method.

The widely used techniques to repair deepwater pipeline are clamp repair, on-bottom repair and surface-lift repair.

2 CLAMP REPAIR

When the pipeline suffers a small damage, clamp can be used to make a permanent repair. Clamp repair is a mature technique which has been widely used in offshore oil and gas fields and has a high success rate. The widely used clamp in this technique is the hydraulic drive clamp which is operated by the ROV under water. The merit of clamp repair is its sample procedure. It only needs the dynamic positioning vessel which supports the ROV (Liu,2004).

The procedure of clamp repair is as follows:

Detect leak and isolate section.

Install Pipe Lift Frames (PLFs) on either side of damage. Then the ROV operate the PLFs to lift pipeline from the seabed.

ROV install clamp over damage and test the tightness of the clamp.

ROV operate the PLFs to lower pipeline back to seabed.

Recover PLFs to surface.

3 SURFACE-LIFT REPAIR

When the pipeline suffers a big damage and triggers big oil and gas leak the surface-lift repair and on-bottom repair are chosen to repair the pipeline damage because the clamp cannot repair the damage effectively (Liang,2009).

The surface-lift repair is a mature technique with a high reliability. The procedure of surface-lift repair is as follows:

(1) Detect leak and isolate section, or flood pipeline.

(2) Install PLFs on either side of damage section to be removed. Then the ROV operate the PLFs to lift pipeline from the seabed.

(3) ROV attach rigging to the damaged section to be removed.

(4) ROV operate the Diamond Wire Cutting Module (DWCM) to cut the damaged section and then recover the damaged section to the surface.

(5) Attach pipeline recovery tools to cut-end of pipe, and dewater if necessary.

(6) Remove and recover PLFs, the lifting equipment on the support ship lift the pipe to surface.

(7) Weld bend-hub with surface-lift sled to recovered pipe.

(8) Lower surface-lift seld with upward-looking hub back to seabed.

(9) Repeat the steps (5) to (8) to attach bend-hub assembly to other cut-end.

(10) Perform submarine metrology, fabrication and install the bend-hub assembly.

(11) Demobilize the support ship, repair euipmrnts and the repair team.

4 ON-BOTTOM REPAIR

The on-bottom repair is chosen when the big damage is on the intersection of several pipes or it is near to submarine slippery pry where the pipe cannot to be lift to the surface. It is also chosen when there is no heavy support ship available to carry out the surface-lift repair.

There are two on-bottom repair methods. One methods use a grip and seal hydraulic connector (GSHC) to repair the pipeline, the other one use the gas metal arc welding (GMWA) technique to repair the damaged pipeline.

The procedure of the first method is as following:

(1) Detect leak and isolate section.

(2) Install PLFs on either side of damage section to be removed. Then the ROV operate the PLFs to lift pipeline from the seabed.

(3) ROV attach rigging to the damaged section to be removed.

(4) ROV operate DWCM to cut the damaged section and then recover it to the surface.

(5) Install Guide Frame over cut-end.

(6) Install Gantry sled onto Guide Frame.

(7) Attach GSHC-bend-hub assembly.

(8) Lock the GSHC-bend-hub assembly onto the Gantry sled, remove and recover the Gantry Frame from the Gantry sled, recover the PLFs.

(9) Repeat the steps (5) to (8) to attach bend-hub assembly to other cut-end.

(10) Perform submarine metrology, fabrication and install the bend-hub assembly.

(11) Demobilize the support ship, repair euipmrnts and the repair team.

The procedure of the second on-bottom repair method is as following:

(1) Detect leak, isolate section.

(2) Land the welding robot system.

(3) Clean up the protective coating over the pipe, such as the concrete weight coating and the fusion bonded epoxy coating.

(4) ROV operate the DWCM to cut the damaged section and then recover the damaged section to the surface.

(5) Land the PLFs and a bend-hub assembly with a sleeve connected onto it to the seabed.

(6) Align the centerline of sleeve to the centerline of pipeline. The bend-hub assembly is connected with the pipeline through sleeve and sealed by ROV, and then recover ROV.

(7) Land welding robot and operate fillet weld on the end of the sleeve using GMAW. Make sure the bend-hub assembly and pipeline are connected and sealed. Recover the welding robot.

(8) ROV operate the PLFs to lower the pipeline, and then recover the PLFs.

(9) Repeat the steps (3) to (8). Carry out GMAW on the other end of the pipeline.

(10) Demobilize the support ship, repair euipmrnts and the repair team.

All of the repair jobs are carried out by ROV.

5 EQUIPMENTS OF SUBSEA PIPELINE REPAIR

From the repair procedure mentioned above, the repair system mainly includes the following equipments:

Pipeline Lift Frame (PLF) which lift the damaged section from the seabed and provide enough space for ROV to carry out the operation; Diamond Wire Cutting Module (DWCM) which cut the damage pipeline; End Prep Tools which clean the ends of pipeline cut by the DWCM; the pipeline recovery tools; the Guide Frame; the Gantry Frame; the Gantry sled; the clamp and the connectors et al. During the repair process the support ship and ROV are needed to complete the repair procedure.

6 ROUTINE MAINTENANCE OF DEEPWATER PIPELINE

Pipeline damage not only have a severely effect on oil and gas production but also generate high fees for repairing. To reduce the possibility of the pipeline damage and keep the deepwater pipeline in a good condition, a pefect routine maintenance should be carried out in every offshore oil and gas field (Zhao et al, 2009). Using the detecting instruments and ROV to check up on the performance of the pipeline, inculding buckling, bending curvature, ellipticity, nonlinearity, dent and flaw caused by corroding. It is more important to prevent the damage than to repair the damage.

7 MAINTENANCE AND REPAIR OF SUBSEA PRODUCTION TREE

Subsea production tree are the full equipments which installed in the subsea wellhead. It is the heart of the production. It controls and manages the oil and gas production and it also provide support for the workover job (Wang, 2009).

Subsea production trees are widely used in deepwater oil and gas fields, this is due to the following merits: the less initial investment, the reusability of the equipments, the flexibility of the well pattern and saving the plateform space. As a sophisticated equipment it requires the system have a high reliability. Due to the high water depth the repair operation of subsea production tree is expensive and time-consuming (Vossen, 2003).

The subsea production trees are controled by the control station on the producing platform and it also controled by ROV. There are a lot of valves on the body of the subsea production trees, ROV can control the subsea production trees by open and close the valves. The routing maintenance is carried out by ROV, it inculding the appearance inspection, corrosion detection, valve test and the leak detection. ROV also responsible to change the SCM module, the tree cap and the sealing ring et al..

8 MAINTENANCE AND REPAIR OF SUBSEA BOP

The subsea BOP is the key equipment during drilling and it also important for the production. Its function is to close the annulus and seal the wellhead to prevent blowout. The subsea BOPs are controlled by the remote controlling system and the maintenance and repair opration is carried out by ROV. The routing maintenance inculding the appearance inspection, corrosion detection, valve test, the leak detection and especially the check of control system. The control system is very important for the BOP. In the blowout happened in the Gulf of Mexico, the BOP did not respond to the control signal is one reason to cause the blowout accident. So the check of effectiveness of valves on the body of subsea BOP and the control system is necessary.

9 THE PROSPECT

The huge amounts of oil and gas resources in deepwater is a treasure for people. The harshest environments put forward a high request to the equipment used in subsea. After the blowout accident happened in the Gulf of Mexico, emergency repair techniques became the key research field. Nowadays, ROV play an important rule in the procedure of routing maintenance and emergency repair. The operator can see status of the subsea equipments by the camera system installed in ROV and ROV can use some tools do the maintenance and repair job. But the ROV is controled by the surface control station, so it is affected by the sea state. When the sea state is bad it is difficult to launch and recover ROV, that may delay the emergency repair. So a "deepwater space station" equipped with ROV and the necessary emergency repair tools which can stay in the seabed for a long time maybe the future of the emergency repair system. When an accident is happened the expert can take the space station to survey the scene. According to the actual situation of the accident the expert chould make a quick decision and take step immediately.

CONCLUSION

There are sveral maintenance and repair methods for deepwater pipeline, subsea production tree and BOP. For the deepwater pipeline different damage should take different mesaures. ROV is a necessary equipment in the maintenance and repair system, but the use of it obstructed by the sea state, it maybe delay the chance to repair. There is still a lot of room to develop in the deepwater repair system.

REFERENCE

Killeen J, Taconis T, Whipple J, et al. 2006. Large Diameter Deepwater Pipeline Repair System. DOT.

Liu Chunhou, Pan Dongmin, Wu Yishan. 2004. Description of Maintenance and Repair Method of Subsea Pipeline [J]. China offshore oil and gas,16(1):59 – 62.

Liang Fuhao, Li Aihua, et al. 2009. Research Progress and Discussion on Deepwater Pipeline Repair System[J]. China offshore oil and gas,21(5):352 – 355.

Vossen R, Moore T. 2003. Subsea Tree Installation, Lessons Learned on a West Africa Development[C], OTC 15371.

Wang Wei, Sun Liping, Bai Yong. 2009. Subsea Production System[J]. China offshore Platform,24(6):41 – 45.

Wei Zhongge, Qi Yaru, et al. 2002. Maintenance Technology for Submarine Pipeline [J]. Petroleum Engieering Construction,28(1):30 – 32.

Zhao Hua, Gong Bentao, Han Yuhui. 2009. Study on Submarine Pipeline Leak and Its Preventive Measure[J]. Environmental Protection Oil & Gas Fields,19(2):51 – 52.

Quantitative Risk Analysis of Simplified Subsea Chemical Injection System

Yoo Won-woo[1], Park Min-sun[1], Yang Young-soon[2]

[1] Dept. Naval Architecture & Ocean Engineering, Seoul National University, Seoul, South Korea
[2] Dept. Naval Architecture & Ocean Engineering, Research Institute of Marine Systems Engineering, Seoul National University, Seoul, South Korea

Abstract Subsea chemical injection system treats blockage problem in the subsea production system. It is important to treat problem fast, because production delay causes a fatal loss of profit in the subsea production system. Therefore, the subsea industry requires relatively higher reliability level of production system than other industries. In this study, subsea chemical injection system (linked to control system) to inject chemicals into subsea X-mas tree is analyzed. By using FSA (Formal Safety Assessment), risk factors are defined and quantitative risk analysis utilizing FTA (Fault Tree Analysis) and ETA (Event Tree Analysis) is performed. As a result, effectiveness for risk reduction option is evaluated.

Key Words Subsea chemical injection system; Subsea production system; Blockage; FSA; ETA; FTA.

INTRODUCTION

Subsea oil production began in earnest in the 1990s. The ratio of the total oil production has increased by 3% in 2002, 6% in 2007, and 10% in 2012. After 2015, subsea oil production is the only sector to continue to grow. For this reason, subsea production system enabling subsea oil production becomes more and more the center of attention.

The purpose of the subsea production system is to have a safe and fast production of crude oil. Accordingly, technology development for fast production and way to reduce production delay in given conditions should be considered. The purpose of this study is a reliability improvement of subsea chemical injection system to treat blockage causing production delay in the subsea production system. Blockage occurs at anywhere in the subsea production system. But this study limits the target to subsea X-mas tree for case study. The analysis for risk reduction option is performed using FSA used widely in the shipbuilding industry.

Hydrates, main cause of blockage, are crystalline compounds formed by the physical combination of water molecules and certain small molecules in hydrocarbons fluid. It is easily formed when the hydrocarbon gas contains water at high pressure and relatively low temperature (Fig. 1).

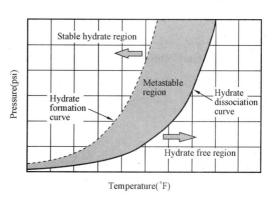

Fig. 1 Hydrate Formation and Dissociation Regions

Solution to de-hydrates consists of heating & depressurization the fluid moving to hydrate free region by increasing temperature or reducing pressure, pigging running through pipelines for cleaning hydrates, or chemical injection interfering with formation of hydrate crystals or agglomeration of crystals into blockages and so on. This study considers only chemical treatment.

1 QUANTITATIVE RISK ANALYSIS OF SIMPLIFIED SUBSEA CHEMICAL INJECTION SYSTEM BASED ON FSA

1.1 FSA (Formal Safety Analysis)

FSA (Fig. 2) is a structured and systematic methodology, aimed at enhancing maritime safety, including protection of life, health, the marine environment and property by using risk analysis and cost benefit assessment. It consists of following five steps.

- Preparatory step
(1) HAZID (HAZard IDentification);
(2) Risk Analysis;
(3) Risk Control Option;
(4) Cost Benefit Assessment;
(5) Recommendations for Decision Making.

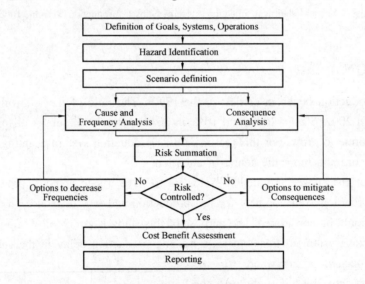

Fig. 2　Framework of FSA (Formal Safety Analysis)

1.2 Preparatory Step

Before performing risk analysis, the problem domain should be defined properly in the preparatory step. The following configuration of subsea chemical injection system (Fig. 3) is considered for FSA application.

Injection target of subsea chemical injection system is chosen at subsea X-mas tree. When detecting blockage at sensor, the information is transmitted to host facility. Then, chemical injection unit at Topside supplies chemical compound to a subsea X-mas tree. This chemical moves to subsea infra as shown in Fig. 3. Opening CIV (Chemical Injection Valve), chemical is then injected into the subsea X-mas tree.

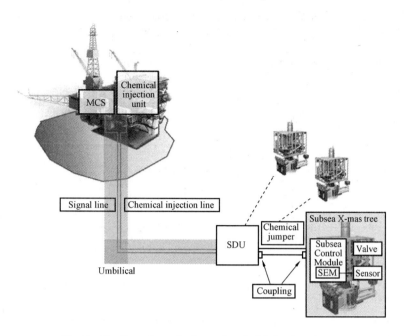

Fig. 3　Configuration of subsea chemical injection system

Subsea chemical injection system is linked to control system in the subsea production system. Therefore, the type of the control system should be defined. In this case, Multiplex electro-hydraulic system (Fig. 4) implemented by computer is selected. The feature of this system is that MCS (Master Control Station) communicates with microprocessor in the SEM (Subsea Electronic Module) and many wells can be controlled via one simple umbilical. It could affect a configuration of subsea chemical injection system.

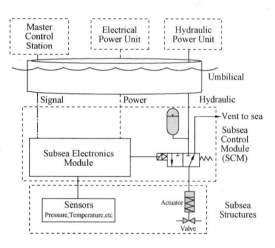

Fig. 4　Multiplex electro-hydraulic control system

Here, the chosen field is Xikomba field in Angola; its operating depth is 1,300m, since it is appropriate for the subsea production system. This field consists of 9 wells (4 production wells, 1 gas injection well and 4 water injection wells). Max. Oil production is 90MBOPD (Thousand Barrels of Oil per Day), gas processing is 115MMscfd (Million standard cubic feet per day).

2　HAZID (HAZARD IDENTIFICATION)

HAZID step makes a list of all relevant accident scenarios with potential causes and outcomes. This step is implemented based on OREDA (Offshore Reliability Data). Table 1 shows the list of risk elements. Very low risk elements, such as signal line connection point, may be absent from the list. Table 2 shows the failure mode of the list showed by Table 1. Other/unknown failure mode means all failure modes except from No. 1 to No. 11.

Table 1 A list of risk elements in the subsea chemical injection system

1	Sensor	6	Chemical line
2	SEM(Subsea Electronic Module)	7	Chemical injection coupling(SDU)
3	Signal line	8	Chemical injection coupling(Tree)
4	MCS(Master Control Station)	9	Chemical jumper
5	Chemical injection unit	10	Valve open system

Table 2 Failure modes

1	Abnormal instrument reading	7	Short circuit
2	Combined/common cause	8	Spurious operation
3	Control/signal failure	9	Transmission failure
4	Erratic output	10	External leakage- process & utility medium
5	Fail to function on demand	11	Plugged/choked
6	Insufficient power	12	Other/unknown failure mode

2.1 Risk Analysis

Risk analysis step evaluates risk factors. The definition of risk is the product of frequency and consequence. There are risk for human, risk for environment, and risk for property. This study considers only risk for property related to chemical treatment of blockage. FTA (Fault Tree Analysis) and ETA (Event Tree Analysis) are used for risk evaluation. Failure rate estimates presented in OREDA used as statistical data are based on the assumption that failure rate function is constant and independent of time, in which case $z(t) = \lambda$, i.e. the failure rates are assumed to be exponential distributed with parameter λ. Using this, failure probability (exponential distribution) could be calculated.

In ETA, it is assumed that the probability of the initial event (the occurrence of blockage) might be half of total flowline blockage probability due to lack of data meeting the conditions. As total flowline blockage probability is 0.00367, it is assumed that probability of initial event is 0.00184. Following the process of subsea chemical injection system to treat accident caused by initial event, event tree (Fig. 8) is formed. The chemical injection failure process might occur when one of any component shows malfunction in the following subsea chemical injection system; Sensor-SEM-Signal line-MCS-Chemical injection unit-Chemical line-Chemical injection coupling (SDU)-Chemical jumper-Chemical injection coupling(Subsea tree)-Valve open (Refer Fig. 3). From the scenario viewpoint of developing ETA, various branching point might be considered as follows: Detecting at sensor-Processing information-Information transmission (Subsea to Host)-Command center-Chemical injection-Chemical delivery (Host to Subsea)-Connection to subsea infra-Chemical delivery (Subsea to Tree)-Connection to tree-Valve open. Valve open system, last branching point in the event tree, is independent from subsea chemical injection system. Using component failure of valve open system, failure probability of valve open should be calculated by FTA as shown in Fig. 5.

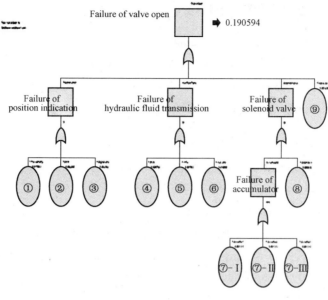

	Components	Failure rate	Probability
1	Valve position sensor	2.01	0.017453492
2	SEM	4.93	0.042267531
3	Signal line	0.27	0.002362405
4	MCS	11.62	0.096781876
5	HPU	0.9	0.007853003
6	Hydraulic line	1.61	0.014004610
7	Accumulator	0.15	0.001313137
8	Solenoid control valve	0.76	0.006635487
9	Utility isolation valve	0.37	0.003235953

Fig. 5 FTA of Valve Open System

From FTA (Fig. 5), it is calculated that failure probability of valve open system is 0.190594. This value is used as a chance of valve open scenario case, last branching point in the event tree. For the consequence evaluation, a property loss related to production delay is only considered as mentioned earlier. So, the consequence is the product of repair time of components and production of a single well (bbl/h) and oil price (2012 3/1 WTI). Table 3 shows annual risk at each branching point presented as the product of frequency and consequence.

Table 3 Annual risk at each branching point ($/a)

Branching point	Risk	Branching point	Risk
Detecting at sensor	484.97	Chemical delivery (Host→Subsea)	68.21
Processing information	125.29	Connection to subsea infra	0.74
Information transmission (Subsea→Host)	0.38	Chemical delivery (Subsea→Tree)	0.22
Command center	141.15	Connection to tree	17.95
Chemical injection	0.59	Valve open	282.01

3 RCO (RISK CONTROL OPTION)

Risk control option step devises regulatory measures to control and reduce the identified risks. To find RCO, RCT (Risk Contribution Tree), representing total risk is usually constructed by integrating fault tree and event tree. In a fault tree, changing the configuration of fault tree, RCO reducing failure probability can be found. In an event tree, changing the probability of branching point, RCO reducing consequence can be found. Through implementation of finding RCO using RCT, it is known from Table 3 that risk contribution of sensor is found to be higher than any others. To find its effectiveness of sensor, application of redundant sensor is carried out at cost benefit analysis, which is a next step in the FSA.

3.1 Cost Benefit Assessment & Recommendations for Decision Making

Cost benefit assessment step determines a cost effectiveness of each risk control option and recommendations for decision making step provides information about the hazards, their associated risks and the cost effectiveness of alternative risk control options. It judges the effectiveness of redundant sensor application. In the shipbuilding industry, FSA generally uses GCAF (Gross Cost of Averting a Fatality) and NCAF (Net Cost of Averting a Fatality) related to human. But, as this study considers only risk for property, benefit-cost ratio (Eq. 1) is used for evaluating risk for property. Table 4 shows a final risk based on 20 - year operating period.

$$R_{b-c} = B/C \tag{1}$$

where, B means Benfit; C means Cost.

Table 4 Sensor risk before & after application of redundancy ($)

	Before application of redundancy	After application of redundancy	B/C
Risk($)	9,699.37	1,950.33	4.84315

Benefit is calculated as 7,749.04 ($). Cost of sensor is assumed as 1,600 ($), based on market research and interview of the expert. Benefit-cost ratio can then be calculated as 4.84315 which is bigger than 1. Therefore, application of redundant sensor is considered to be effective for reducing the risk of subsea chemical injection system. Through an implementation of cost-benefit analysis, the number of sensor in subsea X-mas tree can be determined.

Fig. 6 shows, when operating more than 3 sensors, the cost-effectiveness is not good. Fig. 7 also shows that operating 2 sensors is cost-optimal safety level, namely more effective than any others.

CONCLUSION

This study performs risk assessment of subsea chemical injection system to treat blockage chemically following FSA. As a result, operating double sensor to detect blockage in subsea X-mas tree is more effective than operating single sensor. FSA application in this study has some advantages for handling risk problem in a systematic manner. Through this process, we could come up with various RCOs and evaluate them. It can be applied to other systems in the subsea production sys-

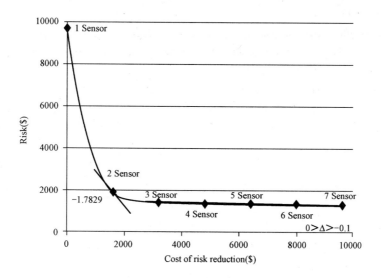

Fig. 6　Cost of risk reduction-Risk Graph

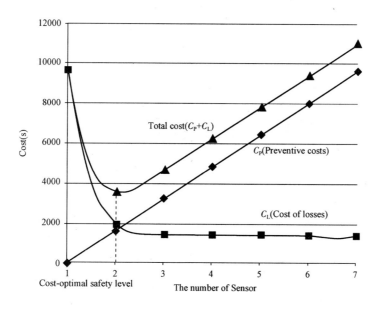

Fig. 7　No. of sensor-cost graph

tem. However, due to lack of data, there seems to be a limitation on accuracy of probability of initial event based on prediction. If taking a practical field data, more accurate risk assessment of the subsea production system could be carried out.

REFERENCES

Brimmer A R . 2006. Deepwater Chemical Injection Systems: The Balance Between Conservatism and Flexibility [C]. *Eni Petroleum*, OTC.

Bai Yong, Bai Qiang. 2010. Subsea Engineering Handbook [M]. - Chap. 7 Subsea Control, Elsevier, USA.

Bai Yong, Bai Qiang. 2010. Subsea Engineering Handbook [M]. - Chap. 15 Hydrates, Elsevier, USA.

Jerry Banks, John S Carson, Barry L Nelson, David M Nicole. 2001. Discrete-Event System Simulation Third Edi-

tion[M]. Prentice Hall, USA.
Mohammad Modarres. 2006. Risk Analysis in Engineering[M]. Taylor & Francis Group, Boca Raton.
Moon Kyung-tae. 2010. The Study on Quantitative Risk Assessment Methodology of Ship Machinery System [D]. Seoul: Seoul National University.
OREDA Participants. 2009. OREDA Handbook, Volume 2 -Subsea Equipment 5th Edition, DNV, Norway.
Svein Kristiansen. 2005. Maritime Transportation-Safety Management and Risk Analysis[M]. Elsevier, UK.

Influencing Factors on Subsea Tree Selection

Lisong Song[1], Mingxuan Gong[2], Menglan Duan[2]

[1] HSE Department, CNOOC Headquarter, Beijing, China
[2] Offshore Oil & Gas Research Center, China University of Petroleum, Beijing, China

Abstract Subsea tree is a wellhead control device for the upper part of underwater oil and gas well, which plays a role in the control and regulation of oil and gas production. With technology of offshore oil and gas drilling and production advances, the subsea tree also has developed various forms. During the process of underwater oil and gas field development, subsea tree selection is becoming increasingly important. Through analysis of influencing factors of subsea tree selection, different effect forms and degrees of influence of different influencing factors on subsea tree selection have been determined. Every influencing factor should be considered in engineering practice to make a safer, more economical and better choice.

Key Words Subsea tree; Wellhead control device; Influencing factors

INTRODUCTION

With onshore oil reserves dwindling and demands increasing, oil exploration is developing to marine. Since 1947, the world's first offshore well drilled successfully in Avery Lake in United States, with advancement of science and technology, offshore oil and gas exploration has developed gradually from shallow waters to deep waters (100 to 500) m, then deep-sea (500 to 1,500) m, even ultra-deep sea (1,500) m. Compared to onshore oil exploration, offshore oil exploitation is more complex, call for higher technical. Subsea tree is a production system placed on the seabed subsea wellhead, which is composed with valves, pipes, connectors and accessories. Subsea tree system plays the role of water injection, gas injection, the control of oil and gas production and chemical reagent injection in the subsea production, controls the well production accidents by the emergency shut-off of the Christmas tree valves to at the same time. As an essential link in the deep-water oil and gas development, technology of subsea tree is developing rapidly (Fig. 1). In the process of subsea oil and gas field development, subsea tree selection is becoming increasingly important.

1992

1994   1996—1998

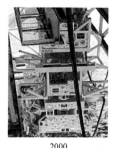
2000

Fig. 1 Shape figure of various periods of Christmas tree

1 CLASSIFICATION AND INTERNAL STRUCTURE OF SUBSEA TREE

Subsea tree is divided into two structural forms, vertical Christmas tree and horizontal Christmas tree (Fig. 2). Selection of the two structures is mainly based on form of well, fluid pressure, fluid property, wellhead form, sleeve shape and size, control system, etc. Subsea horizontal Christmas tree is installed after installation of wellhead system and drilling, while before installation of completion tubing and tubing hanger. Vertical Christmas tree can be recycled without completion.

Advantages of vertical Christmas tree:

(1) The completion can be completed without moving the blowout preventer, avoid need of killing and install shielding;

(2) The swabbing shielded is gate valve;

(3) The Christmas tree can be recycled without completion;

(4) The riser is not directly connected with the Christmas tree.

Disadvantage of vertical Christmas tree:

(1) Lack of fast and effective two holes installation / used in deepwater for 15,000psi workover riser;

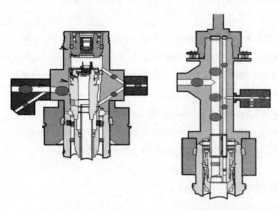

Fig. 2 Partial sectional schematic diagram of the horizontal Christmas tree and vertical Christmas tree

(2) Weight of two holes riser inside the water pipe is beyond of affordability of compensation;

(3) The positioning of tubing hanger installation requires completion guide;

(4) The positioning of slim diameter requires BOP nails;

(5) A Christmas tree device is not compatible with a variety of wellhead device.

Advantage of horizontal Christmas tree:

(1) Single-hole installation of riser can be achieved;

(2) Positioning, preset and testing of tubing hanger can be completed in the factory;

(3) Control the well by limited drilling system (BOP& marine riser);

(4) The 15,000 psi installation tools and SSTT program are very effective;

(5) The tubing hanger is located inside the Christmas tree;

(6) It can move the production tubing without removing the Christmas tree body;

(7) The Christmas tree body can be installed directly without installation of the tubing hanger;

(8) It can short-range drill through the Christmas tree;

(9) The bit must go though the catheter located between Christmas tree and wellhead while drilling, the catheter is also called connecting reducer. The catheter will have small displacement relative to the wellhead head, to limit hydraulic and make tubing hanger be placed on the shoulder conveniently.

(10) The nominal diameter size of the wellhead can be smaller than the horizontal Christmas tree.

Disadvantage of horizontal Christmas tree:

(1) Washing, drain and log of well must be completed within the drilling riser;
(2) The swabbing shield is the plug;
(3) Borings of riser are deposited on the upper portion of the tubing hanger;
(4) The high weight drilling riser sits on tree;
(5) The Christmas tree can be recycled until the completion completed.

Vertical and horizontal Christmas tree spent different workover time, comparison of workover time in water depth of 6,000 feet is shown in Fig. 3.

Seen from Fig. 3, during the wellhead workover in water depth of 6,000 feet, Time spent on tubing installation and recycling is very close, only need four or five days. The completion time is about 20 days. But in the Christmas tree recycling, Vertical Christmas tree needs much less time than the horizontal

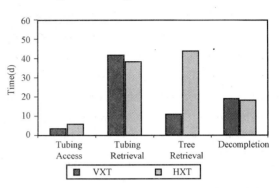

Fig. 3 Diagram of workover time comparison of the two forms of Christmas tree

Christmas tree. As maintenance of deepwater subsea tree requires specialized vessels, and the daily cost is very high, vertical Christmas tree has certain advantages in terms of economic effect.

2 RESEARCH ON INFLUENCING FACTORS OF SUBSEA TREE SELECTION

2.1 Subsea Wellhead Category

The subsea wellhead category has a certain influence on the choice of Christmas tree. Generally, three factors should be considered, pressure, production and fluid properties.

2.1.1 Pressure

Design specifications of subsea tree and its component are shown in Table 1.

Table 1 Design specification of underwater Christmas tree and its components

Name \ Pressure	Rated Pressure Value (psi)				
XT assembly (pressure and control)	5,000	10,000	15,000	—	—
Tubing hanger	7,500	12,500	17,500	—	—
Subsea wellhead	5,000	10,000	15,000	—	—
Conventional mud line equipment	Standard rated working pressure of the conventional mud line equipment does not apply				
Hydraulic control unit	1,500	3,000	In accordance with the manufacturers written specification		

The maximum pressure Christmas tree withstands is the minimum of its various components withstand, exiting Christmas tree rated pressure value is 5,000psi, 10,000psi, 15,000psi. Christmas tree of the rated pressure value can be selected according to the magnitude of the pressure of the exploitation of oil fields.

2.1.2 Fluid Properties

Fluid Properties of different oilfields are different. Viscosity, temperature and media of oil are different too. Therefore, the degree of corrosion on the Christmas tree is different. The corrosion is divided into internal corrosion and external corrosion. Internal corrosion is often due to the electrolyte such as organic acids, carbon dioxide, chloride or hydrogen sulfide. It includes: the mass loss corrosion and chloride or hydrogen sulfide stress cracking corrosion.

External corrosion means oxygen corrosion due to the oxidation when the steel is exposed to the atmosphere or corrosive gases. The severity of corrosion is determined by the temperature, the erosion of the metal surface, the type of corrosion properties of the product, the surface film, the presence of electrolytes. Saltwater can speed up corrosion rate. The main anti-corrosion method are applying appropriate alloy metal preservative, the application of special surface coating or the proper use of cathodic protection devices. Christmas tree material requirements are shown in Table 2.

Table 2 Material requirements

Material category	The minimum requirements of the material	
	Body, hat and flange	Control lever, stem and spindle hanger
AA-General use	Carbon steel and low alloy steel	Carbon steel and low alloy steel
BB-General use	Carbon steel and low alloy steel	Stainless steel
CC-General use	Stainless steel	Stainless steel
DD-Acidic environment	Carbon steel and low alloy steel	Carbon steel and low alloy steel
EE-Acidic environment	Carbon steel and low alloy steel	Stainless steel
FF-Acidic environment	Stainless steel	Stainless steel
HH-Acidic environment	Corrosion Resistant Alloys	Corrosion Resistant Alloys

Different environments are not identical to the material of the various components of the Christmas tree. In short, materials of the Christmas tree and components should be selected according to the requirements of the environmental factors of the target area and the Christmas tree design conditions. At the same time, the corrosion resistance and anti-corrosion measures of the Christmas tree equipment should be analyzed, to meet the requirements of the deepwater engineering.

2.2 Yield and Reserves

Reserves and mining period of different oilfields are different. The impact of reserves and mining period of Christmas tree should be taken into account during the selection of Christmas tree. For oil fields of large reserves, short development time and high daily output, we can choose Christmas tree with relatively large diameter and strong production capacity. At the same time, the mining period also affects the choice of Christmas tree.

2.3 Wellhead Properties

Caliber of Christmas tree tubing hanger and subsea wellhead should be compared when selecting the Christmas tree, to make the installation more convenient. Vertical Christmas tree is faster than Horizontal Christmas tree in workover, especially in the aspect of the recovery of the Christmas tree. Therefore, Workover requirements need to be considered when selecting Christmas tree. At the same time, Affordability of the subsea wellhead need to be considered when selecting Christmas tree too. Christmas tree now usually weighs 40t to 70t, a small Christmas tree less than 40t have been produced too. According to the affordability of pressure of wellhead, choose Christmas trees of different weight. See Fig. 4.

2.4 Undersea Environment

Undersea environment includes seabed temperature, water depth and other factors. In different underwater depths, temperatures are also different. Affordabilities of temperature of the various components of Christmas tree are different. The impact of temperature on various components near the sea floor need to be considered when selecting the Christmas tree. The Christmas tree is divided into shallow Christmas tree and deepwater Christmas tree, Christmas tree in shallow water is more simple and small modular compare to deepwater Christmas tree. General definition in the international of deep water is more than 500m water depth. Therefore, Christmas tree should be selecting Based on the water depth.

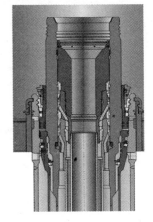

Fig. 4 Wellhead section view

2.5 Connection of the Pipeline

Pipeline and connection of the Christmas tree is a factor to be considered in the selecting of Christmas tree. The connection joints serve to connect the fixed, protection and control the flow.

Fig. 5 shows the cross-over connection of underwater Christmas tree and underwater manifold, usually through a rigid cross takeover connect.

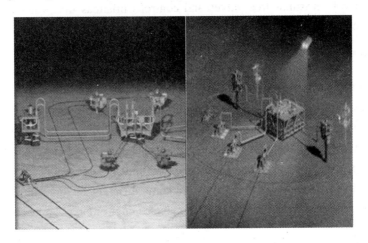

Fig. 5 Subsea production system manifold

Cross-over connection on Christmas tree, you need to consider in the selection process of the Christmas tree. Rigid cross-over and Christmas tree usually connect with four connection methods (Shown in Fig. 6).

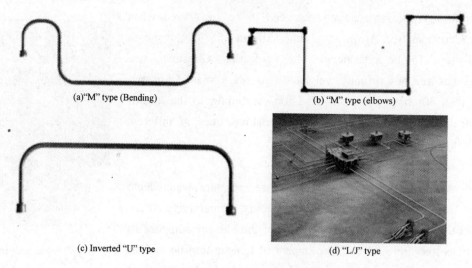

Fig. 6　Form of oil extraction pipeline

3　SCM

SCM is the "command center" of underwater facilities, it receives command from the MCS to guide the hydraulic fluid to operate the underwater valve, it collects information from the underwater sensors, and sends it to the upper facilities.

Usually SCM need to be considered when selecting the subsea tree. Installation of the Christmas tree in shallow water is relatively simple, SCM can be placed up the surface of water to control the installation of Christmas tree. In deep water, SCM is installed along with the Christmas tree and control the connection of the Christmas tree and umbilicals cooperating with ROV, at the same time, SCM can control Christmas tree valves and control Christmas tree water injection, gas injection and other functions.

CONCLUSIONS

(1) Subsea tree has made considerable development, with the continuous development and progress of technical level of the offshore oil and gas drilling and production. The subsea tree is divided into two structural forms of vertical Christmas tree with horizontal Christmas tree. the two forms of the Christmas tree have advantages and disadvantages in all aspects. Relatively speaking, horizontal Christmas tree is more in line with the development trend.

(2) Selection of Christmas tree is impact with some different factors, such as subsea wellhead category, yield and reserves, wellhead properties, undersea environment, connection of the pipeline, SCM, etc.

REFERENCES

DaCosta M, H J. 1998. Hartley Mensa Project. Subsea Tree System [R]. OTC 8579.

Guan Zhichuan, Su Kanhua, Su Yinao. 2010. Numerical Simulation of Subsea Wellhead Stability for Deepwater Drilling[R]. SPE 130823.

Wang Lizhong. 2006. On the Status Quo and Development of China's National Offshore Oil Engineering[J]. China offshore Platform, 21(4): 9 - 11.

Research on the Replacement of Subsea Christmas Tree Connector Sealing Ring

Ying Jiang[1], Li Ma[2], Lei Guo[2], Menglan Duan[2], Chenggong Sun[2]

[1] Offshore Oil Engineering Co., Ltd., Tianjin, China
[2] Offshore Oil and Gas Research Center, China University of Petroleum, Beijing, China

Abstract It is known that subsea Christmas tree is an indispensable part of submarine system. Due to various reasons, subsea tree appear seal failure, structural damage, structural failure, corrosion, control failures and so on. According to the different water depth, we do research on the replacement of easily damaged connector sealing ring of Christmas tree, the selection of different equipment and tools according to different depth. The results show that install with special tools can improved replacement operation convenience and accuracy. We can also use ADS auxiliary repair in 500m water depth.

Key Words Subsea Christmas Tree; Seal ring failed; ADS; Connector

INTRODUCTION

In recent years, with the rapid development of offshore oil fields, underwater accidents happen frequently. "Deep water Horizon" drilling platform 01 well which located in Mexico bay blowout on April 20, 2010, resulted in 11 deaths, 17 injuries, a large area of sea area was seriously polluted (Fig. 1). The most serious accident in recent 50 years offshore of USA led China to study deep water emergency repair technology.

Foreign marine oil developed earlier, the corresponding emergency repair method is relatively better. Especially in the Gulf of Mexico, Brazil, West Africa and North Sea and other oil and gas area, underwater emergency repair technology which dominated by the developed countries is more mature. At present all sorts of underwater repair equipment and technical ability all over the world were basically monopolized by Western company. The technology not only includes equipment and repair process, but also includes management technology. Because the underwater operation is huge investment, and beyond the individual's capacity, and as a result of underwater equipment at the beginning of design, fault risk factors is minimum, the underwater accident itself is low, so there were lots of repair system jointed by many companies. These repair systems are high-

Fig. 1 Deepwater horizon oil spill

ly complex, including the most advanced technology. But due to the underwater accident types are varied, almost every accident has its own characteristics, it is difficult to find common ground, so,2010 in the Gulf of Mexico oil spill accident failed to control timely and effectively, and caused great pollution, it made the owners suffered huge economic losses. Obviously, it is vital to further improve the emergency capability of under water, even for developed western company.

China offshore oil development technology overall relatively advanced countries is relatively backward, mainly reflected in the deepwater oil and gas fields development, only in the initial stage. Underwater repair technology for deep water oil and gas development as the key technology, it did not get corresponding research. For the technology blockade, foreign companies only provide key equipment leasing and technology services. With the social and environmental protection consciousness enhancement, the treatment after pollution, improve the management side edge method have been eliminated, need oil company in mining or job before the accidents that may occur for adequate estimation, and formulate contingency plans, related to repair equipment, repair capabilities must be in a state of readiness. This requires emergency repair technology to advance to, or at least in synchronization with the other development technology research.

1 SUBSEA CHRISTMAS TREE

Subsea Christmas tree has the following functions:
(1) suspension of all weight of the tubing string.
(2) sealing the annular space of tubing and casing.
(3) control and regulate the production of oil wells.
(4) ensure downhole operation safe.
(5) manometry, wax and other daily production management.

The underwater equipment is typically installed in a separate base, because the different base subsidence causes the relative displacement between the device, make the jumper suffering stress. Sealing ring of connector are the weakest link, most likely to damage. It caused serious pollution due to leakage of seal ring, so it is necessary to carry on the replacement operation. Common type accident of Christmas tree is shown in Table 1.

Table 1 common type accident of Christmas tree

Type	Position	Result	Solution
Seal failed	Connection of pipeline oil circuit valve	Leakage of oil and gas, great economic loss and environmental pollution	(1) Shut-in well, inspect leakage, replace seal ring (2) Failed to shut-in well, plugging downhole
	Seal position	Break closed environment, water enter into the tree internal, affects the service life	Replace the seal

Type	Position	Result	Solution
Structural damage	Oil pipeline	Oil and gas leak due to submarine vibration or fatigue damage	Same as seal failed
	Surface of tree	Impact of the external activities such as falling object and fishing caused the structure, make the tree internal lose protection	Take protection measures and repair damaged parts timely
Structural failure	Electrical submersible pump	Cannot production	Replace
	Hydraulic unit	Unable to remove the normal installation	Remove and replace
	Top blocking	Cannot connect with oil circuit, seal failed, affect other part of the tree	Shut in well; Clean inner of the tree; Replace new top blocking
Corrode	Inside	Corroded by oil and gas lead to structure damage	Same as seal failed
	Outside	Corroded by sea creature	Same as seal failed
Hydras	Inside	Affect production	Mechanical and chemical cleaning plugging

2 NECESSARY EQUIPMENT AND TOOLS OF REPLACE SEALING RING

At present, lack of ability of operation let us import, develop and equipped with more advanced equipment and tools to slip the leash of the operating depth bound. The deepwater large equipment, will enable China to have in the deepwater area maritime operation ability. ROV, ADS, HOV and related auxiliary equipment can also used to repair device of subsea. See Table 2 to Table 5.

Table 2 The maximum depth of operation

Device	Maximum depth of operation(m)
ROV	> 6,000
ADS	Between 300~600m, even 700m
SDS	<457

Table 3 Function of operation

Device	Function of operation
SDS	Best
ADS	Normal
ROV	Worst

Table 4 Personal safety

Device	Personal safety
ROV	Safe
ADS	Normal
SDS	Worst

Table 5 Cost of operation

Device	Cost of operation
SDS	Expensive
ROV	Normal
ADS	Cheap

In view of the different depth of water and accident types, we can use different repair equipment and tools.

2.1 Installation Vessel

The vessel must equipped with winch / crane (with heave compensation) which can work more than 1,500 meters water depth, accurate positioning ROV through the dynamic positioning system, and guarantee the operation accuracy.

2.2 ROV

ROV (Remote Operated Vehicles, ROV) is a robot used in underwater for limited operation, it can dive into the water instead of people to complete some operations. People diving depth is limited because of the underwater environment. So the underwater robot has become the important tool for development of marine. It works by the mother ship staff, powers by umbilical, observes through underwater television, sonar and other special equipment, and operates by the mechanical hand.

Specific requirements for ROV to repair the seal ring of the connector of subsea trees as follows:

(1) localization: positive buoyancy, holding function;

(2) mechanical arm: one with five degrees of freedom for holding, one with seven degrees of freedom for operation;

(3) with more than 40L/m hydraulic output, ensure the operation time, with quick connector (flexible joint);

(4) a sonar system (collision avoidance);

(5) observation device, camera and lighting system;

(6) movable in a horizontal plane, a vertical axis, moving and rotating four degrees of freedom above;

(7) all the design accords with API standard;

(8) center of gravity design, ensure the operation is stable;

(9) water depth is not less than 1,500 meters;

(10) suitable for four degree sea condition

2.3 ADS

An atmospheric diving suit or ADS (Fig. 2) is a small one-man articulated submersible of anthropomorphic form which resembles a suit of armor, with elaborate pressure joints to allow articulation while maintaining an internal pressure of one atmosphere. The ADS can be used for very deep dives of up to 2,300 feet (700m) for many hours, and eliminates the majority of physiological dangers associated with deep diving; the occupant need not decompress, there is no need for special gas mixtures, and there is no danger of decompression sickness or nitrogen narcosis. Divers do not even need to be skilled swimmers.

Fig. 2 ADS

The ADS has variously been referred to as a Winnie the Pooh suit (because of its large head), armored diving skirt, articulated diving suit, Iron Duke, Iron Mike, and "deep-sea diving robot". Atmospheric diving suits in current use include the Newt suit and the WASP, both of which are self-contained hard suits that incorporate propulsion units. The Newt suit is constructed from cast aluminum (forged aluminum in a version constructed for the US Navy for submarine rescue), while the WASP is of glass-reinforced plastic (GRP) body tube construction. The upper hull is made from cast aluminum. The bottom dome is machined aluminum.

Specific requirements for ADS to repair the seal ring of the connector of subsea trees as follows:

(1) Umbilical provide air, body can resist external water pressure;
(2) The propulsion system has at least four degrees of freedom such as movable in a horizontal plane, a vertical shaft and rotates;
(3) Buoyancy can keep standing in the deep water;
(4) Joint activities with ease, sealing reliable;
(5) With two manipulator;
(6) work not less than 500 meters.

2.4 SEAL RING REPLACEMENT DEVICE

It can open the connector locking device, hoop seat between connector and connector can be separated, and maintain the relative position between the two, separate volumes meet space requirements for replacing the sealing ring.

3 REPLACE SEALING RING PROCESS

If AX gasket seal failed, special tools can be used to remove and replace the gasket without retrieving the jumper.

3.1 PREPARATIONS

Check the ADS umbilical cable connections are intact, communication is expedite, lighting can work, and carrying tools are complete. Preparatory work should be done very full, can satisfy a dive can finish all work, reduce unnecessary reciprocating submersible, the divers enter the ADS, used a winch to repair position. Pay attention to attitude adjustment in decentralization process, and keep in touch with real time monitoring center. When approaching a tree 5 meters away, decelerate to close tree to avoid collision. On the process of closing tree, continue to take adjustment of speed and direction, slow to arrive at best location to repair easily; After ADS dive down to fault tree, report the surrounding environment, the fault status to the control center. And then close the main valve, cut off the power. Used a winch to lower sealing ring replacement to the fault tree.

3.2 Lower the Install Tool

This tool is lowered on the down line and located above the connector that required gasket replacement. After the tool is position directly above the connector, it is lowered the position and landed on the receiver structure of connection harp. The running tools soft landing plate is lowered under the connector. As the soft landing plate is lowered down, the latch that secured the tool to the receiver structure are activate and locked by the bottom of soft landing plate of the install tool. The latch that made the tool to the connector top plate is also engaged. The lash red ring moved down under the connector. The latch of tool activate seal ring made the connector activate seal ring. Now the connector were ready to be unlocked. See Fig. 3.

3.3 Unlock the Connector

To unlocked connector, the activate seal ring moves uppers over the finger to released melding harp. The hydraulic pressure moves the soft landing plate and connector up, the connector clear the harp. See Fig. 4.

Fig. 3　Lower the install tool

Fig. 4　Unlock the connector

3.4 Replace the Connector

The ADS brings the gasket remover and replacement tool, loaded with a new AX gasket and placed on the harp. The new AX gasket was located on the bottom of gasket remover and replacement tool, the top section of the tool retrieve the old gasket from connector, the connector is lowered under the gasket remover and replacement tool, as the connector is lowered, a slight ring on the top gasket remover and replacement tool disengaged the pins that retained the gasket within connector. The hydraulic pressure moves the soft landing plate and connector up, the connector clear the gasket remover and replacement tool. See Fig. 5.

3.5 ADS Remove the Replacement Tool

The ADS remove the gasket remover and replacement tool loaded with old gasket from the harp. The gasket remover is complete, as the gasket remover and replacement tool remove from the harp; the new gasket is left on the harp. See Fig. 6.

Fig. 5　Replace the connector

Fig. 6　ADS remove the replacement tool

3.6 Lock the Connector

Now the connector is ready to be locked again. The soft landing operation mates the connector's body with a harp. Final a lamb of the connector and the gasket is provided by feature incorporated the body and the harp. Now the connector is ready to be locked. The red ring slides down and locked the connector, the red ring moved down were over the finger sets 40 grade temper and settlement fingers pinvite to grass melding harp. This locks and preload the connector. The pressure allows two lash red ring latched to be disconnect from lash red ring of the connector. In the caters of the outside of the tool show when the latch are disengaged. See Fig. 7.

3.7 Retrieve the Install Tool

With the pressure still applied on the circuit, the lash red ring of the tool moves up and away from the connector of lash red ring. As the lash red ring moves up, the lash in the cater are captured by the soft landing plate of the tool. This prevent the lash from skiding and accidently snagging on the connector when the tool is removed. The pressure allows the two soft landing latches to be disconnected from the top plate of the connector. The hydraulic pressure is diverted to cylinder and the soft landing ring moves up. The lash and the soft landing ring of the tool clear of the connector and the tool release from the receiver structure. The tool retrieve to the surface. Seal ring replacement is complete. See Fig. 8.

Fig. 7　Lock the connector

Fig. 8　Retrieve the install tool

CONCLUSIONS

(1) Strengthening to inspect underwater equipment periodically Deep underwater emergency maintenance method provides efficient solutions for offshore oilfield subsea equipment emergency maintenance. But it's all about remedical measures after the accident, and cannot completely eliminate negative influence on the environment and economic. Therefore, we should strengthen regular inspection.

(2) The establishment of a professional equipment maintenance team. In view of the condition of underwater equipment servicing in china, the establishment of a professional submarine test equipment team is necessary.

REFERENCES

Lin Xiujuan, Xiao Wensheng, Bi Jingbo. 2011. Subsea Wellhead Mechanical Stability Analysis after Placing Tree in Position[J]. Journal of Information and Computational Science, 8(10):1919-1927.

MacKay S. 2008. Completion Design for Sandface Monitoring of Subsea Wells[C]. SPE Annual Technical Conference and Exhibition.

Selection Method for Subsea X-tree

Yuhong Gu[3], Huanhuan Lu[1], Menglan Duan[1,2], Tong He,[1] Honglin Zhao[2]

[1] Department of Mechanics and Engineering Science, Fudan University, Shanghai 200433, China;
[2] Offshore Oil/Gas Research Center, China University of Petroleum, Beijing 102249, China;
[3] King dream Public Limited Company, Wuhan 430223, China

Abstract Subsea production system has been widely used in deep-water oil and gas field development. Subsea Christmas tree (X-tree) is the key equipment used to control the flow of oil and gas to or from a well. First of all, the paper gives a summarization of the subsea X-tree design parameters since the configuration of subsea X-tree must be based on design parameters. And then, the advantages and limitations of the vertical and horizontal subsea X-tree are discussed, and in the meanwhile, the selection principle of the subsea tree layout is presented. Finally, characteristics and applicability of the components and the control systems of the subsea X-tree are analyzed.

Key Words Subsea X-tree; Subsea tree layout; Selection method

INTRODUCTION

Currently deep-water oil and gas exploration has become the inevitable trend of the world's energy development. There are huge reservoirs in South China Sea. In recent years, a number of deep-water oil and gas fields have been put into service. The subsea production system is widely used as the main mode of deep-water oil and gas exploration.

Subsea X-tree is the key equipment used to control the flow of oil and gas to or from a well. According to its applications, the X-tree can divide into X-tree, gas X-tree and water injection tree and so on.

Fig. 1 shows the subsea X-trees serving in one subsea field. Subsea X-Tree is an important equipment of subsea system attached to the top of subsea well and connected with the flowline to transfer oil and gas to pipeline network (API, 2009). Fig. 2 shows a typical subsea X-tree. In this paper, we will analyze the features of various subsea X-trees and selection considerations of X-tree for oil field. Finally, we give the selection method for the components of X-tree which is widely used at present.

X-tree can be carried out in accordance with the requirements of clients specializing in the design with different types in specific oilfield. In general, from the aspect of tree structure, the subsea X-tree can be divided into the mudline X-tree, vertical X-tree and the horizontal X-tree (ISO, 2011). The mudline X-tree is used in shallow water while the others are used in deeper water. Each has its own advantages and disadvantages, and may also various types. So the selection of X-tree should depend on specific conditions. Fig. 3 shows the subsea X-tree with different types.

Fig. 1　Overall layout diagram of subsea X-tree and other underwater equipment

Fig. 2　Typical deep-water vertical guide lineless X-tree

(a)Mudline X-tree　　　(b)Horizontal X-tree　　　(c)Vertical X-tree

Fig. 3　Different types of subsea X-mas tree

1　SUBSEA X-TREE DESIGN PARAMETERS AND TYPES

In order to design a subsea X-tree for a specific oil field, we need to make clear the basic parameters of the X-tree, which covers the following, production type, service standard, pressure, number of wells, water depth and so on(Otten and Brammer,1987).

(1)type of production.

Oil, gas, or water injection? The production type will affect system material selection, tree configuration, and seal selection. Oil production represents the most common type of subsea tree, and gas production generally encompasses the most critical service designs. Water injection typically represents the simplest tree configuration.

(2)Service standard.

The conditions of service materials for the tree valves, components, and seals will be dictated by the composition of the produced fluid. Partial pressures are used to evaluate the specified H_2S and CO_2 contents against the requirements of NACE MR – 01 – 75, API 6A to choose materials (NACE MR0175,2009).

(3)Pressure.

Pressure classes, shut-in and flowing pressures should be evaluated to determine the most economical materials and seal designs. Standard API 6A pressure classes are used for the wellhead components and ASME B31.3 are used for flowline design.

(4) Number of wells.

Satellite or template trees? The connection between the tree and flowline is different for the two types.

(5) Water depth.

With the water depth increasing, the X-tree operation, installation and oriented position is different from each water depth (Qin, Gu, Luo, Li, Duan, Liu, 2012). Water depth of 100m or other shallow waters fewer than 100m, the X-tree can be installed by jack-up platform with the help of divers. With the water depth increasing, floating platform is used to install more complex X-tree. Water depth among 60m to 200m, the installation can be operated by diver. 200m to 900m, the X-tree installed with the help of guide rope must be monitor by ROV. When water depth is deeper than 900m, dynamic positioning floating ship is used and the X-tree is installed with the technology of guideless rope with the help of ROV. Table 1 shows different X-tree for various water depths.

Table 1 Different types of subsea X-mas tree used for different water depths

Water depth(ft)	Installation ship	X-tree type
<400	Jack-up platform	Mud
200 to 800	Floating ship	DA
800 to 2,500	Floating ship	DL
>2,500	Floating ship	DLL

2 X-TREE LAYOUT OPTIONS

The X-tree design parameters and types have been determined. The next question is the X-tree layout. The key to determine the X-tree layout is to make sure the position of oil and gas flowing line and valve. Fig. 4 shows the typical layout of mudline X-tree, vertical tree and horizontal tree.

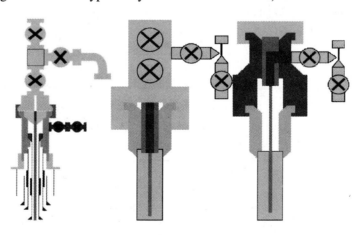

Fig. 4 The configuration of different types of subsea X-mas tree

Mud-wire Christmas tree application is very limited, so we keep our point to discuss the choice of vertical X-tree and horizontal X-tree. The main features of the vertical X-tree cover the following, vertical arrangement of valve, downhole electric and hydraulic lines connected to the hydraulic electric joint of top of the tubing hanger through the bottom of the tree, and tubing and tubing hanger must be installed before the X-tree. The features of the horizontal X-tree are oil export is on the side of tubing hanger and the tree must be installed before the tubing hanger.

Engineer chose vertical or horizontal X-tree for specific oil field should make sure the advantage, disadvantage and the application feature of the tree (Pappas, Maxwell, Guillory, 2005). Table 2 shows the features of the two trees.

Table 2 Comparison between horizontal and vertical subsea X-mas trees

Item	Horizontal X-tree	Vertical X-tree
Delayed completion installation time after well drilling	Fast, quickly lowering speed of the tree	Slow
Completion installation immediately following well drilling	Slow	Fast, drilling and tubing installation at the same time
Test before installation	Downhole function can be test before tubing hanger installation, and install downhole function immediately after the test	Test should be taken after the tubing hanger installation
Well testing	During well testing, the tubing can be recovered immediately if anything break down	During well testing, the tubing we should Shut-in the well and recover the tree and lower blowout preventer(BOP) before recover the tubing if anything break down
Recover X-tree well repairing	Time-consuming, should recover tubing first	Time-saving
Recover tubing well repairing	time-saving	Time-consuming, should recover the X-tree first
Emergency cutout	weak	strong
Applicable reservoir	Complex reservoirs and recover tubing workover frequently	Simple reservoirs and few recover tubing workover
Technical excellence	Electric submersible pumps(ESP) X-tree	High-pressure oil-gas well
Application trends	Oil well, high maintenance rate and well completion to be restored	Gas well, low maintenance rate, well pressure under 10000psi

With the wide application of vertical X-tree and horizontal X-tree, and on the basis of extensive experience, people propose two new designs for problems in production. They are atypical vertical X-tree and atypical horizontal X-tree (Wester, Ringle, 2001).

Atypical vertical X-tree simplifies the design and installation of the tubing hanger and the cap of the tree, and adds tubing head assembly, as shown in Fig. 5.

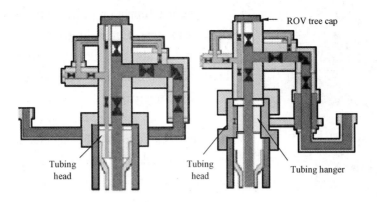

Fig. 5 The comparison of typical and atypical subsea vertical tree

Atypical horizontal X-tree abandons the internal cap of the tree and moves crown plug into tubing hanger obstruction tubing hanger, shown in Fig. 6. The upper seal design still needs the practice of oil field.

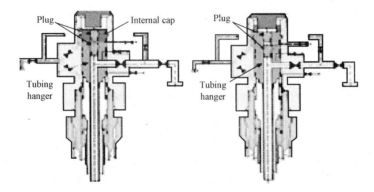

Fig. 6 The comparison of typical and atypical subsea horizontal tree

Many factors come into play in determining the final layout of the X-tree. Such as the size of wellhead tubing/ cannula, type of control system, form and size of flow line, type of well head and so on.

3 COMPONENT SELECTION

The most important job in X-tree design is the selection of component of the tree. The component had better choose the one which had been practical engineering revealed or existing equipment. Different types of X-tree has various selection method in determinate the component (Barnay, Girassol, 2002). Fig. 7 shows the typical components of horizontal X-tree.

The key factors in the selection of the component. cover the flowing, certain types of X-tree, water depth, pressure class, temperature etc (Granhaug, Soul, 1995).

(1) Valve.

Gate valves equipped in the subsea X-tree cover production master valve, annulus valve, annulus master valve, wing valve, and switching valve and those valves should satisfy the requirement of hydraulic driving and long-distance operation. They play a very important role in control sys-

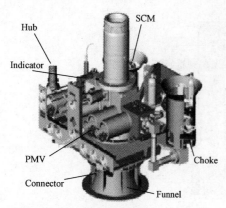

Fig. 7 The typical components of GLL subsea horizontal tree

tem. The determination of the valve depends on pressure class of the tree, output. In general, the master valves are integrated on the tree while the wing valve, and switching valve are connected with flange.

(2) X-tree cap.

X-tree cap protects the top of the tree from corrosion and rock cuttings. The cap can be designed to operate by mechanical or hydraulic means and attached on the external surface of the top of the tree (Juiniti, Roncador, 2001). After installation, the cap should be removed when taking vertical downhole operation. so that the workover system can operate the valves.

(3) Production throttle valve.

Production throttle valve is a flow control device to control the pressure drop and velocity of the production fluid. In general, fault occurs in throttle valve especially easily, so the throttle valve should be designed to be recovered easily. In that case, special operating tool is needed.

(4) Connector.

Connectors include wellhead connector and pipe connector, the former one is to landing on the ground and locking to the subsea wellhead. Providing mechanical and pressure connection as well as positioning the tree and wellhead. The latter connects the underwater pipeline or cross-under pipe to tree. There are two driving modes which can be made by hydraulic drive and mechanical drive, generally, the driving mode need to be equipped with secondary unlock function and can be able to do the seal test. Pipe connector can be divided into vertical connector and horizontal connector, both of them have different features. In addition, link controller connector is to connecting underwater umbilical cord cable and fly line to the X-tree, by the way, this connection is relating to the control system that is chosen.

(5) Tubing hanger.

The type and position of tubing hanger is directly related to the type of X-tree that was chosen. Vertical tree of the tubing hanger is located in internal wellhead, double holes design mostly, the upper configures underground line joint. Horizontal tree of the tubing hanger is located in the interior tree body, be single span design mostly, need to be configured special downhole line through device, and also, the internal need configures block device. The structure size of tubing hanger is decided by pressure rating, production traffic and wellhead size.

(6) Sensor.

Sensor on the X-tree monitors the pressure, temperature, sediment, erosion, multiphase flow etc. Sensor must buy from mature companies whose productions have been strict test validating and oil field practical situation verification.

4 CONTROL SYSTEM SELECTION

Underwater control system can be divided into direct or pilot hydraulic control, pilot electric-hydraulic control, electric-hydraulic complex control and all-electric control, etc. The system needs hydraulic source to realize the drive function, in addition to the new full electric control (Lewis, Money, Kochenower, 2002).

The advantage of directly hydraulic control system is simple and economic structure, wide application, high reliability and the disadvantage is that the distance between tree and ground control station is limited, the control objects should not be too many, what's more, the system is just suitable for controlling the field in the meters of hundreds to one thousand.

Pilot electric liquid control advantage is that it uses electromagnetic valve instead of underwater hydraulic pilot valve, shorts the control command response time, increases the control range, and reduces the hydraulic circuit since it can be connected tree with just a main hydraulic circuit and a cable.

Electric-hydraulic complex control system requires fixing underwater control module to the X-tree and its advantage is shorter reaction time, with remote measurement and control function.

Different control system has a big distinction in tree control element types, number, installation and recovery work, and at the same time, different workover control mode affects the design of tree cap and control module.

PROJECT EXAMPLES

For south China sea LH11-1 offshore oil field (Zhang, Huang, 1998) the water depth is 310m, pressure class is 2250psi. And wells are with great numbers and concentrated as shown in Fig. 8. Crude oil is high viscosity and density.

Carrying out subsea tree selection work base on the features of the LH11 – 1:

(1) Choosing the guide rope assisting the installation of X-tree without divers due to the water depth is close to or exceeds the divers and jack-up platforms limit.

(2) Base disc X-tree is used to reduce difficulty of installation and control due to wells are with great numbers and concentrated.

(3) Submersible electric pump is used due to the features of the LH11-1 oil reservoir

Fig. 8　The oil field development program of SouthSea Liuhua 11 – 1

(Rohloff, Douglas, 1996). Horizontal X-tree at the mean of submersible electric pump has great advantage in reducing time and expense since workover with no need for recovering X-tree.

(4) Finally, select the components of the X-tree based on the of oil field requirement and relevant standards. The X-tree cap has a specially design to meet the oilfield to have cable to connect to downhole. Pipe connector installs vertically to the X-tree to provide convenience for the installation of jumper pipe and flow line. Other components had better choose mature product and take experiment base on LH11-1.

CONCLUSION

(1) First of all, choosing the type and component of subsea production system tree should consider its specific field application. Selection work should be around that tree design need to suitable for oilfield development requirements.

(2) At present, the most commonly used tree pattern is vertical and horizontal, on the basis of preliminary selection, choosing the type need further analysis of two tree of procurement, installation, maintenance and repair, and other comprehensive cost.

(3) This paper summarizes the subsea X-tree technology present situation and the development tendency, compares the different types of tree's features. This paper puts forward the design requirement, determine the types and layout, the tree parts and control system on the selection of subsea X-tree selection method.

(4) At present, the development of each south China sea oil and gas field needs to consider that choose more economic and reasonable type of tree seriously. So this paper has certain guiding significance for the engineering practice.

REFERENCES

API Spec 6A. 2009. Specification for Wellhead and Christmas Tree Equipment.
Gilles Barnay, Girassol. 2002. The Subsea Production System Presentation and Challenges[C]. OTC 14170.
ISO 13628 – 4. 2011. Petroleum and Natural Gas Industries Design and Operation of Subsea Production Systems-Part 4: Subsea Well-head and Tree Equipment.
James M, Rohloff. 1996. Douglas RetcLiuhua 11 – 1 Development -Downhole Completion Program with Electric Submersible Pump and Wet-Mateable Electrical Connectors[C]. OTC 8176.
Lewis M, Money R, Kochenower C. 2002. Subsea System, Production Controls and Umbilicals for Typhoon Project[C]. OTC 14125.
NACE MR0175. 2009 Petroleum and Natural Gas Industries — Materials for use in H_2S-Containing Environments in Oil and Gas Production.
Otten J D, Brammer N. 1987. Equipment Selection Procedure for Subsea Trees Society of Petroleum Engineers. SPE 16847.
Otto Granhaug, John Soul. 1995. The Garden Banks 388 Horizontal Tree Design and Development[C]. Offshore technology conference, OTC 7845.
Pappas J, Maxwell J P, Guillory R. 2005. Tree Types and Installation Method. Northside Study Group, SPE.
Qin Rui, Gu Yuhong, Luo Xiaolan, Li Qingping, Duan Menglan. 2012. Parametric Analysis for the Design of Subsea Christmas tree[J]. The Ocean Engineering, 30(2):116 – 122.
Randy J Wester, Eric P Ringle. 2001. Installation and Workover Time Savings: Key Drivers for Deepwater Tree Selection[C]. OTC 12943.
Ricardo Juiniti Roncador. 2001. Field Development with Subsea Completions[C]. OTC 13259.
Zhang Bo, Huang Changmu. 1998. China South Sea LH11 – 1 Oil Field Subsea Developing Technology[J]. China Offshore Oil and Gas(Engineering), 10(3):38 – 40.

Simple Analysis of Application Prospect For Several Deepwater Wellhead Unit

Haiping Wang [1], Yanjin Ma [1], Ying Jin [1], Yan Liu [2], Haijing Wang [2]

[1] Engineering Department, Offshore Oil Engineering Company, Tianjin, China
[2] Construction Division, Energy Technology & Services-Oilfield Construction Engineering Company, Tianjin, China

Abstract Offshore oil resource potential is tremendous, current offshore oil and gas exploration and development of the range from shallow sea, half a deep sea has extended to the depths of the sea. The wellhead device is the important equipment in oil and gas production, its performance is in relation to oil and gas well whether which have safe security and efficient production. In this paper the general development situation of foreign wellhead device and the structure of each part of the improvement on the basis of current our country, pointed out the shortcomings of the wellhead device. Also recommendations on the future development of our wellhead equipment, presented its preliminary views on the subsea wellhead and Christmas tree type selection.

Key Words The deep sea technology; Oil field development; Engineering technology; The mouth of a well under water; Christmas tree

INTRODUCTION

From the view of ocean engineering, the main differences between deep and shallow water are embodied in the development mode. Deep water technology comes from shallow water, but deep to shallow water is a great cross. At present, the international Marine petroleum engineering in deep water definition is 300 meters water depth. In the 300 - meter depth of less than, can use jacket model development, and 400m to 600m depth can also use fixed platform development. More than 600 meters depth general use the floating platform or underwater system. For the deep water, fixed platform investment is too big, hundreds of meters long jacket cost of steel quantity is very large. And also need to use ultra large barge, marine section installation complex with high risk. In the gulf of Mexico, more than 600 meters depth will have to use tension leg platform (TLP), floating column type platform (SPAR) or underwater tree.

In recent years, China's oil consumption grew rapidly, in 2010 has become the world's second largest oil consumer. External dependence reached 50%. In order to improve the degree of self-sufficiency in oil, in addition to develop onshore oilfield, but also need to fully excavate the deep sea which has the rich oil and gas resources potential: "Ask the deep sea for oil"! Deep water and ultra deepwater oilfield development engineering technology, for our country, still is blank, we need to innovate independently.

1 INTRODUCTION OF OIL AND GAS WELLHEAD DEVICE

Oil and gas wellhead device is an important oil and gas production equipment. Its performance is related to the safe and efficient production of oil and gas wells. Wellhead consists of casing head, tubing head and Christmas tree, which is mainly used to monitor production wellhead pressure and control oil (gas) well flow, also can be used in acid fracturing, water injection, and testing. The development of the petroleum industry constantly put forward higher request to wellhead reliability and controlling, this will prompt and driving the wellhead device also to be improved and developed. At present offshore oil drilling and production wellhead according to its position can be divided into the following categories:

(1) The subsea wellhead is located in underwater submarine, drilling of BOP (Blow Out Preventer) and production of XT (X-Tree) are installed at the bottom of the well head, completion and in case of emergency, the surface ship can evacuate easily.

(2) The underwater wellhead installed in the water floating body, floating body and water platform with flexible cable connected. BOP weight bore by floating body. Completion and in case of emergency, water platform can evacuate timely. But the technology is not mature.

(3) The water head installed on the surface, the maintenance of BOP, XT is convenient, but completion and in case of emergency, the surface of the platform is not easy to leave.

2 WELLHEAD DEVELOPMENT SITUATION IN PRESENT

2.1 Foreign Wellhead Development Overall Situation

At present, overseas production X-tree maximum working pressure is 140MPa. The tree of flange connecting valve maximum working rated pressure is 210MPa. Many foreign companies are constantly developing and to be perfect electric submersible pump wellhead and integrated X-tree. Have produced single tubing electric submersible pump wellhead, double tubing wellhead and double full bore integrated X-tree. They can also according to the oil (gas) field condition and borehole retention fluid to produce land (including offshore oil platform) of wellhead equipment and subsea wellhead device. At present the main producing countries have the United States, Britain, Italy, more than 10 countries, and for the products of technology and capacity, the United States should in a leading position in the world.

2.2 Wellhead Components And Structure

(1) Gate valves. Gate valve as head of the main parts, major foreign manufacturers develop flat gate structure, and inside screwed yoke lever type plate valve has an absolute advantage. The new production of gate valves have FL and FIS type two kinds, working pressure of 13.79 ~ 34.5MPa. The structure characteristic is: the integral gate structure, can prevent pipeline of sediment into the body cavity. High bearing capacity of two thrust bearing is used to absorb gate opening and closing of the load, thus will reduce the handwheel force to the minimum. Special inert material spring-loaded lip type seal, can not only protect the metal sealing surface and can strengthen the low pressure seal performance. FL type gate valves in each seat is using a single lip seal. FIS type gate valves use inside and outside seal. Another kind of JS type gate valves are the character-

istics of valve seat to valve plate and valve body metal to metal seal, special outside diameter and inside diameter of lip type seal not only strengthen the low pressure seal and protect the seat, valve plate and valve body metal sealing surface. JS type gate valve design of the main characteristic is to have a rotary orifice valve plate, it can be to the open position not have to consider the valve stem position. This feature enables the JS type gate valves special suitable for bottom valve of the X-tree device.

(2) Casing head. Casing head including four-way casing head, casing hanger, packing, top wire, gate valve and pressure gauge, is installed in the whole wellhead most of the bottom, the function is to connect the layer casing and seal between the annular casing space. In recent years, the casing head structure constantly improved, mainly to the simple operation, good seal performance (automatic seal), suitable for deep well multilayer casing and integrated development direction. Compact type casing head is multilayer casing head of the shell made of an organic whole, internal each layer hanger transfer weight through mutual load. It has small volume and saving materials and fittings to reduce its weight.

(3) Tubing head. It has the composition of four-way tubing head, tubing hanger, packing, top wire, gate valve, pressure gauge and flange. It's packed in casing head above, mainly to bearing the pressure inside the oil string, bearing oil string and oil (gas) X-trees and other plant loading. The tubing hanger of electric submersible pump tubing head device, uses increase installed wear cable device to realize the electric submersible pump recovery operation. This form of cable joint, according to the voltage, current, working temperature is divided into many kinds of level for the user choice, the voltage from 3,000V to 5,000V, current is respectively 70A, 125A, 140A, 164A, working temperature from 60℃ to 178℃. Italian tubing head have DP70, DP4, DP7 three types. The DP70 is suitable for single tubing suspension. DP4 is suitable for double tubing suspension. In the four-way shell it has additional tight top screw, to control double tube suspension direction and position. DP7 type is suitable for multiple tubing suspension. In the four-way shell it has four tight top screw, to control the hanger installation direction and depth.

(4) Oil (gas) X-tree. Oil (gas) X-tree is refers to the main body part above the wellhead total gate. It looks like a tree. So, it is called oil (gas) X-tree, mainly consists of casing gate, total gate, production gate, paraffin gate, four-way or three-way tubing spool and nozzle. Its function is to control and adjust the flowing production Wells, guide the extrusive oil and gas into the pipeline, guarantee to record oil pressure, casing pressure, oil and gas production, take samples and paraffin removal, etc. In recent years the general trend is integral structure. Marine platform recovery operation because of the limitation of the platform area and drilling cost general first chooses the integral structure.

In addition, foreign wellhead (mainly in seal) now also tends to develop high temperature, high pressure and super high temperature, super high pressure. It can conclude that for wellhead device the movement between the metal sealing element due to the changes in pressure and temperature in after limited number of cycle, can lead to wellhead device failure. Can be further conclude that, in the future wellhead developing, metal seal will play a very important role, so must develop effective technical methods to produce stable sealing environment to deal with the risk of wellhead due to the high pressure.

3 SUBSEA SYSTEM

With the development of offshore oil industry, as well as to reduce the production cost requirements, production wellhead system from the sea surface to underwater, and develop from single well underwater X-tree to more wells underwater system.

Subsea system's main features:

(1) Can make full use of already drilling well and appraisal well production, even if some of the offset well hit area, but also can use these wells to inject water.

(2) For the shallow reservoir of oil field, in the area beyond control of platform, can use subsea X-water the tree to exploit and back to receive platform.

(3) Compared to build platform, subsea system costs less. This makes the deep water area and reserves of smaller marginal oil field can be developed.

(4) Some subsea production system can product the oil and gas, not need to build production platform. This situation not only save money, to production facilities to avoid water fierce storm and the influence of ice also have certain significance.

(5) Late stage of oilfield development in the platform location has filled circumstances, infill wells in order to improve the recovery efficiency can use underwater X-tree mining, and back to receive platform.

(6) After mining, underwater device can easily and economical fishing up, such already save a lot of money for rehabilitation measures, and some equipment can also be repeated use.

3.1 Single Well Subsea Production System

Underwater X-tree from its place of the space state, can be divided into wet X-tree and dry X-tree. Wet X-tree that tree ontology directly immerged in water.

Single well wet X-tree can use the guide frame to link with casing head and tubing from the above water platform the bottom. If there are multiple wells, the wet X-tree shall be installed in several wellhead together of the submarine base plate. If there are satellite wells in the nearby, also can through the pipeline connection part, send the satellite well crude oil to this base plate, along with all the oil recovery. The workover tools can be divided into through the flowline (TFL) and not through the flowline (NON-TFL) two types. Dry X-tree have a anhydrous space. X-tree and its auxiliary equipment are maintained in an atmosphere and water isolation airtight indoor, also known as wellhead chamber. Therefore to be needed a pretty big metal atmospheric chamber, the X-tree can be closed at among them. The person can enter into the space through the butt channel from submarines or work ship. It's easy to operate and safety performance is superior. At the top of the wellhead is the lead nozzle, it's the through hole to draw out pipeline and lead in cable and hydraulic pipeline which is requested to have good sealing device. At the top center is a goblet joint door linking with the repair tank, maintenance personnel can enter the wellhead through chamber of the joint door to carry out maintenance work from water by repair tank sink to the bottom of the sea. Single well underwater X-tree can be directly connected to the floating production device for production, it sets the underwater casing head and total valve set under the sea bed. It's extremely safe from anchor, net or rapids erosion and damage. This X-tree has specialties as following:

(1) Using the extensive used SG underwater casing head, drilling steps is similar to common underwater drilling system.

(2) X-tree height by ordinary 9~10.7m decreased to 3.3m, and can be used well cover protected.

(3) Easy to install and operate from drilling device.

(4) The whole system has the accident relief valve.

(5) Caisson can release joint in total valve group upper, and in case of deep water ice flow after scouring, still can keep on the oil well control ability.

3.2 Multiple Well Subsea Production System

Multiple well production system is that to concentrate several satellite wells type X-tree together through a submarine manifold center to connect to the floating production platform. Sometimes through the underwater chassis to drill a group of oil well, and then back to production platform, it also belongs to a kind of multiple well production system. Multiple well subsea production system characteristic is:

(1) Allow early production, it can quickly get oil and capital recovery.

(2) Each Wells can be independently production. It will not affect the other well production status.

(3) Can monitor and control each well separately from the ground.

(4) All seal, valves and other equipment required regular inspection and maintenance can replace. Production riser can be quickly removed.

Multiple well production system has several key parts as follows:

(1) Chassis (Template). Chassis is a frame placed in the bottom of the sea, and its role is to guide drilling equipment and provide proper well spacing. It's helpful for drilling cluster well on the platform. After drilling completion, the chassis can be installed subsea X-tree for production, also can be back to receive platform X-tree for production. Chassis is applicable for floating drilling ship and fixed platform. Using chassis can drill at the mobile platform at the same time to build production platform, so as to realize the early recovery and reduce oilfield total development cost. Using chassis will make drilling and oil production to become more flexible and convenient. According to the required of the well number, water depth, submarine conditions and installation requirements can choose different types of chassis. Simple distance chassis can be used in the shallow water area. Packaged chassis structure is flexible. To add cantilever component can increase the drilling number. While well number has been set, can use integrated chassis.

(2) Production Riser (Riser) Production riser can provide a dual channel. The crude oil and natural gas be out from the oil well should be delivered to upper platform for separation and processing through the riser. After treatment the crude oil will be transport to the tanker or other storage device. Riser system includes a central core and the oil well flown line around the central core. Riser design and manufacturing process is relatively complex, its main characteristic is:

① Each well all have their own separate flown line, so as not to interfere with other oil well production.

② For floating platform, because chassis fixed on the sea floor, and the platform will float on the surface of the sea, so riser system should be able to provide upper, lower, left and right small expansion changes. The changes can be absorbed by upper goose neck, stretcher and lower the universal coupling joint.

③ If there will be great waves according to the weather forecast, riser system must to be removed from bottom and lifted up to the top. The emergent remove system's operation is more complex, to 24 hours to finish the homework.

4 GAP BETWEEN OUR COUNTRY WELLHEAD AND FOREIGN ADVANCED

Compared with foreign advanced level, our country wellhead still has the disparity, mainly lie in the following aspects: (1) The structure form is single. In the domestic gate valve although now also uses plate gate, but still given priority to wedge gate valve, cant to be according to pressure, hole depth, oil and gas layer reserves and conditions factors to decide on gate valves. (2) The pressure level is too close and temperature rating range is small. (3) Material. Main parts material of our country for oil (gas) wellhead use is less, such as body material. Also is the castings and forgings, the high pressure casting are using ADO refining in the United States, appearance and inner quality is good. (4) Other aspects. The United States manufacture company can choose bearing, spring and seals according to condition to working condition, our country haven't achieved yet. In the machining method, procedure and quality control etc. also has a large gap.

5 SUGGESTIONS TO OUR COUNTRY DEEP-SEA WELLHEAD DEVICE

(1) See from the wellhead installed position, subsea wellhead is recommended first. Because of from technical feasibility analysis, whether workover operation, or maintenance, subsea wellhead all can meet the requirements. From the economic aspect analysis, for single well, the total cost for water well completion project, is about underwater completion of 3.8 to 4.0 times; For multiple well (cluster Wells) is 1.2 to 1.8 times. This is contrasted in the same water depth, the same production period and the same depreciation rate. It shows subsea has great superiority.

(2) From the type of X-tree, the proposal is chosen wet X-tree. Because technically, China had had multiple factory can produce pressure as high as 15,000 lbf/in^2 (103.42MPa) subsea wellhead device (oil tubing head and X-tree). In addition, in the underwater equipment and component of sacrificial anode anticorrosion technology, our country has accumulated certain experience for many years. Then from the economic analysis, according to the front of the same contrast method, the ratio between dry X-tree total engineering cost and wet X-tree, for single well is 5.8 to 6.0 times, multiple well is 6.1 to 6.3 times. Obviously, wet X-tree investment saves cost, more appropriate the promotion.

(3) Should be unified design criteria in the industry. Although API Spec 6A has stipulated the ASME allowable stress and deformation energy theory, test stress analysis design method of bearing housing design, but due to the different of the design method by each production factory adopted, product compatibility is poor, also gives product maintenance more inconvenience. Some manufacturer due to design criterion is not reasonable, so the safe coefficient chose too big in raw materials and

caused great waste in processing. From the bearing parts stress measurement and blasting test, the domestic product safety coefficient is too big commonly.

(4) In the new materials and blank, should speed up the steps of production (gas) wellhead new material research, especially in the packing materials. For high H_2S, CO and chloride condition with seal domestic it's also difficult to produce, still depend on import.

(5) In the new products aspect, as China's oil industry speeding up in reforming and opening up, we should accelerate the deep ocean of oil (gas) wellhead, especially underwater oil (gas) wellhead researching, in order to meet China's ocean petroleum industry development needs.

REFERENCES

Fang Huacan. 1990. Marine Petroleum Equipment And Structure[M]. Beijing:Petroleum Industry Press.

Zhao Zhengzhang. 2005. Foreign Marine Deep Water Oil and Gas Exploration Development Trend And Enlightenment [C]. Exploration Forum.

Subsea Gate Valve Design

Ming Zhou[1], Shulin Li[2], Hu Zhang[2]

[1]DOUSON Drilling & Production Equipment Co. ,Ltd. ,Suzhou 215137,China

Abstract Deepwater oil and gas exploration is the trend of the future resource development. Subsea gate valve is one of the key components of the subsea Christmas tree equipment and manifold assembly. Subsea gate valve design is the basis for design of the subsea Christmas tree equipment and manifold. This paper introduced the working principle of the subsea gate valve, and designed for 3km water depth subsea gate valve by a combination of theoretical calculations and finite element analysis method which meets API 17D requirements.

Key Words Subsea gate valve; Actuator; Finite element method; Hydrostatic pressure

INTRODUCTION

With the development of the global economy, the contradiction is increasing prominent, because of the insufficient supply and the growing demand for energy supply of the world. Human is demanding resources from land to ocean, from the shallow water to the deeper water. Subsea wellhead and Christmas tree equipment will need to move towards for high-reliability, high adaptability, automation, intelligence, and deep-sea areas of development.

The research and development of subsea equipment are rather late in China. There is no subsea wellhead and Christmas tree equipment production capacity of scale currently, offshore oil and gas exploitation usually use imported equipment. The development of subsea wellhead and Christmas tree equipment can not only to fill in the domestic blank, but also is the necessary of the development for China's offshore oil and gas industry. DOUSON Drilling & Production Equipment Co. Ltd. , (hereafter DOUSON) had launched the project of "marine and underwater oil and gas drilling and production of key equipment and technology (Subsea wellhead and Christmas tree) R & D and industrialization" in cooperation with China University of Petroleum in 2011. Subsea gate valve is a key component of the subsea Christmas tree and manifold equipment, playing an important role in subsea equipment.

With the development of the Christmas tree to the deeper water, the subsea gate valve design and performance requirements are facing great challenges. First, the subsea environment is one of the harshest, most corrosive environments in the world. Second, the pressure is very high. Third, the service life is 20 years, so that require high reliability. The high external ambient pressures which occur in deep water, coupled with high specific gravity hydraulic control fluids, are shown to have a significant impact on the performance of the actuators. The design & analysis methods and the verification test procedures which are required to develop and qualify new deep water actuator designs (S. Z. Ali, 1996).

1 SUBSEA GATE VALVE WORKING PRINCIPLE

DOUSON developed subsea gate valve with hydraulic control, mechanical spring return and ROV operations. Subsea gate valve contains both the gate valve and safety valve. The working principle is: when input instructions or there are special circumstances underwater feedback command the telecommunications signal through the cable to pass information to the solenoid valve to open commutation manipulation accumulator, through the control of high-pressure liquid pipeline into the valve of the hydraulic cylinder, the piston down, spring compression, and promote the valve gate down, the valve is opened until position; keep pressure control, that remains open; vent out the pressure of the hydraulic cylinder, spring immediately return, and led the line of the valve plate, the valve is closed until position. The actuators are designed to be 100% failsafe. This means that it should always be possible to operate the valve despite any failure that the system can or has experienced.

The subsea gate valve structure shown in Fig. 1, the subsea gate valves level is U-HH-PSL3G-PR2, main parameters are as follows:

Valve Model: 2 – 1/16
Length: 1.6m
Working temperature: −18 ~ 121℃
Rated working pressure: 10,000psi
Pressure of the control fluid: 3,000psi
Power stroke: 66mm
Maximum working depth: 3,000m

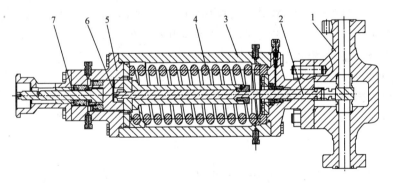

Fig. 1　The subsea gate valve assembly drawing

1—Body; 2—Stem; 3—Spring chamber; 4—Spring; 5—Cylinder; 6—Piston; 7—Override stem

2 SUBSEA GATE VALVE STRUCTURE DESIGN

Valves and actuators must be designed to meet two requirements in accordance with the requirements of API 17D (API 17D 2nd, 2011).

(1) For a fail-closed valve, with the assembly subjected to external hydrostatic pressure of the maximum rated water depth and full rated bore pressure, applied as a differential across the gate, it shall be shown that the valve opens fully from a previously closed position with a maximum of 90% the hydraulic RWP above actual or simulated ambient pressure.

(2) For a hydraulic fail-closed valve, with the assembly subjected to external hydrostatic pressure of the maximum rated water depth and atmospheric pressure in the body cavity, the valve shall be shown to move from a previously fully open position to a fully closed position as the hydraulic pressure in the actuator is lowered to a minimum of 100psi above ambient pressure.

In this paper, the design of underwater valves with hydraulic linear drive, drive design is the focus of the underwater gate valve design.

2.1 Force of Actuator

Designs of hydraulic actuator pistons for subsea valves were sized based on the force required to open the valve under full differential pressure divided by the available control system supply pressure. The return spring was then sized larger than the sum of all of the forces left in the valve and from the surrounding environment, when control pressure is lost, to assure the valve's closure over and above what the valve stem thrust alone can provide. Equations for sizing a fail close valve stem, effective piston area, return spring force, and minimum closing force are (Fowler and Herd, 1976):

$$F_c = F_p - F_r + S_p - S_f + F_o > 0 \tag{1}$$

$$F_{co} = F_p + F_r + S_p + S_f + kh - F_o \tag{2}$$

$$S_p > S_f - F_o \tag{3}$$

$$F_o = B_e - U_e - C_e \tag{4}$$

where, F_c is critical valve closure force; F_{co} is force required to open the valve; F_o is net seawater depth (sea head) force acting on the actuator; B_e is sea chest force; C_e is control line head force; U_e is upper manual override stem force; S_p is preloaded force in actuator's return spring mechanism; S_f is valve stem packing drag; F_p is cavity pressure at actuator; F_{fr} is friction force (between the gate and seat); k is spring rate; h is stroke.

Force calculation:

(1) Bore pressure towards valve stem:

$$F_p = \frac{\pi p D_r^2}{4} \tag{5}$$

where, p is bore pressure; D_r is valve stem diameter.

(2) Valve stem packing drag:

$$S_f = n\psi D_r b p \tag{6}$$

where, n is the number of packings; ψ is coefficient; b is packing width.

(3) Friction force (between the gate and seat)

$$F_r = \frac{\pi}{4}(D_M + 2B_M)^2 p\mu \tag{7}$$

where, D_M is seat inside diameter; B_M is seat seals width; μ is friction factor.

Net required force to open full pressure in bore:

$$F_{co} = F_p + F_r + S_p + S_f + kh - F_o = 155,530N$$

Net required force to full close:

$$F_c = F_p - F_r + S_p - S_f + F_o = 8,360\text{N}$$

hydraulic control pressure:

$$p_C = \frac{F_{co}}{A_p} = 2,697\text{psi} \tag{8}$$

$$p_C = \frac{F_c}{A_p} = 145\text{psi} \tag{9}$$

where, A_p is actuator effective piston area; p_C is hydraulic control pressure.

The results of Eq. (8), Eq. (9) meet the two requirements of API 17D valve and actuator is designed, So the design of the actuator force is reasonable.

2.2 The Design and Calculation of the Key Components of the Actuator

(1) The design of ROV override rod.

The ROV override rod is one of the key components of the actuator, which withstand high pressure. When needed ROV open the valve, override rod bear impetus to 155,530N. The material of the override rod is NES 833, the yield strength is 500MPa, the tensile strength is 635MPa after the heat treatment.

$$\sigma = \frac{4F}{\pi d^2} \leqslant [\sigma] \tag{10}$$

$$[\sigma] = 0.67 S_Y$$

where, F is force of ROV override rod, N; d is ROV override rod minimum diameter, mm; S_Y is 500MPa.

(2) The design of stem.

The stem is one of the key components of the actuator, which withstand high pressure. When the valve is opened and closed, the force of stem is 86,830N. The material of stem is UNS 07718, the yield strength is 825MPa, the tensile strength is 931MPa.

$$\sigma = \frac{4F}{\pi d^2} \leqslant [\sigma] \tag{11}$$

$$[\sigma] = 0.67 S_Y$$

where, F is force of ROV override rod, N; d is stem minimum diameter, mm; S_Y is 825, MPa.

(3) The design and calculation of spring.

The spring is one of the key components of the actuator, mainly rely on the resilient force of the compression spring to close the valve in the work. The working pressure in the valve bore helps the valve closed, consider the worst case, the valve bore pressure does not generate a thrust on the stem, the spring can be successfully return.

$$F_2 \geqslant S_p + \frac{\pi D_r^2}{4} \rho g H \tag{12}$$

where, F_2 is force of spring, N; ρ is density of sea water (1,030kg/m³); H is the maximum water depth rating (3,000m).

In accordance with the requirements of API 17D, the actuator fail-safe mechanism shall be designed and verified to provide a minimum mean spring life of 5,000 the actuator. Selection of hot-rolled spring class Ⅲ according of spring working conditions. The material of spring is AISI 5160, working temperature is −20℃ to 60℃, shear modulus G = 79GPa, the allowable shear stress τ_p = 740MPa (BS 1726 − 1, 2002).

Spring formula:

$$d = \sqrt[3]{\frac{8FD}{\pi \tau_p}} \tag{13}$$

$$C = \frac{D}{d} \tag{14}$$

$$K = \frac{4C - 1}{4C - 4} + \frac{0.625}{C} \tag{15}$$

$$k = \frac{Gd^4}{8nD^3} \tag{16}$$

where, k is spring rate; C is spring index; D is mean diameter of coil, mm; d is nominal diameter of wire, mm; K is the stress correction factor; F is spring force, N; n is number of active coil.

3 FINITE ELEMENT ANALYSIS OF THE MAIN PARTS OF SUBSEA GATE VALVE

3.1 Hydrostatic Pressure Calculation of Body

Table 1 Shows the properties of materials relevant for the calculations.

Table 1 Material properties

Component	Material	S_Y (MPa)
Body	AISI 8630	552

According to API 6A Distortion energy theory (API 6A 20th, 2010).

$$S_E \leq S_Y \tag{17}$$

where, S_E is the maximum allowable equivalent stress at the most highly stressed distance into the wall, computed by the distortion energy theory method; S_Y is specified minimum yield strength.

According to ISO 13628 − 7; D.2.4 Elastic-plastic finite element analysis (ISO 13628 − 7, 2010).

(1) Global strain criteria (Max. Principal strain):

$$\varepsilon_{prin} \leq 0.02 \tag{18}$$

(2) Local failure criteria (Equivalent plastic strain):

$$\varepsilon_{peq} \leq \min\left[0.1; 0.5 \times \left(1 - \frac{\sigma_y}{\sigma_u}\right)\right] \tag{19}$$

where, σ_y is specified minimum yield strength; σ_u is specified minimum tensile strength.

A finite element model of body was created to investigate the stress levels at test pressure of 103.5MPa. A half-symmetry model was utilized to reduce the computational time with acceptable accuracy. The body was fixed at the flange facing and the test pressure of 103.5MPa was applied along all the pressure containing surface. Fig. 2 show the element mesh. The element used for whole model is Quadratic 10 noded tetrahedron, in the local area were mesh subdivision. The number of elements is 78,489, the number of nodes is 117,624, the degrees of freedom is 342,996.

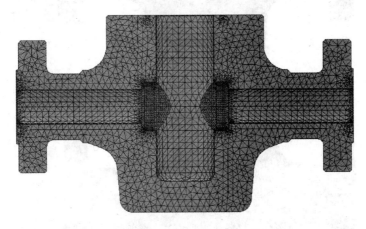

Fig. 2 Element mesh

Fig. 3 show the smallest thickness/radius was found in the bore of body, the maximum Von Mises stress is 533.5MPa, less than the material yield strength 552MPa.

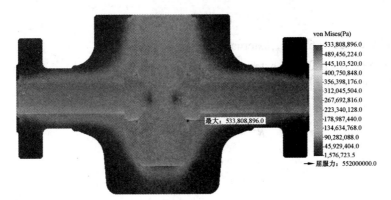

Fig. 3 Von Mises stress

Fig. 4 shows the maximum principal strain is 0.0024≤0.02.

Fig. 5 shows the maximum equivalent strain is 0.0022≤0.01.

Calculated by the above analysis can determine the body designed to meet the strength requirements of API 17D.

3.2 Calculation of the Actual Working Conditions of Body in Underwater

The valve body withstand internal pressure produced by the medium fluid pressure and external pressure produced by the depth of the water in the actual conditions of the underwater. There are two limiting cases of the valve body in the actual conditions of the underwater.

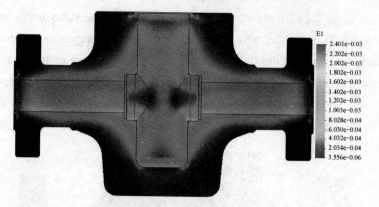

Fig. 4 Principal strain

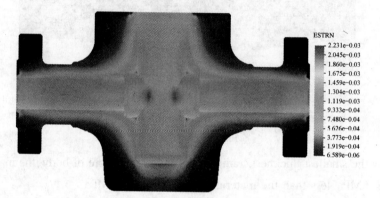

Fig. 5 Plastic equivalent strain

(1) Internal pressure is 69MPa, external pressure is 32MPa;
(2) Internal pressure is 0MPa, external pressure is 69MPa.

According to the Lame formula:

$$\sigma_\theta = \frac{p_i R_i^2 - p_o R_o^2}{R_o^2 - R_i^2} + \frac{(p_i - p_o) R_i^2 R_o^2}{R_o^2 - R_i^2} \cdot \frac{1}{r^2} \tag{20}$$

$$\sigma_r = \frac{p_i R_i^2 - p_o R_o^2}{R_o^2 - R_i^2} - \frac{(p_i - p_o) R_i^2 R_o^2}{R_o^2 - R_i^2} \cdot \frac{1}{r^2} \tag{21}$$

$$\sigma_a = \frac{p_i R_i^2 - p_o R_o^2}{R_o^2 - R_i^2} \tag{22}$$

$$\sigma_{Von} = \sqrt{\frac{(\sigma_\theta - \sigma_r)^2 + (\sigma_\theta - \sigma_a)^2 + (\sigma_r - \sigma_a)^2}{2}} \tag{23}$$

Where, σ_θ is hoop stress; σ_r is radial stress; σ_a is axial stress; p_i is internal pressure; p_o is external pressure; R_i is inside radius; R_o is outside radius; r is calculated stress at the radius.

The maximum Von Mises stress is 95MPa in first case, The maximum Von Mises stress is 105MPa in second case, there are less the than material yield strength 552MPa. So the design of the valve body is safe.

3.3 Calculation of the Hydrostatic Pressure of the Hydraulic Cylinder

The hydraulic cylinder is one of the important parts of the subsea gate valve. The interior of the cylinder will withstand high pressure. The size of the cylinder should be properly determined to ensure that the hydraulic cylinder has sufficient output power, the movement velocity and the effective stroke. Hydraulic cylinder surface should be appropriate with the tolerance level, surface roughness and geometric tolerance level, to ensure that the hydraulic cylinder sealing, motion stability and reliability.

Table 2 Shows the properties of materials relevant for the calculations.

Table 2 Material properties

Component	Material	S_Y, MPa
Cylinder	AISI 4140	517

A finite element model of cylinder was created to investigate the stress levels at test pressure of 31MPa. A half-symmetry model was utilized to reduce the computational time with acceptable accuracy. The cylinder was fixed at the flange facing and the test pressure of 31MPa was applied along all the pressure containing surface. Fig. 6 shows the element mesh. The element used for whole model is Quadratic 10 noded tetrahedron in the local area were mesh subdivision. The number of element is 94533; the number of nodes is 140133; the degree of freedom is 352996, respectively.

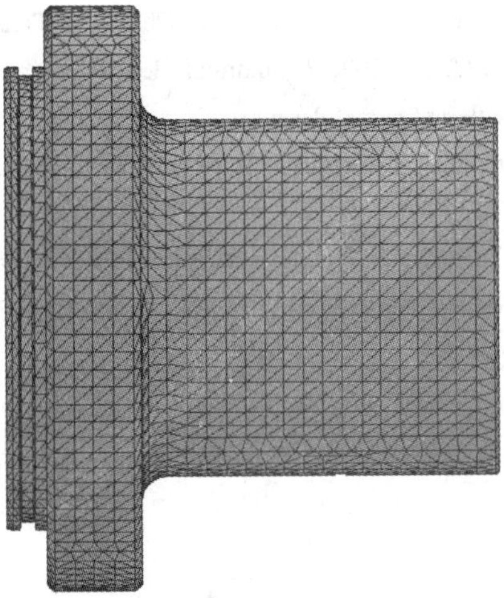

Fig. 7 shows the smallest thickness/radius was found in the bore of cylinder, the maximum Von Mises stress is 165MPa, less than the material yield strength 517MPa. Fig. 8 shows the maximum principal strain is 0.0008≤0.02.

Fig. 6 Element mesh

Calculated by the above analysis can determine the cylinder designed to meet the strength requirements of API 17D.

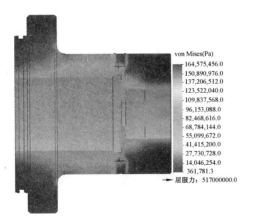

Fig. 7 Von Mises stress

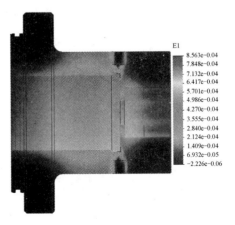

Fig. 8 Principal strain

CONCLUSION

In this paper, the face of the deep-water Christmas tree condition requirements, designed level of U – HH – PSL3G – PR2 2 – 1/16 – 10K subsea gate valve. The theoretical calculation of actuator and the calculation of the key components of the valve and the finite element analysis, the subsea gate valve design meets the relevant requirements of API 17D.

REFERENCES

Ali S Z. 1996. Subsea Valve Actuator For Ultra Deepwater[R]. OTC 8240.

American Petroleum Institue. 2011. Design and Operation of Subsea Production Systems—Subsea Wellhead and Tree Equipment [M]. API 17D 2nd.

API 6A 20th, 2010. American Petroleum Institue, Petroleum and Natural Gas Industries—Drilling and Production Equipment— Wellhead and Christmas Tree Equipment [S].

BS 1726 – 1. 2002. Cylindrical Helical Springs Made Fromround Wire and Bar —Guide to Methods of Specifying, Tolerances and Testing.

Herd D P, McCaskill J W. 1976. How to Make a Valve Which Will Fail-Safe in Very Deep Water. ASME Paper: 76 – 35.

ISO 13628 – 7, 2010. American Petroleum Institue. Petroleum and Natural Gas Industries—Design and Operation of Subsea Production Systems— Completion/Workover Riser and Systems [S].

The Effect of Locking Surface Angles on the Mechanical Advantage of Subsea Connector

Xiaojian Jin[1], Jinlong Wang[2], Menglan Duan[2,3], Kai Tian[3], Biwei Ren[3], Xiaolan Luo[3]

[1] China National Offshore Oil Corporation, Beijing, China
[2] Department of Mechanics and Engineering Science, Fudan University, Shanghai, China
[3] Offshore Oil/Gas Research Center, China University of Petroleum, Beijing, China

Abstract   The devices of subsea production system are all connected by subsea connectors. It's important to analyze the locking mechanism in the basic design. Through the freebody diagrams and simple friction theory, the mechanical advantage related to coefficient of sliding friction and locking surface angles is developed. It show the effect of the different locking surface on the mechanical advantage. It provides significant reference on the design of the locking structure of subsea connector.

Key Words   Subsea Connector; Mechanical Advantage; Freebody Diagram; Friction Theory; Locking Surface Angle

INTRODUCTION

With the development of deepwater oil and gas fields, the subsea production system has become an important development mode. In general, a subsea production system includes subsea trees, subsea manifolds, jumpers, PLEM/PLET, etc. These devices are all connected by subsea connectors. Subsea connector is a significant device of subsea production systems equipment in the offshore oil/gas development. Fig. 1 is a typical subsea manifold and its connector, which is vertical mechanical collet connector.

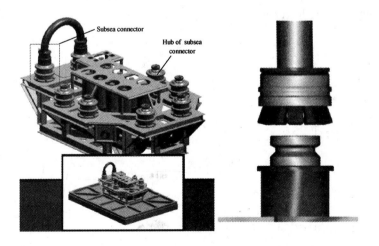

Fig. 1   A typical subsea manifold and its connector

The locking parts in a collet connector include a seal, an upper hub, a lower hub, about 10 to 40 fingers and a collet. Pushing force from the motive source, usually several hydraulic cylinders, makes the collet slide downward and hence the fingers move inward. Fingers clamp the upper hub and the lower hub, resulting in a load on the seal. The force is amplified in the process of a series of transmission, which is realized by conic surfaces on fingers and hubs. To ensure the locking parts to function successfully, it is necessary to analyze the function of each component concerned. A section of the locking parts is shown in Fig. 2.

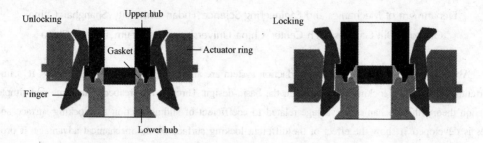

Fig. 2 Cross Section of the connector's locking part

To assist the design of connector locking mechanism it is useful to get the mechanical advantage equation and analyze the influencing factors. Freebody diagram and static analysis will be used to derive the equation. Different sets of friction coefficients and inclination angles are used to demonstrate how they work on the load transferring.

1 MECHANICAL ADVANTAGE OF SUBSEA CONNECTOR

To optimize the design of connector locking mechanism the equation of mechanical advantage should be obtained which is ratio of hydraulic actuating force to sealing force. Kai Tian has researched the mechanical advantage of subsea connector.

As Fig. 3 the actuator ring driven by the hydraulic actuating cylinder, the finger's function is to fasten the upper hub and the lower hub.

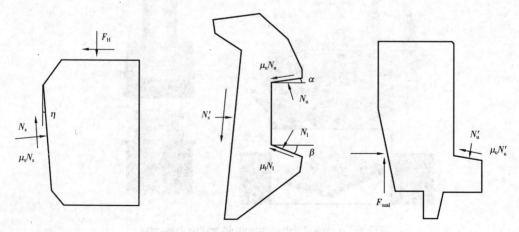

Fig. 3 Freebody diagram of actuator ring, finger, upper hub

To obtain the relation between hydraulic force and sealing force, freebody diagram and static analysis are used below. As a result the mechanical advantage are concludes as:

$$MA = \frac{F_{seal}}{F_H} = \frac{K - M}{N - M}$$

$$K = \frac{\cos\eta - \mu_s \sin\eta}{\sin\eta + \mu_s \cos\eta}$$

where

$$M = \frac{\sin\beta + \mu_l \cos\beta}{\mu_l \sin\beta - \cos\beta}$$

$$N = \frac{\sin\alpha + \mu_u \cos\alpha}{\cos\alpha - \mu_u \sin\alpha}$$

where, η is the slope angle which must be less than 8.5° for self-locking; α,β are the two inclination angles of two locking surfaces separately; μ_u, μ_l are the corresponding friction coefficients; μ_s is the coefficient of friction on the back surface.

2 PARAMETRIC ANALYSIS

To make the design of connector more efficient the mechanical advantage should be large enough to reduce requirements for hydraulic cylinder. As the above model shows, there are two influencing factor: coefficient of friction and the angles of the locking suafaces. the parametric analysis should be done to demonstrate the relation between these influencing factors and mechanical advantage.

The general effects from friction are illustrated in. In this paper, only the effects from three locking surface angles are considered.

To find out the general effects from locking surface angles, a simplify can be made the equation (1). Setting $\alpha = \beta, \eta = 4°$ (it equals 4° for self-locking.) with $\mu_s = \mu_u = \mu_l = \mu = 0.15$, the curve of mechanical advantage is shown in Fig. 4.

It's demonstrated that as the angle increases, the mechanical advantage decreases. A mechanical advantage of 13.9 is indicated at an high efficiency and $\alpha = \beta = 1°$. Most importantly the curve suggests that mechanical advantage drops from 13.9 to 3.3 when the angle increases from 1° to 30°. To be concluded maximizing the efficiency of subsea connector needs to minimize the angles. But in the real process of design, there are some other considerations, so the angles of α,β are determined as 15°.

After the angles of α,β are determined, it is of necessity to know how the effect form the self locking angle η. Fig. 5 plots effect from self locking angle. The mechanical advantage drops from 7.7° to 4.2° when the angle η increases from 0.5° to 8.5°.

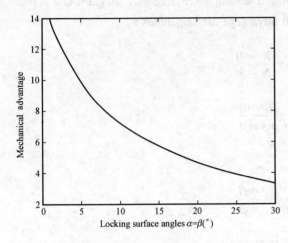
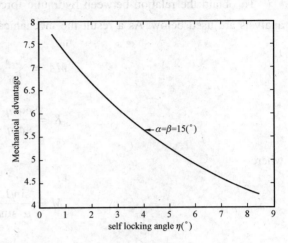

Fig. 4   Plot of MA with equal locking surface angles, $\alpha = \beta$

Fig. 5   Plot of effect from self locking angle

CONCLUSIONS

According to the above analysis, the simple method of subsea connector analysis demonstrated provides a useful physical picture of how connector performance is affected by locking surface angles. The results are as follows:

(1) The mechanical advantage of locking mechanism is mainly related to coefficient of sliding friction and locking surface angles.

(2) The mechanical advantage is highest at the surface angles near zero. Changes of a few range may alter connector clamping force by a significant factor.

(3) Keep the locking surface angles below certain level will help maintian the locking ability of subsea connector.

ACKNOWLEDGEMENTS

This work was financially supported by the National Science and Technology Major Project (2011ZX05026 – 003 – 02 Prototype Development of Subsea Connector for manifolds) and National High Technology Research and Development Program (863 Program, 2012AA09A205 Key Technical Research of Subsea Tree and Complete Sets of Equipment Development).

REFERENCES

Gledhill A, Hart B. 1987. Advances in Underwater Technology, Ocean Science and Offshore Engineering [M]. Springer.

Jacqueline C. 1999. Deepwater High-capacity Collet Connector[C]. Offshore Technology Corference.

James H Owens. 1978. Friction Effects in Marine Hydraulic Connectors[C]. Offshore Technology Conference.

Tian Kai, Duam Menglan, Hong Yi, et al. 2012. Research on the Mechanical Advantage of Subsea Connector [J]. Advanced Materials Research, 524 – 527:1543 – 1547.

Wang Liquan, An Shaojun, Wang Gang. 2011. Study and Design of a Deepwater Sub-sea Pipeline Collet Connector

[J]. Journal of Harbin Engineering University, 32(9): 1103.

Wang Y Y, Duan M L, Feng W, et al. 2011. Investigation on Installation Methods of Deepwater Manifolds and Their Applications to LW3 – 1 Gas Field in South China Sea[J]. The Ocean Engineering, 29(3): 23.

Zhou C F, Jiao X D, et al. 2011. Selection and Design of Jumper Connector[J]. Journal of Petrochemical University, 24(3): 75.